LINEAR INTEGRATED CIRCUITS

APPLICATIONS AND EXPERIMENTS

THEODORE F. BOGART, JR.
UNIVERSITY OF SOUTHERN MISSISSIPPI

JOHN WILEY & SONS
NEW YORK CHICHESTER BRISBANE TORONTO SINGAPORE

ISBN 0 471 87512 0

Printed in the United States of America

10 9 8 7 6 5 4 3 2 1

PREFACE

LINEAR INTEGRATED CIRCUITS: APPLICATIONS AND EXPERIMENTS is a comprehensive survey of modern linear integrated circuits and the practical systems in which they are used. Each of the 12 experiments in the book is accompanied by discussion material that summarizes the important characteristics of a particular type of integrated circuit and covers the theory related to a number of specific applications. The book could be used as a stand-alone text for a predominantly laboratory-oriented course, or as a companion to a textbook in a lecture-laboratory sequence.

All experiments have been successfully performed by students whose background included the traditional two-semester sequence of courses in DC and AC circuit analysis and one course in basic semiconductor device theory. In recognition of the importance of the operational amplifier in so many linear circuit applications, the first two experiments deal exclusively with the characteristics and limitations of these versatile devices. The inexpensive and widely available 741 operational amplifier is used in many of the experiments. Other integrated circuits used in the experiments were similarly selected on the basis of low cost and wide availability. Complete manufacturers' specifications and data sheets are provided in Appendix B for all of the integrated circuits used. No special trainers or unusual equipment is required for any of the experiments, all of which can be constructed and performed in any reasonably equipped electronics laboratory.

Many of the experiments will be found to be too long for a typical scheduled laboratory session. The experimental procedures were intentionally made long to provide an instructor with flexibility in the choice of topic coverage and/or scheduling of successive meetings. The Questions section following each experiment generally requires the student to compare experimental results with values predicted by the theory outlined in the Discussion section. These too may be pruned at the instructor's discretion. Following each Discussion section is a set of Exercises, many of which are similar to the Questions in that they require the student to perform theoretical calculations on the experimental circuits. These have been included because many instructors prefer that students perform theoretical calculations before constructing circuits in the laboratory.

A number of Design Projects are provided at selected intervals after certain of the experiments have been completed. The purpose of these projects is to enable students to gain experience designing linear circuits of the types they have studied and to interface and combine circuit types into larger systems.

Appendix A, Writing Lab Reports, summarizes standard practices for presenting and interpreting experimental results. It emphasizes the importance of quantitative references to the data, the correct method for graphing experimental data, and the expression of concise error analyses and conclusions. Included is a sample lab report that illustrates the major points of good technical writing and the special considerations applicable to linear integrated circuit experiments.

I wish to thank Mr. Bruce Elliot, my laboratory assistant at the University of Southern Mississippi, for his help in constructing and checking all of the experimental circuits in this book and for suggestions that improved the instructional value in many of the procedures. I am also grateful to Dr. C. Howard Heiden for departmental support and encouragement during the preparation of this material.

THEODORE F. BOGART JR.

CONTENTS

LINEAR INTEGRATED CIRCUITS

APPLICATIONS AND EXPERIMENTS

1 Introduction to Operational Amplifiers

OBJECTIVES

1. To learn how to construct operational amplifier circuits that perform inverted and non-inverted voltage scaling.

2. To verify experimentally the theoretical closed-loop voltage gain of inverting and non-inverting operational amplifier circuits.

3. To learn how to construct and to verify experimentally the voltage follower circuit using an operational amplifier.

4. To verify experimentally the virtual ground of an operational amplifier with feedback.

5. To verify experimentally the input impedance presented to a signal source by an operational amplifier with feedback.

6. To learn how an operational amplifier can be operated with a single power supply voltage.

EQUIPMENT AND MATERIALS REQUIRED

1. Dual trace oscilloscope.

2. Sinewave signal generator, 0-10 V peak, 100 Hz - 1 kHz.

3. $\pm$ 15 V DC power supplies.

4. 741 operational amplifier.

5. 10K potentiometer.

6. Resistors: 1K, 4.7K, 10K (2), 22K, 33K, 100K (2).

7. 0.1 μF capacitor.

DISCUSSION

An operational amplifier is a high gain, high input impedance voltage amplifier with two inputs and a single output. One input(labeled + on the amplifier symbol) is called the <u>non-inverting</u> input, and the other (labeled -) is called the <u>inverting</u> input. When the <u>non-inverting</u> input is grounded and a signal is connected to <u>the inverting</u> input, the output is 180° out of phase with the input. If the input connections are reversed (signal applied to non-inverting input) the output is in phase with the input. If signals are applied to both inputs, the output is proportional to the <u>difference</u> of the inputs.

The amplifier is called operational because in many applications it is used to perform mathematical operations (summation, integration, etc.) on voltages connected to its input (s). In one such application, a resistor is connected between the output and the input to provide voltage feedback. See Figure 1.1.

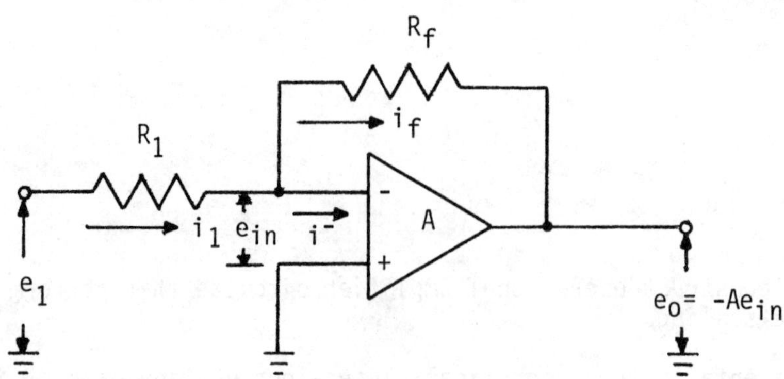

Figure 1.1 An operational amplifier connected to perform voltage scaling. $e_o/e_1 = -R_f/R_1$, where the minus sign denotes phase inversion.

Applying Kirchhoff's current law at the - input, we have

$$i_1 = i_f + i^- \tag{1}$$

Since the input impedance of the operational amplifier is very large (typically several megohms or greater), i^- is negligibly small and equation (1) becomes

$$i_1 = i_f \tag{2}$$

But $i_1 = \dfrac{e_1 - e_{in}}{R_1}$ and $i_f = \dfrac{e_{in} - e_o}{R_f}$.

$$\text{Hence, } \frac{e_1 - e_{in}}{R_1} = \frac{e_{in} - e_o}{R_f}. \tag{3}$$

Since $e_o = -Ae_{in}$, where A is the open loop gain of the amplifier (the gain when the feedback resistor is removed), we have $e_{in} = -e_o/A$, which when substituted in (3) yields

$$\frac{e_1}{R_1} + \frac{e_o}{AR_1} = \frac{-e_o}{AR_f} - \frac{e_o}{R_f} \tag{4}$$

Solving (4) for e_o/e_1 we find

$$\frac{e_o}{e_1} = \frac{-R_f}{R_1}\left(\frac{A}{A + 1/\beta}\right) \tag{5}$$

where $\beta = R_1/(R_1 + R_f)$ = the feedback ratio and the minus sign denotes the 180° phase inversion. In the above derivation we used the fact that $e_{in} = -e_o/A$. Since A is very large, we see that e_{in} is very small. It is in fact so close to zero in most cases that the -terminal is essentially at ground potential and is called virtual ground. For this reason, $i_1 \simeq e_1/R_1$ and we see that the impedance seen by the signal source is effectively R_1.

Since an operational amplifier has a very large value of open loop gain A, typically 10,000 to 1,000,000, we have $A + 1/\beta \simeq A$, and equation (5) may therefore be written

$$\frac{e_o}{e_1} = -\frac{R_f}{R_1} \tag{6}$$

The significance of this result is that the closed-loop gain, R_f/R_1, of the circuit depends only on the ratio of the resistors and not on precise values of the amplifier characteristics. This operation is called scaling, since the output equals the input multiplied by the scale factor R_f/R_1 (and, of course, inverted.)

Figure 1.2 shows how the operational amplifier may be connected for non-inverted scaling.

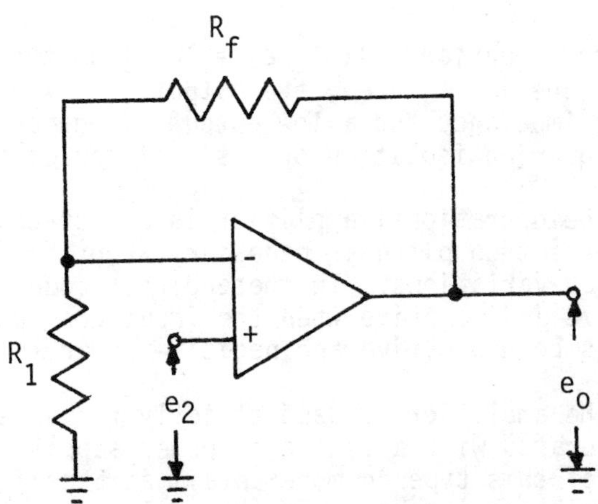

Figure 1.2 An operational amplifier connected to perform non-inverted voltage scaling. $e_o/e_2 = (R_f + R_1)/R_1$.

For the configuration of Figure 1.2 it can be shown that

$$\frac{e_o}{e_2} = \frac{1}{\beta}\left(\frac{-A}{1/\beta + A}\right) \tag{7}$$

where A and β are the same as previously defined. Again, since $A \gg 1/\beta$, equation (7) reduces to

$$\frac{e_o}{e_2} = \frac{1}{\beta} = \frac{R_f + R_1}{R_1} \tag{8}$$

Another frequently used operational amplifier configuration is the voltage follower, shown in Figure 1.3.

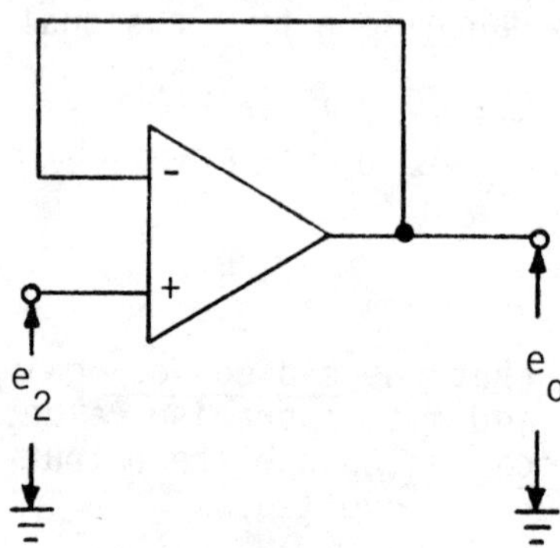

Figure 1.3 A voltage follower. $e_o/e_2 = 1$.

In Figure 1.3 $R_f = 0$, so from equation (8), $e_o/e_2 = 1$. Thus the output voltage is the same (in magnitude and phase) as the input, i.e., the output follows the input. The voltage follower has a very high input impedance and a low output impedance, and is therefore used as a buffer in applications requiring isolation of a signal source from a load.

In many applications the operational amplifier is direct-coupled to the signal source that provides its input. Direct-coupling is necessary when the input signal is a dc voltage or has very low frequency variations. In these direct-coupled situations, the output voltage must go both positive and negative when the input goes positive and negative. Therefore, the amplifier requires both positive and negative power supply voltages.

In applications where the amplifier is used strictly for ac signals and dc levels are not important, it may be operated with a single dc power supply voltage. We will investigate a typical application of this type in more detail in Experiment 4, "Audio Amplifiers." When a single power supply voltage is used a dc (bias) level must be added to the output of the amplifier. The ac signal variations then cause the output to vary above and below this dc bias level. Figure 1.4 shows the input and output waveforms of a unity-gain amplifier whose input is a 2 volt peak sine wave and whose output has a 5 volt dc level.

Figure 1.5 shows an operational amplifier circuit in which a dc level is added to the output by connecting a dc voltage to the non-inverting input. The potentiometer is used to adjust the magnitude of the dc output level. Note that the (ac) input signal connected to the inverting input must be capacitor-coupled. The ratio of the ac signals e_o/e_1 is still given by: $e_o/e_1 = -R_f/R_1$.

When the operational amplifier is used with a single (positive) power supply, as shown in Figure 1.5, the terminal used for the negative supply is grounded. Note that the dc level in the output will influence the maximum peak-to-peak ac voltage that can appear at the output without distortion. For example, if the positive supply voltage is 10 V DC and the dc output level is 7 V DC, then the output is limited to 3 V peak (6 V pk-pk), since the output can only rise 3 V (from 7 V to 10 V). Similarly, if the output dc level is 4 V, then it can only decrease 4 V (from 4 V to 0 V), so the maximum output would be 8 V peak-to-peak. For optimum operation, that is, to permit the maximum peak-to-peak output signal variation, the dc level in the output should be set to one-half the positive supply voltage.

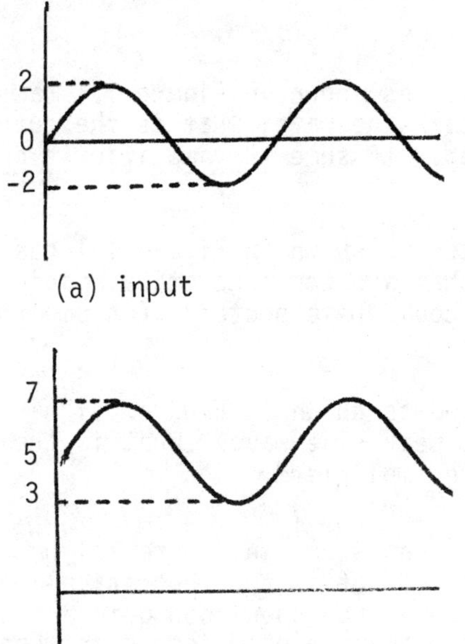

Figure 1.4 A 4 V pk-pk input producing a 4 V pk-pk output having a 5 V DC (offset) level.

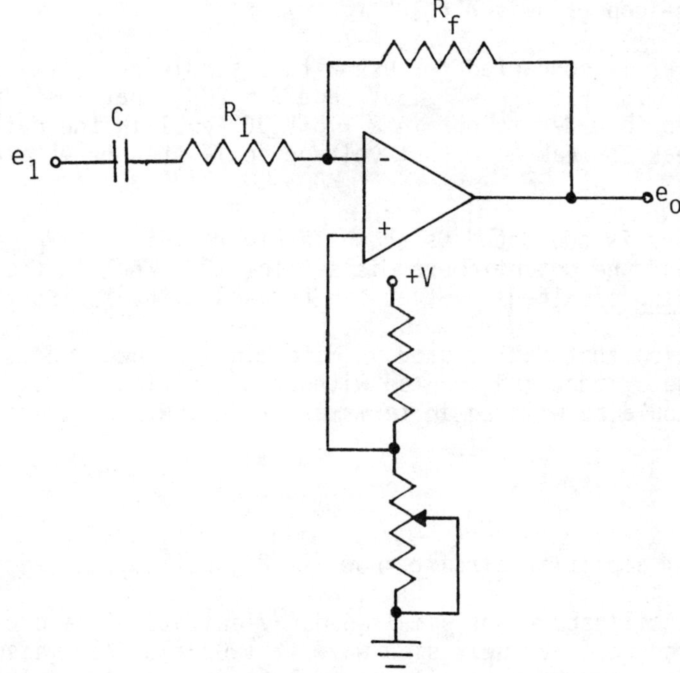

Figure 1.5 A dc level is added to the output using the non-inverting input. The amplifier is operated with a single positive power supply.

EXERCISES

1. An operational amplifier connected as shown in Figure 1.1 has Rf = 33K and R_1 = 10K. If the input e_1 is a 2.5 volt peak sine wave, what is the output? Repeat, if the values of Rf and R_1 are interchanged. (Assume the amplifier's open-loop gain A is extremely large.)

2. An operational amplifier connected as shown in Figure 1.1 has Rf = R_1 = 10K. If these resistors have 10% tolerances, what are the theoretical minimum and maximum values of closed-loop amplifier gain that could be expected? (Assume the amplifier's open-loop gain A is extremely large.)

3. An operational amplifier connected as shown in Figure 1.1 has an open-loop gain of A = 2×10^5. The output is a 5 volt peak sine wave. What is the peak value of the voltage e_{in} at the inverting input to the amplifier?

4. An operational amplifier connected as shown in Figure 1.1 has an open-loop gain of A = 2×10^5. If Rf = 47K and R_1 = 10K, what is the theoretical (exact) closed-loop gain of the amplifier (taking into account the open-loop gain and the feedback ratio β)? Compare this result with the approximate closed-loop gain when it is assumed that A + 1/β = A. Repeat for A = 2×10^3. What is your conclusion, as regards the validity of calculating the closed-loop gain by the ratio R_f/R_1?

5. An operational amplifier connected as shown in Figure 1.2 has R_f = 100K. The input e_2 is a 0.5 volt peak sine wave. Assuming the open-loop gain A is extremely large, what should be the value of R_1 in order that the output be a 5.5 volt peak sine wave?

6. Using the values calculated in exercise 5, what is the exact theoretical peak value of the output if the open-loop gain is A = 10^4?

7. An operational amplifier is connected for use with a single positive power supply voltage as shown in Figure 1.5. If V_{CC} = + 15 V, and R = 22K, what resistance should the potentiometer be set to in order to obtain a + 5 V DC level in the output of the amplifier? What maximum peak-to-peak ac output voltage could then be obtained at the output without distortion?

8. An operational amplifier is connected as shown in Figure 1.5. If V_{CC} = + 15 V, R = 10K, R_1 = 4.7K, R_f = 10K, and the potentiometer is set for 10K, what is the peak value of the maximum sine wave input voltage e_1 that can be used without distorting the output?

9. Derive a general equation that can be used to determine the peak value of the maximum sine wave input voltage e_1 that can be used without distorting the output in Figure 1.5. Your equation should be written in terms of R_1, R_f, R, and the potentiometer resistance R_1 in Figure 1.5.

PROCEDURE

1. Connect the operational amplifier circuit shown in Figure 1.6.

2. Connect a dual trace oscilloscope for simultaneous viewing of e_o and e_1. Adjust the signal generator to produce a 1 V peak sine wave at 100 Hz. Then measure and record the peak value of e_o for each of the following values of Rf: 1K, 4.7K, 10K, 33K, and 100K. Note the phase of e_o with respect to e_1.

3. With R_f = 100K and e_1 = 1 V peak as in step 2, measure the voltage at pin 2. (Recall that this point is at virtual ground.)

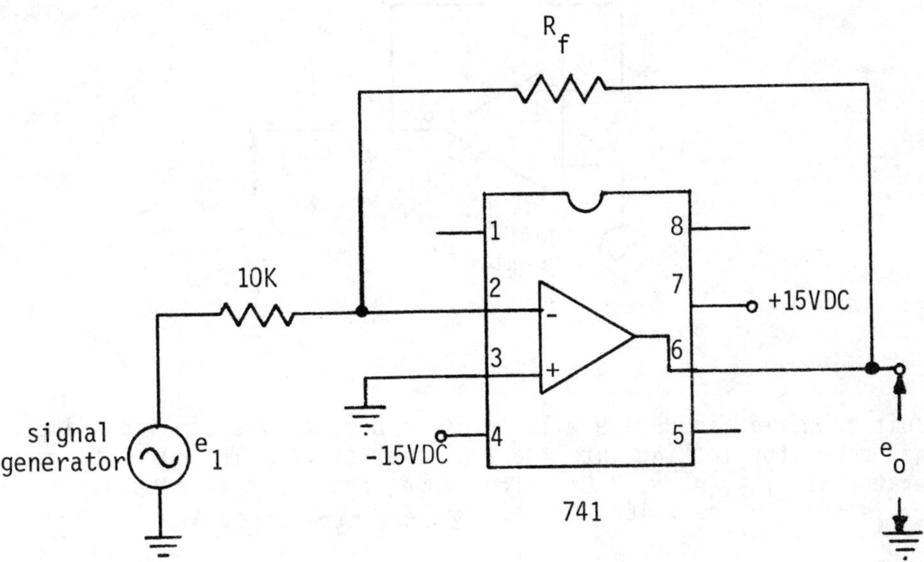

Figure 1.6

4. Now connect the circuit shown in Figure 1.7. (Pin numbers are shown for the 741 amplifier output and inputs; all other pin connections are the same as in Figure 1.4.)

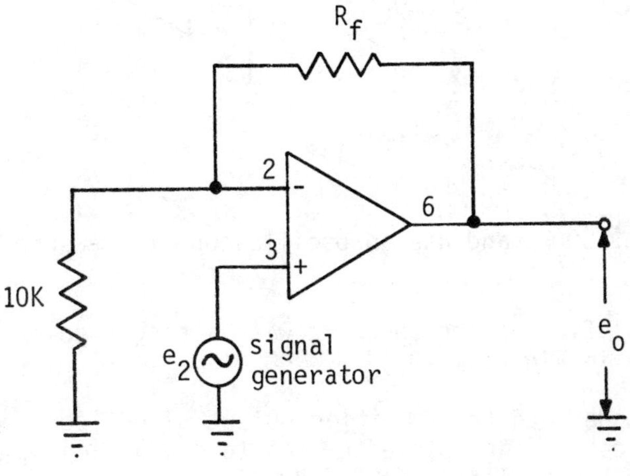

Figure 1.7

5. Set e_2 to 1 V peak at 100 Hz and measure and record the peak value of e_o for R_f = 1K, 4.7K, 10K, 33K, and 100 K. Note the phase of e_o with respect to e_2.

6. Connect the circuit shown in Figure 1.8.

7. Using a dual trace oscilloscope, measure and record the peak value of e_o for each of the following signal generator settings: 1 V peak at 100 Hz, 5 V peak at 500 Hz, 10 V peak at 1 kHz. Note the phase of e_o with respect to e_2 in each case.

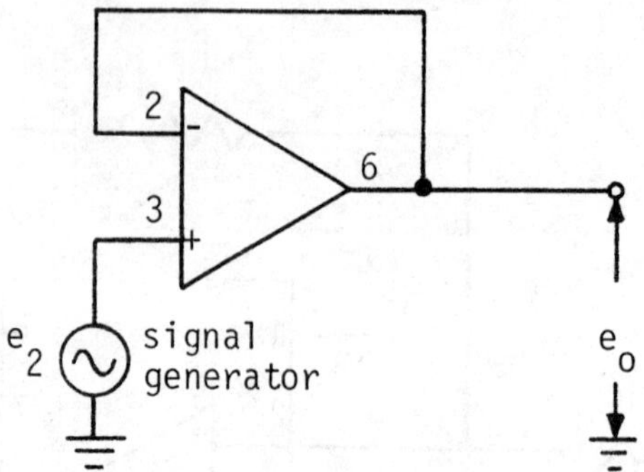

Figure 1.8

8. The circuit shown in Figure 1.9 will be used to determine the input impedance seen by the signal generator looking into the amplifier stage. The current drawn from the signal generator is $i_1 = (e_1 - e_2)/10K$. Therefore, the input impedance seen by the signal generator is $e_1/i_1 = (e_1 \times 10^4)/(e_1 - e_2)$. Connect the circuit.

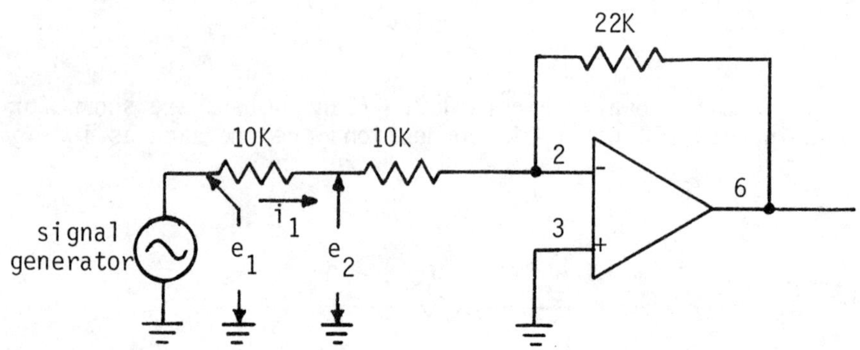

Figure 1.9

9. Set e_1 to 10 V peak at 100 Hz and use an oscilloscope to measure and record the peak values of e_1 and e_2.

10. To verify that the amplifier can be operated with a single power supply voltage, connect the circuit shown in Figure 1.10.

11. With the input e_1 grounded and the amplifier output direct-coupled to the oscilloscope (i.e., with the oscilloscope input selector set to "dc"), adjust the 10K potentiometer until the dc output of the amplifeir is 7.5 V.

12. Now connect the input e_1 to a signal generator adjusted to produce a 2 V pk sine wave at 1 kHz. Record the waveform observed on the oscilloscope, which should still be direct-coupled to the amplifier output. Note in particular the minimum and maximum values of the output waveform.

13. Increase the input signal level to 10 V pk and repeat your measurements. Sketch the waveform observed on the oscilloscope.

14. Remove the signal generator and again ground the input e_1. Adjust the potentiometer to achieve the lowest possible dc level in the amplifier output. Measure and record this level.

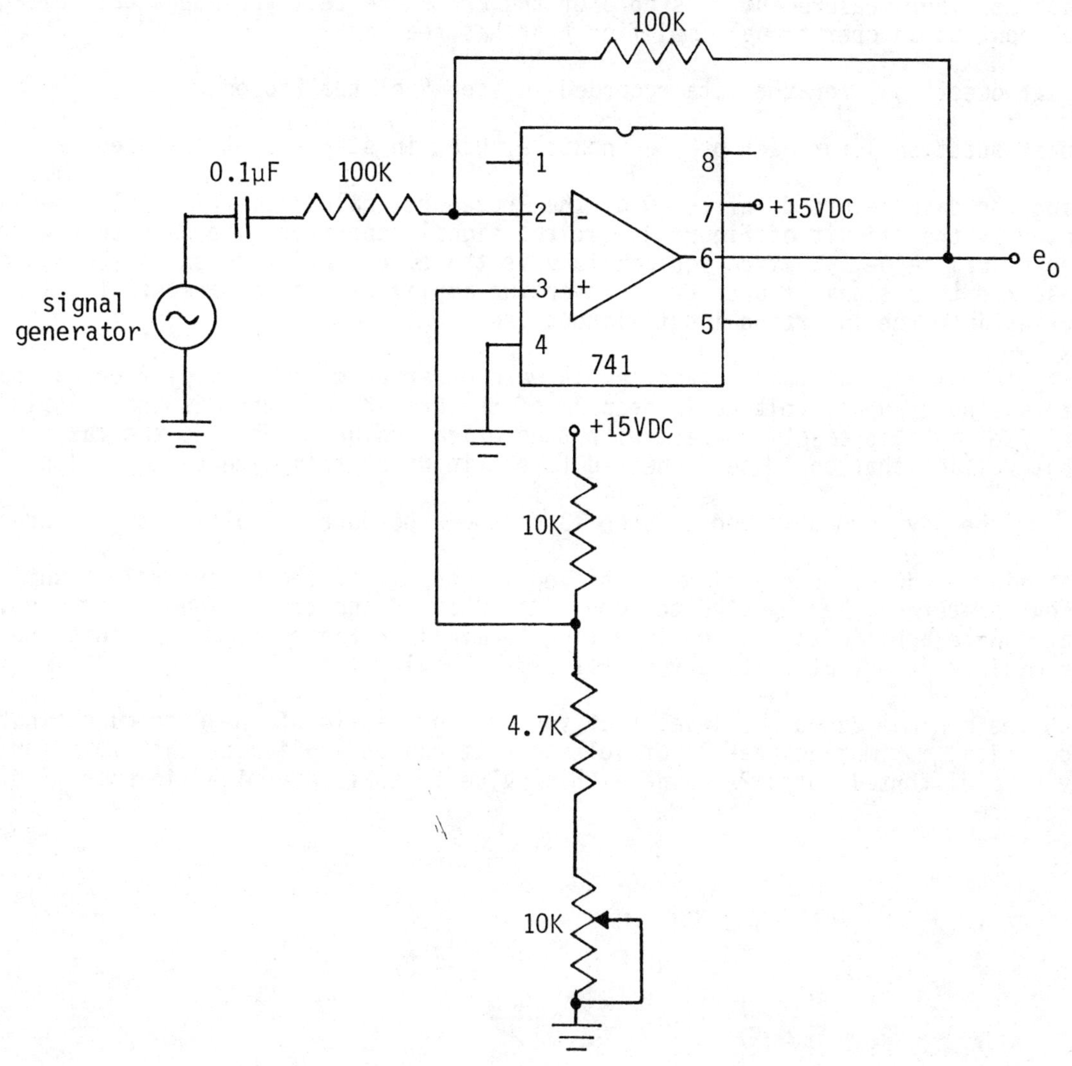

Figure 1.10

15. Reconnect the signal generator and adjust its amplitude to achieve the maximum possible (ac) output from the amplifier that is displayed without distortion by the oscilloscope. Measure and record the peak value of the ac input signal when the maximum undistorted output is achieved.

QUESTIONS

1. Calculate the theoretical closed-loop voltage gain of the circuit in Figure 1.6 for each of the values of R_f used in step 2 of the Procedure. Compare these with the voltage gains determined from the experimental data recorded in step 2. Report your results in a data table that shows all measured values, results of theoretical calculations, and percent differences between theoretical and experimental results. Account for any significant differences between theoretical and experimental results. What do your phase angle observations confirm about this circuit?

2. What does your measurement in step 3 of the Procedure tell you about the voltage at the input to an operational amplifier that has feedback?

3. Repeat question 1 for the data recorded in step 5 of the Procedure.

4. Repeat question 1 for each of the inputs e_2 used in step 7 of the Procedure.

5. Using the data recorded in step 9 of the Procedure, calculate the input impedance presented by the circuit of Figure 1.9 to the signal generator. Compare this with the theoretical value. What do you conclude is the effect of feedback on the impedance presented to a signal source by an inverting amplifier, as compared to the impedance looking into the inverting input without feedback?

6. What minimum and maximum voltage levels were observed when the amplifier was operated with a single supply voltage in step 12 of the Procedure? What is the ac gain of the amplifier in this case? Compare with theoretical values. What is the maximum ac peak input voltage that could be connected to the input of this single-supply amplifier?

7. Sketch the waveform observed in step 13 of the Procedure. Explain its appearance.

8. What minumum dc output level was achieved in step 14 of the Procedure? Assuming the potentiometer can be adjusted to zero ohms, what is the theoretical minimum voltage that can be applied to the non-inverting terminal in Figure 1.10? Is this the same as the minimum dc output voltage you measured? Should it be?

9. With the minimum dc output level that you set in step 14 of the Procedure, what is the theoretical maximum ac peak input voltage that can be applied to the amplifier without having a distorted output? Compare this value to that determined in step 15 of the Procedure.

2 Op-Amp Limitations: Bandwidth, Slew Rate, Offsets

OBJECTIVES

1. To learn through experimental observation how the bandwidth of an operational amplifier decreases when its closed-loop gain is increased.

2. To verify experimentally that the gain-bandwidth product of an operational amplifier is constant.

3. To learn how to measure the slew rate of an operational amplifier and to observe the effect of driving the amplifier with a signal that causes the slew rate to be exceeded.

4. To learn how to measure and interpret input offset voltage and input offset currents.

5. To verify experimentally the theoretical output offset voltage due to input offsets.

6. To learn circuit techniques for reducing or eliminating offset.

7. To learn how to measure open-loop gain and how to calculate common mode gain given the common mode rejection ratio (CMRR).

EQUIPMENT AND MATERIALS REQUIRED

1. Dual-trace oscilloscope

2. Sine/square wave signal generator, adjustable to 1 MHz, 15 V peak.

3. dc millivoltmeter.

4. ± 15 V DC power supplies.

5. 741 operational amplifier.

6. Resistors: 1K (2), 10K (2), 33K, 100K (2), 220K, 330K, 470K, 1M (2).

7. 10K potentiometer.

8. 0.1 μF capacitors (3).

DISCUSSION

The practical operational amplifier has certain limitations that adversely affect its performance in some applications. The user or designer of operational amplifier circuits must be aware of these limitations so that he or she can select an amplifier whose specifications meet the requirements of the application, or, in some cases, so that he or she can provide circuitry necessary to compensate for the limitations. Failure to do so may result in errors, in the sense that the output voltage may not conform to that which is predicted by the theory. Broadly speaking, amplifier limitations that may give rise to erroneous outputs can be classified as being due to either the dc or ac characteristics of the amplifier.

An important ac limitation of an operational amplifier is its frequency response. Since operational amplifiers are used down to zero frequency (dc), the bandwidth (BW) of the amplifier is equal to the cutoff frequency f_C, where the output level is 3 dB down from (or .707 times) its dc value. It is an important fact, expressed by equation (1), that the gain-bandwidth product of an operational amplifier is constant:

$$G \times BW = \text{constant} \tag{1}$$

The actual constant depends upon the particular amplifier. If the amplifier is used in its open-loop configuration ($R_f = \infty$, a relatively rare case), its gain is exceptionally high and its bandwidth is therefore quite small. In closed-loop applications where R_f is large compared to R_1, that is, in applications requiring high values of gain $G = R_f/R_1$, we must be aware of the reduced frequency range that equation (1) implies as a consequence.

Another ac limitation of the operational amplifier is its specified slew rate. This refers to the maximum rate at which the output voltage can change in large signal applications, and is given in V/sec (or V/μsec). If the amplitude and frequency of a signal are both large, the rate at which the voltage must change may exceed the specified slew rate, even though the frequency of the signal may be within the bandwidth of the amplifier. The maximum frequency of a sinusoidal signal with peak value E_p that an amplifier can accommodate, due to the limitation of its slew rate S, is given by inequality (2):

$$f \leq \frac{S}{s\pi E_p} \tag{2}$$

For accurate operation, the output of an operational amplifier should be zero volts when both inputs are at zero volts. The dc, or bias, characteristics of a typical amplifier are such that this is rarely the case. Unless the amplifier is used strictly for its ac characteristics, some provision must be made to compensate for, i.e., eliminate or reduce, the output voltage, called the output offset voltage, that exists when the inputs are zero.

There are two causes of output offset. One of these is the fact that the internal voltage drops at the inverting and non-inverting inputs are not identical, due to slight differences (mismatches) in the input devices. This gives rise to a different voltage, called the input offset voltage, V_{io}, that is amplified and therefore contributes to output offset. Consider the configuration shown in Figure 2.1. Resistor R_2 in this figure is used to compensate for offset and will be discussed subsequently. In Figure 2.1, it can be shown that the output offset voltage V_{os} due to input offset V_{io} is

$$V_{os} = \frac{V_{io}}{\beta} \tag{3}$$

where $\beta = R_1/(R_f + R_1)$, as before. Input offset V_{io} is usually given in manufacturers' specifications and is typically in the range from 1 to several millivolts.

The other source of output offset is the fact that both inputs must draw bias currents (I^+ and I^- in Figure 2.1). These currents may be slightly different and flow through different values of external impedance, again giving rise to a difference voltage. Manufacturers generally specify an input bias current I_B which is the average of I^+ and I^-:

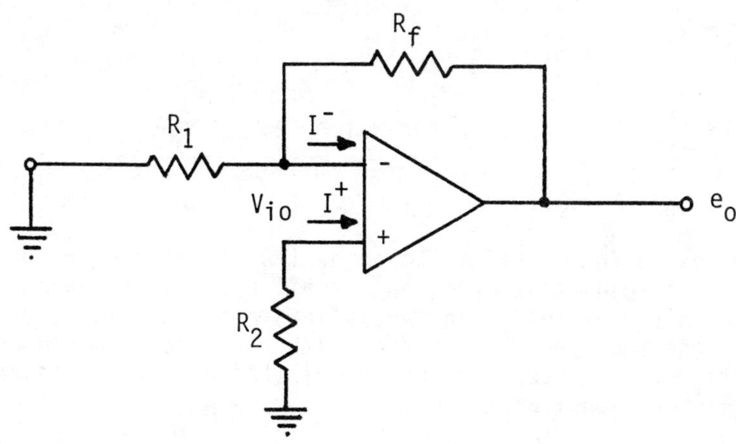

Figure 2.1 An operational amplifier circuit showing the input offset voltage V_{io} between the + and - inputs and the input bias currents I^+ and I^-.

$$I_B = \frac{I^+ + I^-}{2} \tag{4}$$

Since I^+ and I^- are approximately equal, we usually have

$$I_B \simeq I^+ \simeq I^- \tag{5}$$

I_B is typically in the nanoamp or picoamp range. It can be shown that the component of output offset due to I^+ and I^- is

$$V_{os} = I^- R_f - I^+ R_2/\beta \tag{6}$$

One way to reduce this component of offset is to set R_2 in Figure 2.1 equal to the parallel equivalent of R_1 and R_f:

$$R_2 = \frac{R_1 R_f}{R_1 + R_f} \tag{7}$$

When this is done, equation (6) becomes:

$$V_{os} = I_{io} R_f$$

where I_{io} is the input offset current, and is equal to the difference between I^+ and I^-. Clearly I_{io} is much less than I^+ or I^- and manufacturers typically specify it as a few nanoamps or picoamps.

The total output offset voltage is the sum of its components due to input offset voltage and input offset current:

$$V_{os} = \frac{V_{io}}{\beta} + I^- R_f - \frac{I^+ R_2}{\beta} \tag{8}$$

The worst case occurs when $R_2 = 0$, for which (8) becomes

$$V_{os} = \frac{V_{io}}{\beta} + I^{-}R_f \simeq \frac{V_{io}}{\beta} + I_B R_f \tag{9}$$

The best case occurs when $R_2 = R_1||R_f$, for which (8) becomes

$$V_{os} = \frac{V_{io}}{\beta} + I_{io}R_f \tag{10}$$

Another way to compensate for offset is to connect a potentiometer to one of the inputs in such a way that a small adjustment produces a small positive or negative input that is just sufficient to null out the offset. Many amplifiers are designed so that a potentiometer may be connected across two special pins and, with the wiper arm connected to a supply voltage, the potentiometer may be adjusted to null the output. This is the case, for example, with the 741 operational amplifier.

If the same signal is simultaneously connected to both the inverting and non-inverting input, then theoretically the output should be zero. However, due to small internal differences in the characteristics of the inverting and non-inverting sections of the amplifier, there will in fact be a non-zero output voltage. When the same signal is applied to both inputs, the amplifier is said to be operating in the <u>common mode</u>, and we can define the common mode gain A_C as the ratio of the output voltage to the common input voltage. Obviously the common mode gain should be much smaller than the open loop gain A, which we have discussed previously and which we distinguish from A_C by referring to it as the <u>difference mode</u> gain. The <u>common mode rejection ratio</u> (CMRR) is defined as the ratio of the difference mode gain A to the common mode gain A_C:

$$\text{CMRR} = A/A_C \tag{11}$$

The CMRR should be as large as possible and is frequently expressed in dB in manufacturers' specifications.

$$\text{CMRR(dB)} = 20 \log_{10} \frac{A}{A_C} \tag{12}$$

EXERCISES

1. An operational amplifier connected as shown in Figure 2.2 has $R_f = 100K$. If the gain-bandwidth product is 2.5×10^5, what is the bandwidth of the amplifier? What should be the value of R_f in order to obtain a bandwidth of 500 kHz?

2. An operational amplifier has a slew rate of 0.5 V/μsec. Its maximum output voltage is ± 15 V. If it is driven by a square wave having negligible rise time, how much time is required for the output to change from -15 V to + 15 V? What is the maximum frequency at which it can be driven by a 12 V peak sine wave without exceeding the slew rate?

3. When an operational amplifier is driven to its maximum output by a square wave, the output waveform that results is shown in Figure 2.3. If $\Delta V = 20$ V and $\Delta t = 25$ μsec, what is the maximum frequency at which it can be driven by an 8 V peak sine wave without exceeding the slew rate?

4. An operational amplifier connected as shown in Figure 2.1 has $R_1 = 22K$ and $R_f = 47K$. If the input offset voltage is 8.4 mV, what is the output offset voltage due to the input offset? What value should R_1 be changed to in order to reduce this offset by one-half?

5. The currents I^+ and I^- in Figure 2.1 are measured to be 45 and 60 nA, respectively. What would be a typical manufacturer's specification for the input bias current, based on these values? If $R_1 = R_f = 1M$, and $R_2 = 100K$, what would be the output offset voltage due to the bias currents?

6. In the circuit of Figure 2.1, $R_1 = R_f = 1M$ and the input offset current is 5 nA. What should be the value of R_2 to minimize the output offset voltage due to input bias currents? What would be the value of the output offset voltage due to bias currents in this case?

PROCEDURE

1. Connect the circuit shown in Figure 2.2.

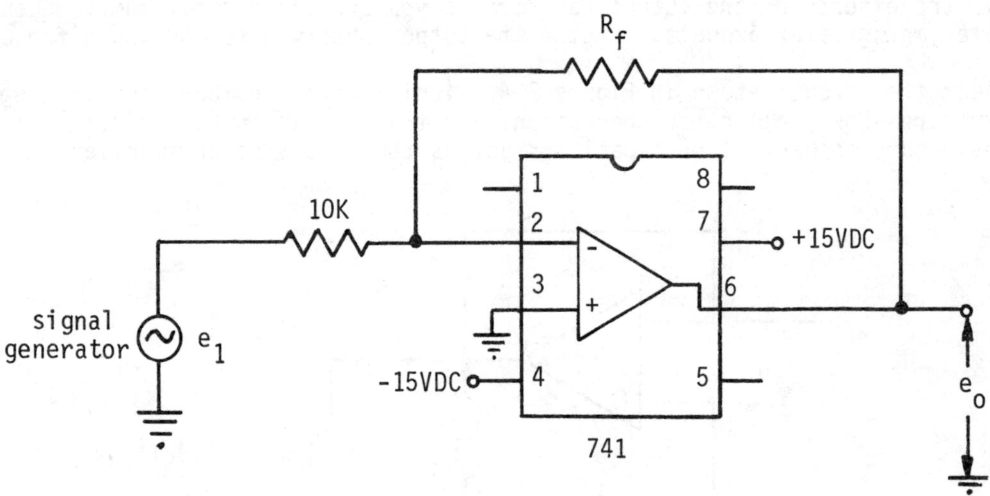

Figure 2.2

2. Connect a dual trace oscilloscope for simultaneous viewing of e_1 and e_o. With R_f = 33K, set the signal generator output to .5 V peak at 100 Hz. Measure and record the peak value of e_o. To determine the bandwidth of the amplifier, increase the frequency until the peak value of e_o is .707 times its value at 100 Hz. Record this frequency.

3. To verify that the gain-bandwidth product is constant, repeat step 2 for R_f = 100K and R_f = 220K.

4. With R_f = 10K, apply a square wave input that alternates between equal positive and negative voltages at 100 Hz. Adjust the amplitude of the square wave until the output e_o viewed on the oscilloscope just reaches + 15 V. To determine the slew rate of the amplifier, it will be necessary to measure the change in voltage ΔV which occurs in a measured time interval Δt. See Figure 2.3. Expand the horizontal sensitivity of the oscilloscope as necessary to obtain a display similar to Figure 2.3. Then measure ΔV over a convenient interval of the linear portion of the display. Also measure the time interval Δt over which the voltage change ΔV occurs, being careful to take into account any horizontal expansion factor used.

5. Calculate the slew rate $S = \Delta V/\Delta t$. Now connect a 10 V peak sinewave e_1 to the 10K input resistor in Figure 2.2 (with R_f = 10K). Increase the frequency of the sinewave

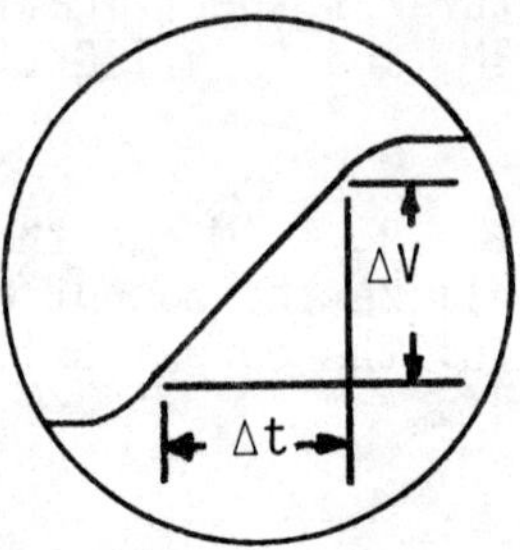

Figure 2.3

until it exceeds the maximum value specified by equation 2, $S/2\pi E_p = S/20\pi$ Hz. Observe the effects on the output waveform as you exceed this maximum frequency by greater and greater amounts. Sketch the output observed for one such frequency.

6. Connect the circuit shown in Figure 2.4. (Only the pin numbers for the inputs and output are shown; all other connections are the same as in Figure 2.2.) If oscillations occur, connect .1 μF capacitors across the $\pm$ 15 V power supplies.

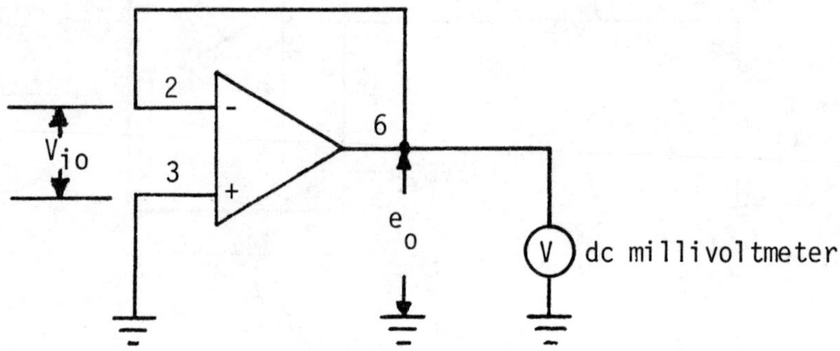

Figure 2.4

Note that Figure 2.4 is a voltage follower and therefore e_o measured by the millivoltmeter equals the input offset voltage V_{io}. A very sensitive meter will be required for this purpose, since V_{io} will be on the order of one millivolt. A dc connected oscilloscope may also be used for this measurement, if it has a sufficient vertical sensitivity and a X1 probe is used. Measure, or estimate as best as possible, and record the value of $e_o = V_{io}$.

7. Connect the circuit shown in Figure 2.5. Measure and record e_o. Note that $I^- = (V_{io} - e_o)/1M$.

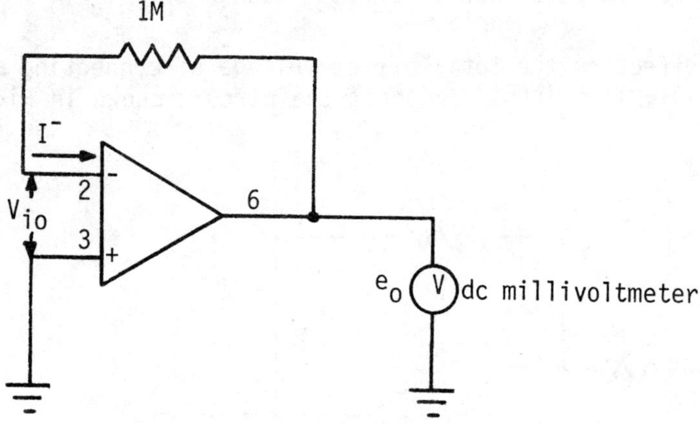

Figure 2.5

8. Connect the circuit shown in Figure 2.6. Measure and record e_o. Note that $I^+ = (e_o - V_{io})/1M$.

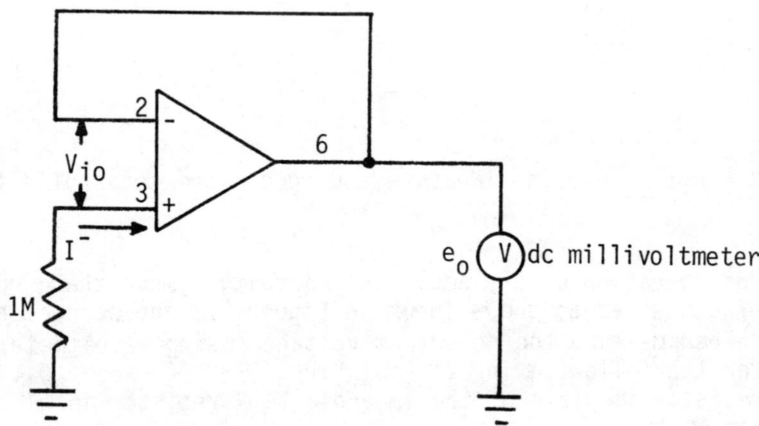

Figure 2.6

9. Connect the circuit shown in Figure 2.7.

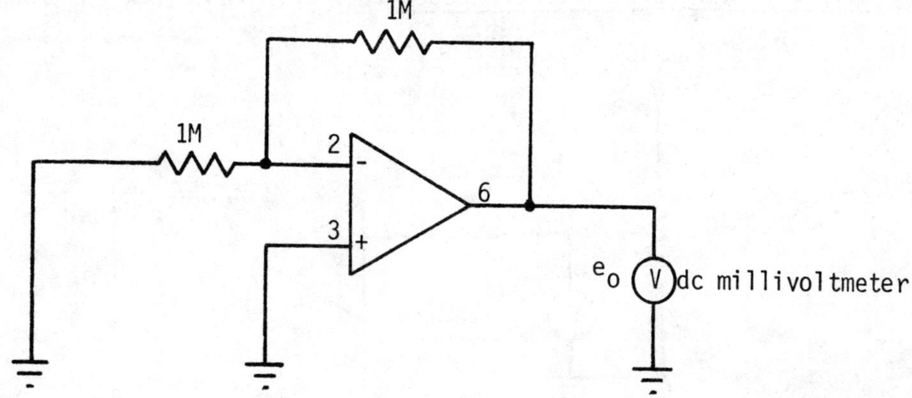

Figure 2.7

Measure and record the total output offset voltage $e_o = V_{os}$.

10. To observe the effect on the total offset voltage of connecting a compensating resistor to the non-inverting input , connect the circuit shown in Figure 2.8.

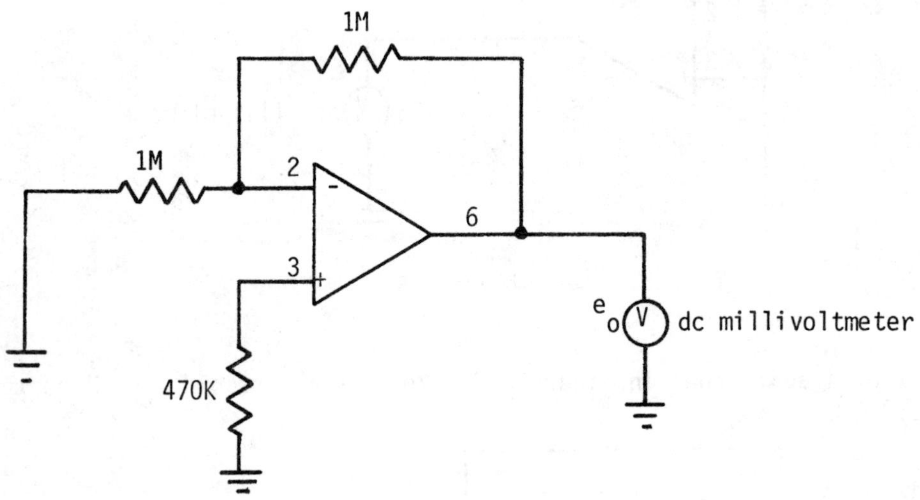

Figure 2.8

Note that 470K $\approx R_1||R_f$ = 500K. Measure and record the total offset voltage $e_o = V_{os}$.

11. To verify the dc operation of the amplifier circuit, remove the ground connection on the 1M resistor connected to the - input in Figure 2.8 and connect a dc power supply* in its place. Then measure the dc output voltage (using a voltmeter set to an appropriate scale) for the following inputs: + 1 V DC, -2 V DC, and + 5 V DC. Repeat, using a 470K input resistor in place of the 1M and a 330K resistor in place of the 470K connected to pin 3.

*If a third power supply is not available, an adjustable voltage-divider network employing a potentiometer may be connected across the existing $\pm$ 15 V supplies. Adjust the potentiometer to supply the dc voltages required in this step by measuring the voltage at the 1M input resistor with the voltage divider connected to the 1M resistor, as shown below:

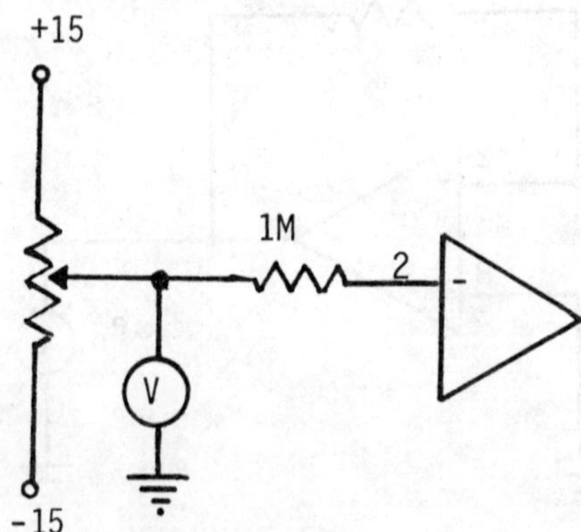

12. Connect the circuit shown in Figure 2.9. Be certain that the wiper arm of the 10K potentiometer is connected to -15 V DC and not + 15 V DC.

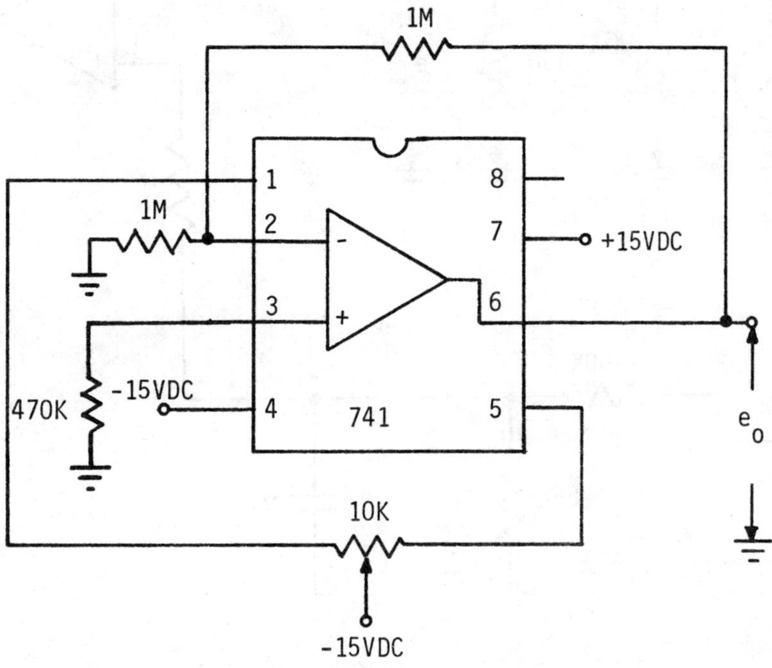

Figure 2.9

Monitor e_o on a millivoltmeter and adjust the 10K potentiometer until the output is as close as possible to zero volts. Then repeat step 11 (referring to Figure 2.9 rather than 2.8).

13. To determine the open loop voltage gain of the amplifier, connect the circuit shown in Figure 2.10.

The network connected to the + terminal is used to compensate for offset. Since the open-loop gain is very large, very small changes in the amplifier input voltages will radically affect the output voltage. Notice that the DC voltage applied from the potentiometer is divided down by a factor of 1000 in the 100 Ω - 100K divider network. Even so, the amplifier will be extremely sensitive and very small changes in the potentiometer setting may drive the amplifier output to plus or minus 15 V. Therefore, the potentiometer must be adjusted very slowly and very carefully. If possible, use a precision multi-turn potentiometer. (The total potentiometer resistance is not critical.)

Similarly, the signal generator voltage e_s is divided down by a factor of 1000 before its application to the - input. Note that

$$e_1 = \left(\frac{100}{100K + 100}\right)e_s \simeq e_s/1000$$

Consequently, the ac open loop gain is

$$G = \frac{e_o}{e_1} = 1000\,\frac{e_o}{e_s}$$

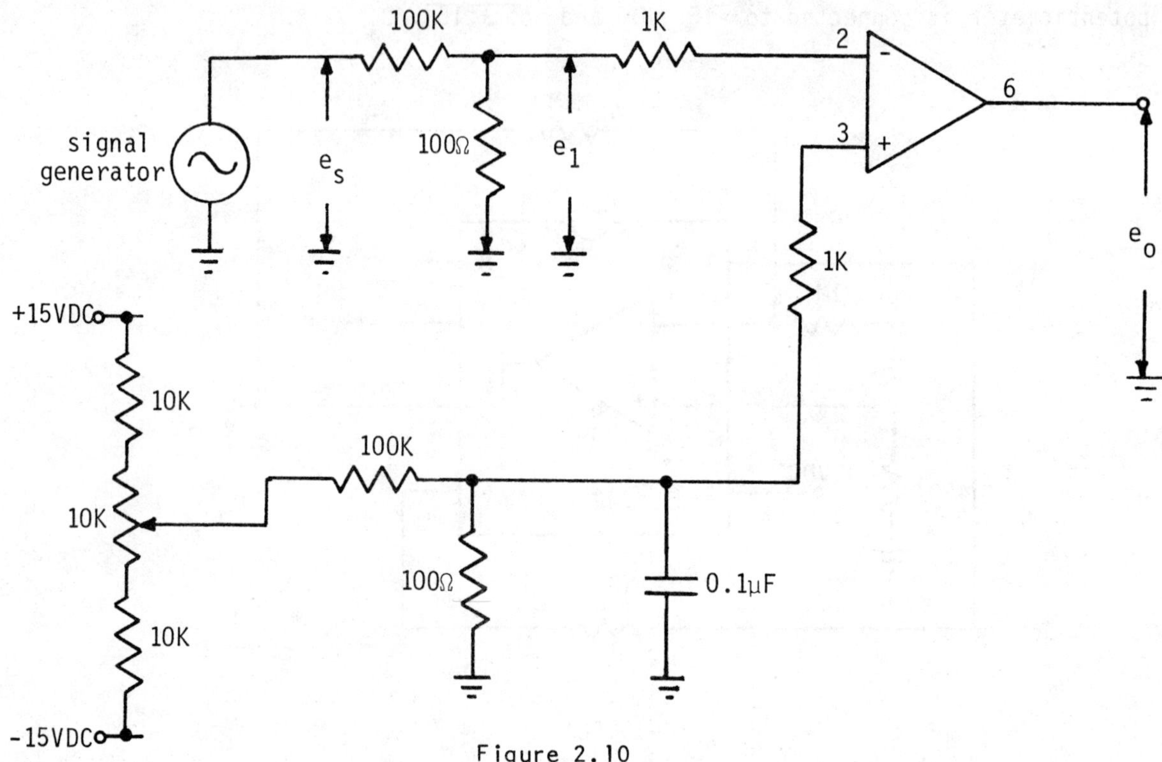

Figure 2.10

Connect an oscilloscope for simultaneous viewing of e_s and e_o. With e_s grounded, carefully adjust the potentiometer until the amplifier's dc output is as close to zero as possible. (Small fluctuations in output level may occur and are acceptable; the purpose of the 0.1μF capacitor is to dampen these out as much as possible.) Set the signal generator output e_s to approximately 50 mV peak at 1 Hz. Increase or decrease the peak value of e_s as necessary to obtain an amplifier output of about 10 V peak. Then record the peak values of e_s and e_o.

QUESTIONS

1. Using the data obtained in steps 2 and 3 of the Procedure, compute the gain-bandwidth product for each of the values of R_f used. Is this product (essentially) constant? Should it be?

2. Using the gain-bandwidth product obtained in question 1, estimate the bandwidth of the circuit in Figure 2.2 when R_1 = 22K and R_f = 150K.

3. Using the data obtained in step 4 of the Procedure, show your calculation of the slew rate S. How does this result compare with the manufacturer's specified value of slew rate?

4. What is the maximum frequency at which the amplifier can be driven when the input signal has a peak value of 10 V, based upon the slew rate you determined in question 3? Describe the appearance (include a sketch) of the amplifier output when the input frequency exceeded this value.

5. Compare the value of input offset voltage measured in step 6 of the Procedure with

the manufacturer's specification.

6. Based upon your measurement in step 7 of the Procedure, what is the value of I^-? Compare this value with the manufacturer's specification for the input bias current I_B. Should you expect these two values to be exactly equal?

7. Repeat question 6 for I^+ and step 8 of the Procedure.

8. Compute the average value of I^+ and I^-, using the values of I^+ and I^- determined in questions 6 and 7. Should this value equal I_B?

9. Compute the input offset current I_{io} using the values of I^+ and I^- determined in questions 6 and 7. How does this value compare with the manufacturer's specification?

10. Calculate the theoretical output offset voltage V_{os} for the circuit of Figure 2.7, using input offset values determined in the preceding questions. How does this value compare with the value of V_{os} measured in step 9 of the Procedure?

11. Repeat question 10 for Procedure step 10.

12. Compare the DC output voltages measured in step 11 of the Procedure with their theoretical values. How would these output voltages change if offset compensation had not been incorporated in the circuit?

13. Repeat question 12 for Procedure step 12.

14. Using the data obtained in Procedure step 13, calculate the open-loop gain of the amplifier. Show your calculations. Find the open-loop gain in dB and compare this value with the manufacturer's specification.

15. Using the manufacturer's specified value for the CMRR of the amplifier and the value of open-loop gain determined in question 14, calculate the common mode gain A_C. Also express A_C in dB.

16. In this experiment, you used three different techniques to compensate for offset. List relative advantages and disadvantages of each, based on your experience in constructing and using them.

3 Voltage Regulators

OBJECTIVES

1. To gain familiarity, through experimental observation, with the capabilities and applications of monolithic voltage regulators.
2. To learn how to connect a fixed (three terminal) voltage regulator in a circuit for proper operation.
3. To learn how an adjustable regulator may be connected with external components for the purpose of adjusting output voltage and setting current limits.
4. To learn how to measure and interpret percent load regulation and percent line regulation.
5. To learn through observation of load voltage and load current the difference between the current sensing and current foldback methods of overload protection.
6. To verify experimentally the equations and specifications given by device manufacturers for certain regulator parameters.
7. To gain experience designing an adjustable voltage regulator circuit to meet certain specifications.

EQUIPMENT AND MATERIALS REQUIRED

1. Adjustable dc power supply, 0-18 V DC.
2. dc voltmeter, 0-10 V, 0-20 V.
3. dc ammeter(s), 0-10mA, 0-20 mA, 0-100 mA, 0-500 mA, 0-1A (or similar).
4. 7805 voltage regulator.
5. 723 voltage regulator.
6. Resistors: 15 Ω 2W, 22 Ω 1W (2), 33 Ω ½W, 47 Ω 1W, 100 Ω ½W, 220 Ω ½W, 680 Ω, 1K, 2.2K, 2.7K, 5.6K.
7. Potentiometers: 1K, 10K.
8. Capacitors: 100 pF, .001 μF.

DISCUSSION

A voltage regulator is used to maintain the output voltage from a dc voltage source at an (ideally) constant value, independent of the amount of load current drawn from it. Without voltage regulation, the output of a voltage source decreases as the current drawn from it increases, due to the increasing voltage drop across its internal resistance. The voltage regulator monitors the output voltage and through a feedback process automatically adjusts some parameter in the source to compensate for any tendency of the output to change.

One method that is widely used in integrated circuit regulators to maintain a constant output is to compare the output with an internally generated reference voltage, using a high gain operational amplifier. The output of the amplifier is proportional to the difference between the output and reference voltages and is used to control the bias on a transistor, which in turn directly affects the output voltage level. Whether this transistor, called the pass transistor, is connected in series with the load (a series regulator) or in parallel with it (a shunt regulator), its bias is changed in such a way that it will boost the output voltage whenever the output drops below the reference level and reduce the output when it rises above the reference. The unregulated output voltage of a DC source is called the input voltage to the regulator, while the constant voltage produced by the regulator is called its output voltage. In order for the regulator to operate properly, its input voltage must be greater than its regulated output by an amount (several volts) that is usually given in the manufacturer's specifications.

Voltage regulators may be classified as being of the fixed or of the adjustable type. As these names imply, the fixed regulator provides the single value of output voltage for which it is designed, while the output level of the adjustable type may be changed by the user over some limited range, using external components connected to the regulator. In the adjustable type, the internally generated reference voltage is brought out through one of the device's terminals, and this reference can then be modified externally, for example, by the use of a potentiometer. The modified value of reference voltage is then fed back into the regulator for comparison with the output level, resulting in the output being maintained at a different level.

A practical voltage regulator cannot maintain an absolutely constant output voltage for all changes that may occur in its input voltage nor for all values of load current drawn from it. A measure of the voltage regulator's ability to maintain a constant output is called its percent voltage regulation. To refine this measure we define percent load regulation and percent line regulation. Load regulation refers to the extent to which the regulator is able to maintain a constant output when changes in load occur, and is defined by

$$\%\text{ load regulation} = \frac{V_{NL}-V_{FL}}{V_{NL}} \times 100\% \qquad (1)$$

where V_{NL} = no-load voltage (output voltage with zero load current)

V_{FL} = full-load voltage (output voltage when maximum rated load current is supplied)

Note that the ideal regulator has zero percent regulation since, ideally, $V_{NL} = V_{FL}$, making the numerator of (1) equal to zero. In practice, manufacturers generally specify percent regulation over some limited range of load current, not necessarily between the extremes of no-load and full-load. Also, the specification may be given as percent of output voltage. This method of specification is essentially the same as equation (1).

Another measure of the regulator's ability to maintain a constant output during load current changes is its output impedance, or output resistance, R_o By definition,

$$R_o = \frac{\Delta V_L}{\Delta I_L}\ \Omega \qquad (2)$$

where ΔV_L is the change in voltage that occurs for a given change in load current, as for example from I_{NL} to I_{FL}. Again, for the ideal regulator R_O is zero Ω, since ΔV_L should ideally be zero. We may think of R_O as the Thevenin equivalent resistance of the regulator, and since this resistance is in series with the output, greater load currents cause greater internal voltage drops across it, resulting in lower output voltages.

Line regulation (also called input regulation) refers to the ability of the regulator to maintain a constant output when changes in its (unregulated) input voltage occur. Manufacturers generally specify line regulation as the percent change in output voltage that occurs when the input voltage changes over some stated range.

A good voltage regulator incorporates some means of load current limiting and/or short-circuit protection. Current limiting refers to the ability of the regulator to prevent the load current from exceeding a certain preset value, regardless of how small a resistance is connected across its terminals. Current limiting may be used for protection of the regulator, protection of the load, or both. In some versions, the current limit can be set by the user through selection of the size of an external resistor, called the current sense resistor, R_{SC}. The voltage across this resistor increases as the load current increases, and when that voltage reaches a certain limit, it changes the bias of the pass transistor in such a way that no additional load current can flow. The normal feedback control function of the regulator is thus interrupted to restrict current flow. In other versions, a fixed sense resistor is located inside the integrated circuit chip and may be used to sense a short-circuit condition.

Two methods of current limiting used in voltage regulators are called current sensing and current foldback. The current sensing technique maintains load current at a preset value when an overload condition occurs, while dropping the output voltage to near zero. The current foldback technique reduces both output voltage and load current when an overload condition occurs.

One type of fixed voltage regulator is known as a three terminal device. This integrated circuit requires only three connections: input (unregulated) voltage, output (regulated) voltage, and ground. Because of their simplicity and small size, these devices are widely used to provide local regulation, that is, to supply a regulated voltage to one part of a larger electronic system, for example, to one printed circuit board. Local regulation results in cost savings because the percent regulation required in one portion of a system can be provided to that portion only, rather than to the whole system.

A popular example of three terminal devices is the 7800 and 7900 series regulators, supplied by a number of different manufacturers. These devices are available with fixed output voltages ranging from 5 to 24 volts, both positive and negative. The 7800 series produce positive output voltages, whose values are specified by the last two digits of the number. The 7805, for example, is a + 5 V regulator. The 7900 series produce negative voltages, with the last two digits again indicating the (negative) output voltage value.

An example of the adjustable voltage regulator is the versatile 723 device. This is available in a 10-lead TO5 or a 14 pin DIP case. Through external connections, the user may set the value of regulated voltage between 2 and 37 volts, may select a current limit, and may choose the current sensing or current foldback technique for overload protection. Figure 3.1 shows a block diagram of the device. Note in Figure 3.1 the operational amplifier (labeled "error amplifier") that is used to detect differences between reference level and output voltage. The inverting and non-inverting inputs are accessible to the user via external terminals, so a variety of connections can be made, to produce a corresponding variety of controlled outputs, as discussed previously. Note also that the internally generated reference voltage V_{REF}, typically about 7 V, is brought out to an external terminal. The current limit (CL) and current sense (CS) terminals are used for external connections to components that determine the type and value of current limiting. Note that these terminals are the base and emitter terminals of a transistor whose

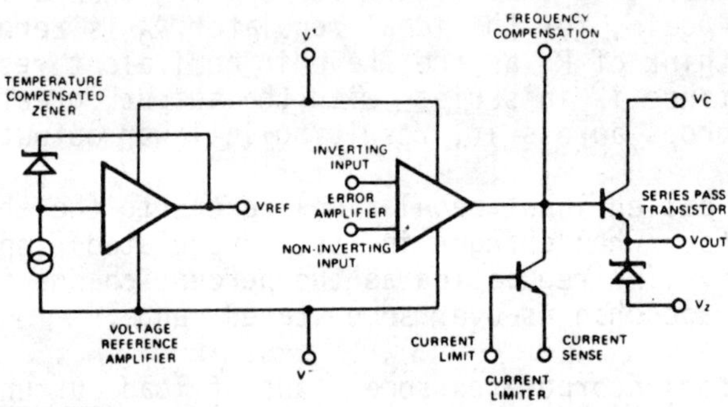

Figure 3.1 Block diagram of the 723 voltage regulator.

collector is connected to the base of the pass transistor. Thus the voltage across these base-emitter terminals controls the bias of the pass transistor. The output voltage V_O is taken from the emitter of the pass transistor, which acts as an emitter follower, so this device is of the series regulator type. The frequency compensation terminal is used for connecting external capacitance to supress oscillations in this high gain circuit.

Figure 3.2 shows one way the 723 will be connected in this experiment for tests of its performance.

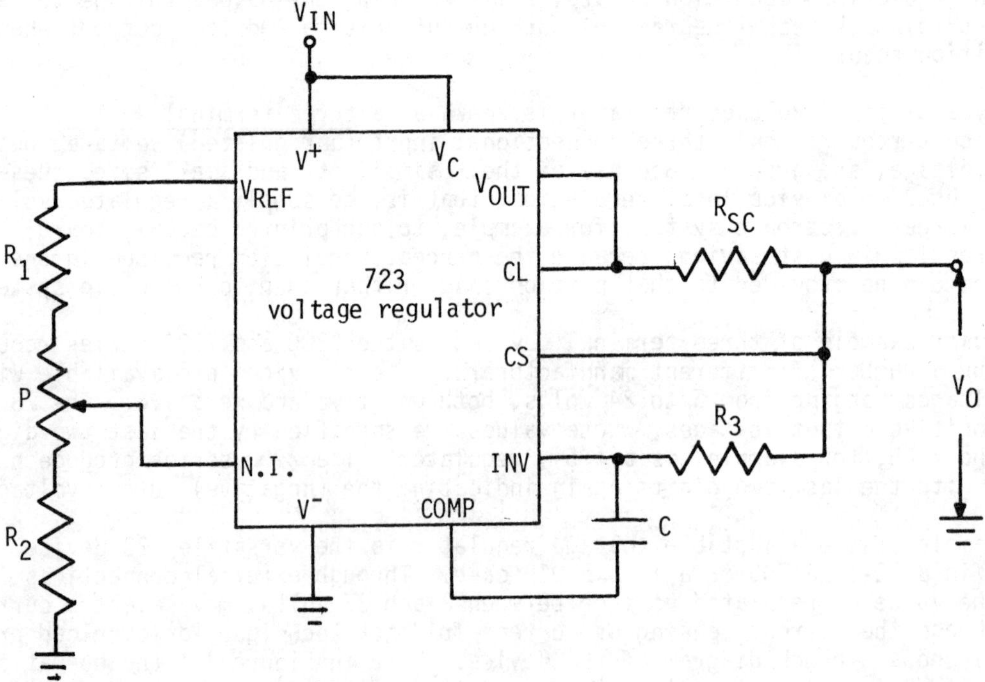

Figure 3.2 The 723 voltage regulator connected to maintain an output voltage that is determined by the voltage at N.I.

Note in Figure 3.2 that the reference voltage V_{REF} is divided down by the R_1, potentiometer P, and R_2 combination and fed into the non-inverting (N.I.) input of the error amplifier. Since the inverting input (INV) is connected through R_3 to the output, the voltage divider determines the value of regulated output voltage. In the absence of the potentiometer P,

$$V_{OUT} = \frac{R_2}{R_1 + R_2} V_{REF} \tag{3}$$

The potentiometer is used to make small adjustments in the value of the regulated output voltage, V_{OUT}.

Note also in Figure 3.2 that load current flows through R_{SC} and that R_{SC} is connected directly across the CL and CS terminals. Therefore, the voltage across the base-emitter junction of the transistor controlling the pass transistor (see Figure 3.1) depends directly on the load current. The current limit, I_{LIMIT}, is determined by

$$I_{LIMIT} = \frac{V_{SC}}{R_{SC}} \tag{4}$$

where V_{SC} is the voltage across R_{SC} when current limiting occurs. V_{SC} is typically about 0.6 V, the forward-biasing base-to-emitter voltage required to turn on the transistor it controls, and the voltage that is therefore necessary to reduce conduction of the pass transistor.

We will also investigate foldback current limiting using the 723 regulator. In this case the CL input is taken from a voltage divider, as shown in Figure 3.3. (Other connections are the same as in Figure 3.2.)

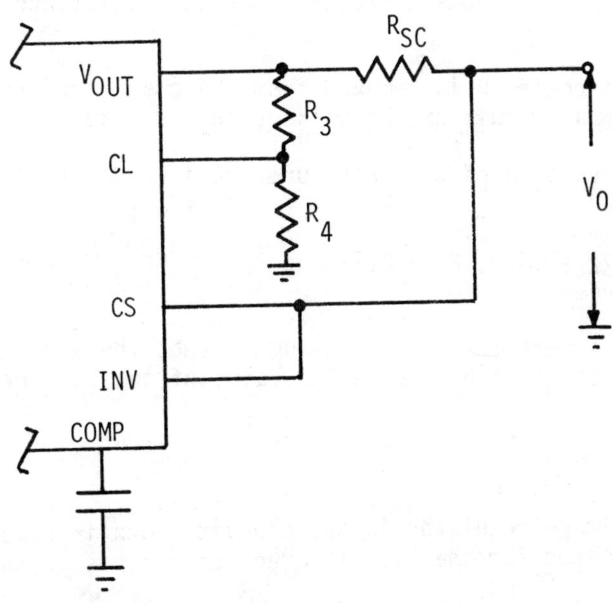

Figure 3.3 The 723 voltage regulator connected for foldback current limiting.

A detailed description of the theory of operation of this current limiting technique may be found in Young, Linear Integrated Circuits (Wiley, 1981), pp. 120-121. According to manufacturer's specification sheets,

$$I_{SHORT\ CIRCUIT} = \frac{V_{SC}}{R_{SC}} \frac{(R_3 + R_4)}{R_4} \tag{5}$$

The knee current, where current reduction due to foldback first begins to occur, is

$$I_{KNEE} = \frac{V_{OUT}R_3}{R_{SC}R_4} + \frac{V_{SC}(R_3 + R_4)}{R_{SC}R_4} \tag{6}$$

The manufacturer's specification sheets for the 723 regulator (see the Appendix) contain a large number of addional ways the regulator may be used for a variety of applications, including negative voltage outputs, floating outputs, switching regulation, and the use of external pass transistors.

EXERCISES

1. The open-circuit output voltage of a certain voltage regulator is 5.12 V DC. When the regulator supplies its rated full-load current of 500 mA, its output voltage is 4.97 V DC. What is its percent voltage regulation? What is its output resistance?

2. A 723 voltage regulator is connected as shown in Figure 3.2. V_{REF} = 7 V, R_1 = 680 Ω, R_2 = 2.2K, and the potentiometer has a total resistance of 1K. If the potentiometer wiper arm is set exactly in the middle, what is the output voltage V_o of the regulator? What are the theoretical minimum and maximum values of V_o that can be obtained by rotating the potentiometer from one extreme to the other?

3. Repeat exercise 2 if the potentiometer has total resistance 5K, and all other values are the same.

4. Assume the potentiometer P is removed from the circuit of Figure 3.2. If R_1 = 680 Ω and V_{REF} = 7 V, what should be the value of R_2 to make V_o = 6.3 V?

5. What should be the value of R_{SC} in Figure 3.2 in order to limit the output current to 18 mA?

6. In Figure 3.3, R_{SC} = 33 Ω, R_3 = 2.7K and R_4 = 5.6K. What current flows when the output (V_o) is grounded?

7. Using the values in exercise 6, and assuming that the input to the "N.I." terminal of the 723 regulator is 4.5 V, at which value of output current does foldback limiting begin?

PROCEDURE

1. Connect a 7805 voltage regulator in the circuit shown in Figure 3.4. Note how the terminals are numbered for the two different case types in which the 7805 is packaged.

2. With the power supply voltage V_{in} set to + 10 V DC, measure and record the load current I_L and load voltage V_L when R_L has each of the following values: 220 Ω (¼W); 100 Ω (½W); 47 Ω (1W); 22 Ω (1W); 15 Ω (2W); 22 Ω||22 Ω = 11 Ω (each resistor 2W). Wattage ratings are the minimum that should be used.

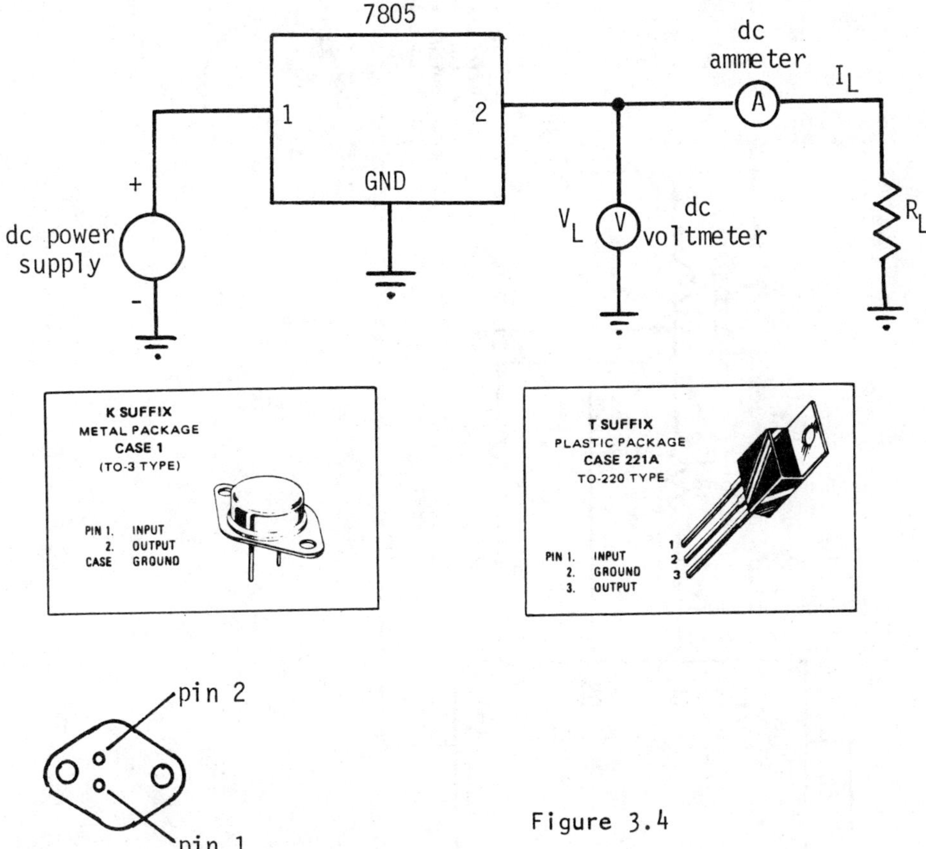

Figure 3.4

3. To determine line regulation, with the two 22 Ω parallel resistors still connected, measure and record V_L for each of the following values of V_{in}: + 7 V DC, + 12 V DC, + 18 V DC. (Caution: do not leave the higher voltages connected for longer than is required to make a measurement; watch for excessive heating.)

4. With V_{in} set to + 10 V DC, momentarily short the output terminal of the regulator to ground with a piece of heavy wire. Observe the short circuit current that flows. After removing the short circuit, verify that the regulator still operates properly by restoring R_L to 100 Ω (½W) and 15 Ω (2W) and observing the output voltage.

5. Connect the 723 regulator in the circuit shown in Figure 3.5. **Note how the terminals** are numbered for the two different case types. Note also that V_Z is not available in the 10-lead metal can. Ground the V_Z terminal on the 14-pin case.

6. With the power supply voltage V_{in} set to + 10 V DC, measure and record V_{REF} with respect to ground. With load resistance R_L removed from the circuit (output of the regulator open), measure the minimum and maximum output voltages attainable by rotating the 1K potentiometer through its full range.

7. Now adjust the 1K potentiometer until V_O is + 5 V DC. Measure and record the voltage between the wiper arm of the 1K potentiometer and ground.

8. Connect R_L (the 10K potentiometer) and adjust it until a load current of I_L = 1 mA flows. Record the value of V_L. Repeat for the following values of load current:

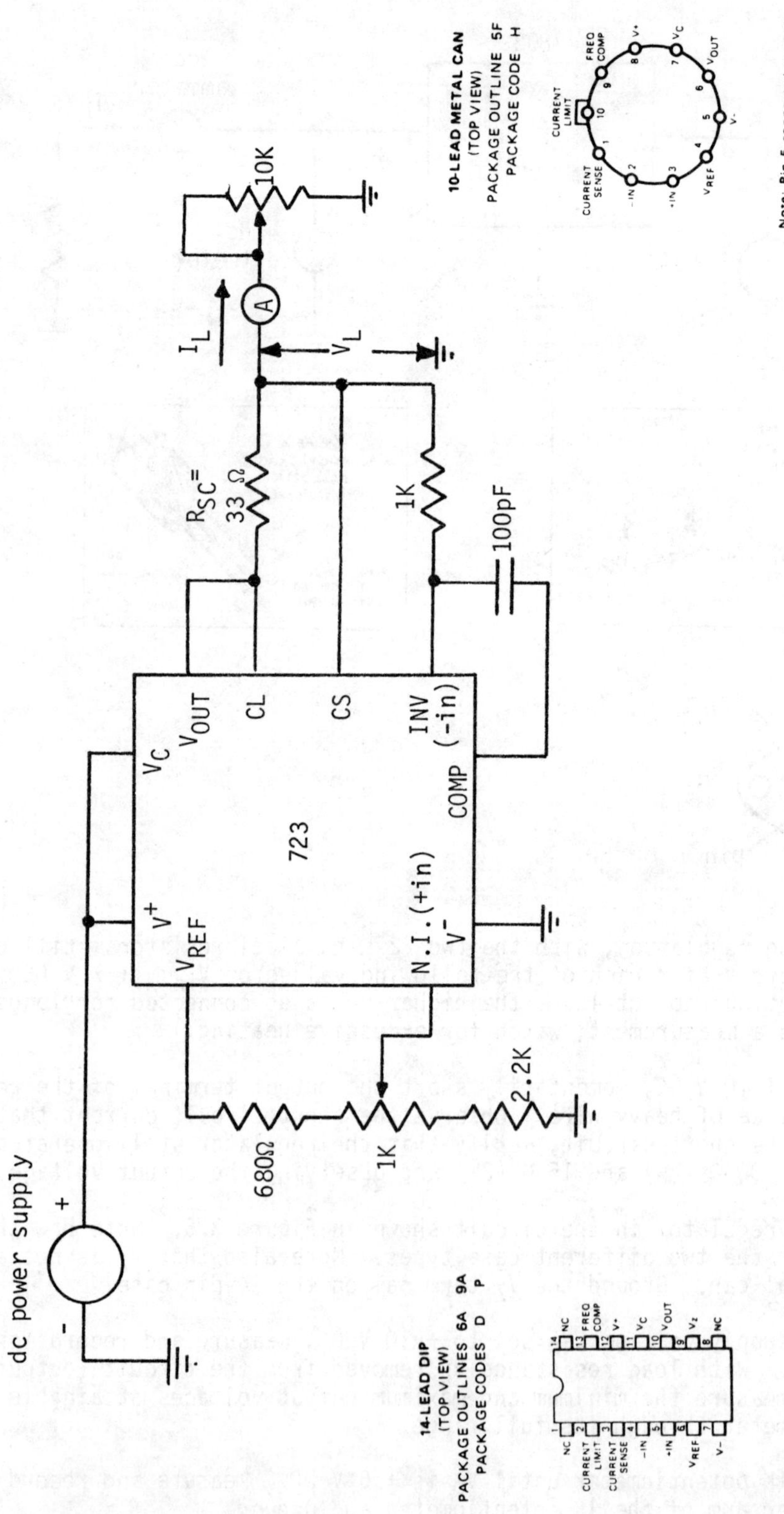

dc power supply
+
-
723
V^+
V_{REF}
V_C
V_{OUT}
CL
CS
INV
(-in)
COMP
N.I.(+in)
V^-
680Ω
1K
2.2K
R_{SC}=
33 Ω
1K
100pF
I_L
V_L
A
10K
14-LEAD DIP
(TOP VIEW)
PACKAGE OUTLINES 6A 9A
PACKAGE CODES D P
10-LEAD METAL CAN
(TOP VIEW)
PACKAGE OUTLINE 5F
PACKAGE CODE H
Note: Pin 5 connected to case.

Figure 3.5

5 mA, 10 mA, 12 mA, 15 mA, and 18 mA. For the higher load currents you may replace the 10K potentiometer with a 1K potentiometer to improve your ability to set the currents accurately.

9. Gradually increase the load current beyond 18 mA. You should observe that the load voltage suddently decreases somewhere in the vicinity of 18 to 20 mA of load current, when the voltage across the sense resistor R_{SC} becomes great enough to begin current limiting. Measure and record a few values of load current and load voltage in this vicinity, both below and above the current limiting point.

10. Replace R_L with a short-circuit and measure the load current that flows. Also measure the voltage drop across R_{SC} in this condition.

11. Replace R_{SC} with a 22 Ω resistor, and repeat steps 9 and 10, except take your measurements in 5 mA increments of load current until you reach the vicinity of the current limiting point.

12. Replace R_{SC} with a short circuit (R_{SC} = 0). With V_{in} = 10 V, measure and record I_L and V_L in 5 mA increments of I_L: 5 mA, 10 mA, ..., up to I_{LMAX}, where I_{LMAX} is 5 mA greater than the short-circuit current measured in step 11. That is, $I_{LMAX} = I_{SC}$ (R_{SC} = 22 Ω) + 5 mA. CAUTION: do NOT short circuit the output of the regulator; make very fine and careful adjustments in R_L as you approach I_{LMAX}. You may want to connect a fixed 100 Ω resistor in series with R_L to guard against accidentally shorting the output.

13. With R_{SC} still equal to zero, adjust R_L for a load current I_L of 1 mA. To determine the line regulation of the 723, measure and record V_L for each of the following values of V_{in}: 10 V, 15 V, ..., up to 35 V in 5 V increments.

14. Set V_{in} to + 6 V and measure and record V_L. Also measure and record the voltage with respect to ground at the wiper arm of the 1K potentiometer connected to V_{REF}.

15. Connect the current foldback circuit shown in Figure 3.6. Refer to Figure 3.5 for pin numbers.

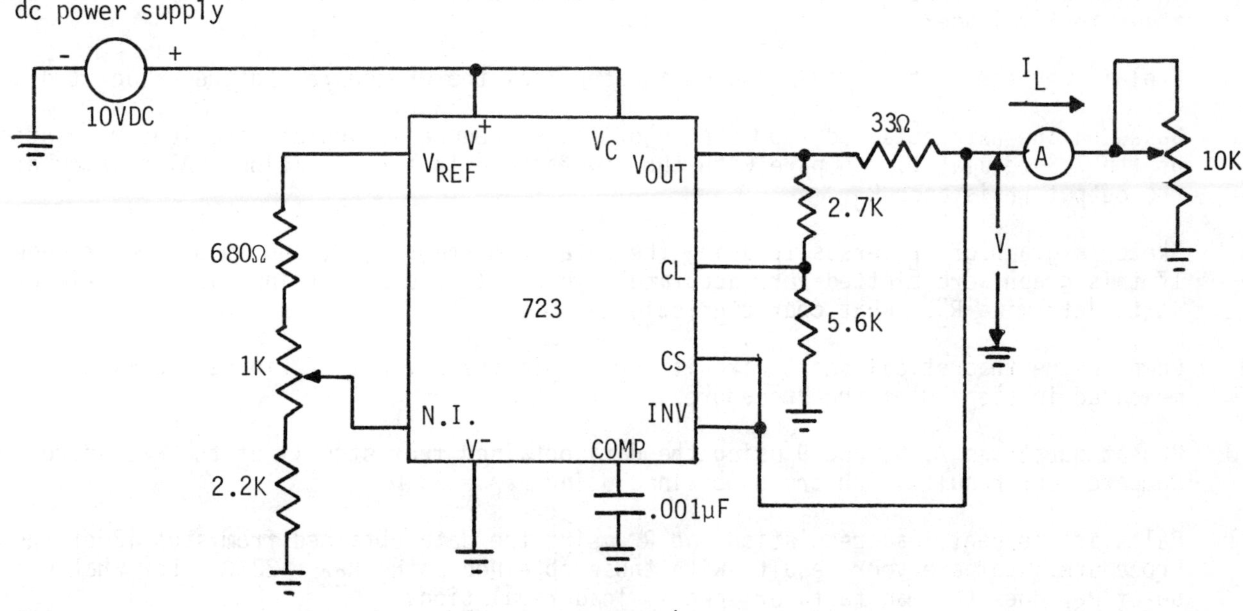

Figure 3.6

By adjusting the load potentiometer R_L, gradually increase the load current I_L until you observe a sudden reduction in V_L and I_L. The point where this rapid drop first begins is the knee of current foldback. Carefully adjust R_L to obtain and record I_L and V_L values in the vicinity of the knee, both before, after, and at the onset of foldback. Measure and record V_{SC} across R_{SC} at the knee.

16. Replace R_L with a short-circuit and record the short-circuit load current. Also measure and record V_{SC} across R_{SC} in this condition.

17. Using the 723 regulator, design a voltage regulator circuit to produce an output of + 4.3 V $\pm$ 10%. Do not use a potentiometer to divide V_{REF}. Use only standard resistor values. NOTE: The total current drain on V_{REF} should not exceed 10 mA. Use the current sensing method for overload protection. Short circuit current should be 20 mA $\pm$ 10%. Record all component values used in your design. Then connect your circuit and perform tests to verify its performance. Record all test results so they may be compared with the requirements given.

QUESTIONS

1. Calculate the percent load regulation of the 7805 regulator using the data obtained in step 2 of the Procedure. Compare your calculated value with the manufacturer's specification.

2. Calculate the output resistance R_o of the 7805, based on the data from step 2 of the Procedure. Compare your value with the manufacturer's specification.

3. Calculate the line regulation (percent change in output voltage) using the data obtained in step 3 of the Procedure. Compare your calculated value with the manufacturer's specification.

4. What value of short-circuit current did you measure in step 4 of the Procedure? Did the regulator operate properly after the short circuit was removed? Do you believe the 7805 incorporates internal short-circuit current limiting?

5. What range of output voltages were you able to set by rotating the 1K potentiometer in Figure 3.5? What is the theoretical range through which you should be able to set the output voltage?

6. Explain why the voltage you measured in step 7 of the Procedure had the value it did.

7. Based on the data obtained from step 8 of the Procedure, calculate the load regulation of the 723 regulator. Compare with the manufacturer's specification. Also calculate the output resistance R_o.

8. Sketch a graph of V_L versus I_L using the data from steps 8, 9, and 10 of the Procedure. If this graph were plotted very accurately you could use a certain characteristic of it to determine R_o. What characteristic?

9. Compare the theoretical short-circuit current in the circuit of Figure 3.5 with that measured in step 10 of the Procedure.

10. Repeat questions 7, 8, and 9 using the data obtained from step 11 of the Procedure. Compare your results with those obtained using R_{SC} = 33 Ω.

11. Calculate percent load regulation and R_o using the data obtained from step 12 of the Procedure. Compare your results with those obtained using R_{SC} = 22 Ω. For what value of R_{SC} does the manufacturer specify load regulation?

12. Calculate percent line regulation using the data obtained in step 13 of the Procedure and compare with the manufacturer's specification.

13. Why did the 723 regulator fail to hold the output at or near 5 V in step 14 of the Procedure? Justify your answer in terms of the other voltage measured in that step and by reference to manufacturer's specifications.

14. Sketch a graph of V_L versus I_L in the vicinity of the knee of current foldback, using the data obtained in steps 15 and 16 of the Procedure. What differences do you notice between the current limiting behavior of the 723 using current foldback and using current sensing? Base your comparison on previous results obtained with R_{SC} = 33 Ω.

15. Calculate the theoretical values of I_{KNEE} and $I_{SHORT\ CIRCUIT}$ in current foldback (based on measured values of V_{SC}) and compare with the values you observed in steps 15 and 16 of the Procedure.

16. Draw a schematic diagram of the 723 voltage regulator you designed in step 17 of the Procedure. Show calculations you used to determine component values. List performance data based on the tests of your design and describe how well your circuit met the specified design criteria.

4 Audio Amplifiers

OBJECTIVES

1. To learn how to connect and use a monolithic audio amplifier in a circuit.

2. To investigate the effect of coupling capacitor sizes on the lower cutoff frequency of an amplifier, and to verify experimentally the cutoff frequency predicted by theory.

3. To learn how to use log-log and semilog graph paper to plot a frequency response characteristic.

4. To verify experimentally the rates at which amplifier gains fall off in dB per decade and dB per octave beyond an amplifier's cutoff frequency.

5. To hear and see the effect of clipping on an audible tone when an amplifier is overdriven.

6. To measure stereo channel separation and investigate the effects of power supply impedance on separation.

EQUIPMENT AND MATERIALS REQUIRED

1. Dual trace oscilloscope.

2. 15 V DC power supply.

3. Sine wave signal generator, 0-100 kHz, 0-7 V RMS.

4. Speakers (2).

5. 377 dual audio amplifier. (A 378 may be used in place of the 377.)

6. Resistors: 10 Ω 2W, 10K (2), 1M (2).

7. Capacitors: .01 μF, 0.1 μF (2), 1.0 μF (2), 10 μF, 100 μF (3), 500 μF.

DISCUSSION

An audio amplifier is an ac amplifier designed to amplify signals whose frequencies fall in the audible range: from about 20 Hz to 20 kHz. Although an integrated circuit operational amplifier may be the basis for the design of such an amplifier, its characteristics are different to the extent that the requirements of the application are different. For example, the typical audio amplifier is capable of delivering several watts of power to a low impedance load (such as a speaker). Furthermore, the presence of a dc output offset voltage is not a problem in the design, since the load is usually capacitor coupled to the amplifier. Indeed, the typical integrated circuit audio amplifier requires but a single power supply voltage, and the dc output level is one-half that supply voltage. Thus the ac output signal is superimposed on a significant dc level which must be blocked from the load by a coupling capacitor. The input to the amplifier is also capacitor coupled from a signal source or preamplifier.

The range of frequencies over which the gain of the amplifier is essentially constant (independent of frequency) is called the midband range. As the signal frequency is progressively decreased, we find that we eventually reach a frequency where the gain begins to diminish, or "fall off," as evidenced by the output level decreasing rapidly with decreasing frequency. The frequency at which the output level is .707 times its midband value, that is, the frequency at which the amplifier gain is 70.7% of its midband value, is called the lower cutoff frequency. Since 20 $\log_{10}(.707) = -3$ dB, this frequency is often called the (lower) 3 dB frequency. (it is also called the lower break frequency and the half-power frequency.) Similarly, if the frequency is progressively increased to a point beyond its midband range, we find once again that gain falls off. The upper cutoff frequency is the frequency where the gain has once again fallen to .707 times its midband value. Figure 4.1 is a sketch showing the output voltage of a typical ac amplifier versus frequency. Note in this example that the output level is 10 V in the midband range and .707 x 10 = 7.07 V at the cutoff frequencies.

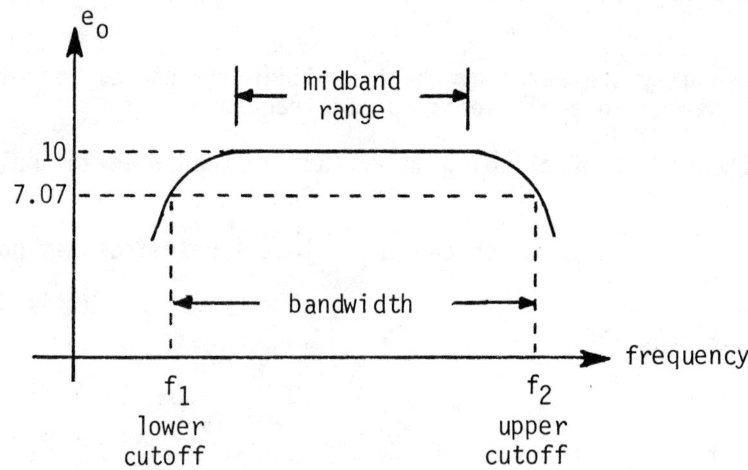

Figure 4.1 Output voltage of a typical ac amplifier versus frequency.

The bandwidth of the amplifier is by definition

$$BW = f_2 - f_1 \tag{1}$$

where f_1 = the lower cutoff frequency,
and f_2 = the upper cutoff frequency. (See Figure 4.1.)
Figure 4.1 is an example of a frequency response characteristic. This response is said to be "flat" in the midband region.

The input and output coupling capacitors have a significant effect on the value of the lower cutoff frequency f_1. The larger these capacitors, the lower the value of f_1. Figure 4.2 shows the equivalent input and output circuits of a capacitor-coupled amplifier.

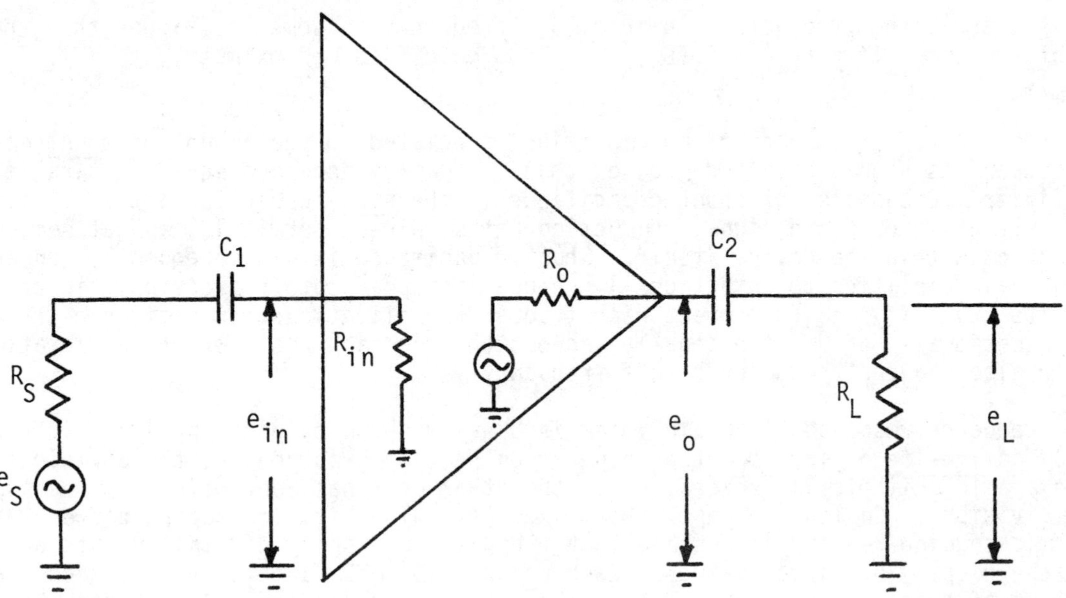

Figure 4.2 The equivalent circuit of a capacitor-coupled AC amplifier.

Since C_1 is in series with the amplifier, the ac voltage drop across it constitutes a loss of signal, in the sense that a portion of the source voltage e_S does not reach the amplifier input. In other words, R_{in} and C form a voltage divider, and the greater the drop across C the less signal voltage appears at the input of the amplifier, resulting in a reduction in the overall gain from source to load. The reactance X_C of the capacitance is inversely proportional to frequency, $X_C = 1/(2\pi fC)$ ohms, so the lower the frequency f, the greater the voltage drop across C. The lower cutoff frequency $f_1(C_1)$ due to capacitor C_1 occurs at that frequency where X_{C_1} equals $R_{in} + R_S$, namely,

$$f_1(C_1) = \frac{1}{2\pi(R_{in} + R_S)C_1} \text{ Hz} \tag{2}$$

Similarly, C_2 is in series with the load R_L and the output resistance of the amplifier R_o, so the lower cutoff frequency due to capacitor C_2 is

$$f_1(C_2) = \frac{1}{2\pi(R_L + R_o)C_2} \text{ Hz} \tag{3}$$

If $f_1(C_1)$ is neither equal nor close to $f_1(C_2)$, then the actual lower cutoff frequency is the larger of the two. If $f_1(C_1)$ and $f_1(C_2)$ are reasonably close in value (less than an octave apart), then the actual lower cutoff frequency is somewhat higher than the larger of the two. If $f_1(C_1) = f_1(C_2)$, then the cutoff is approximately $f_1 \simeq 1.5f_1(C_1) = 1.5f_1(C_2)$.

The upper cutoff frequency of an amplifier is related to the amount of shunt capacitance between its signal path and ground, that is, capacitance connected in parallel with the amplifier. One source of shunt capacitance is the stray capacitance that always exists between external ground and signal conducting paths, wires, terminals, and at semiconductor junctions within the device itself. Shunt capacitance is also frequently connected to a high gain amplifier to intentionally reduce its upper cutoff frequency, in order to improve its stability and to prevent high frequency oscillations. Capacitance that appears, intentionally or unintentionally, across the feedback resistor of an operational amplifier also reduces the upper cutoff frequency.

The frequency response of an amplifier is usually shown by plotting its gain versus frequency on log-log or semilog graph paper. On semilog graph paper, the divisions along one axis are logarithmically spaced, while the other axis has conventional linear spacing between divisions. On log-log paper, both axes have logarithmic spacing between divisions. Logarithmic spacing results in a scale that expands the display of small values and compresses the display of larger values. Each decade (10 to 1 range of values) occupies the same amount of space on the graph. Study the graph paper in Figure 4.3. Note that all decades have printed scale values ranging from 1 to 10, but that actual scale values are assigned as needed (1-10, 10-100, 100-1000, etc.). Note also that each decade occupies the same amount of space.

One advantage of logarithmic spacing is that a larger range of values can be shown in one plot without losing resolution in the smaller values. For example, if we wished to plot frequency values between 10 Hz and 100 kHz on 100 divisions of linear paper, each division would represent approximately 1000 Hz, and it would be impossible to plot values in the decade between 10 Hz and 100 Hz. On the other hand, using log paper the decade between 10 and 100 Hz would occupy the same space as the decade between 10 kHz and 100 kHz. Log-log and semilog graph paper is specified by the number of decades it contains. Each decade is called a cycle. For example, 2 cycle by 4 cycle (or simply "2 by 4") log paper has two decades on one axis and four on the other. The number of cycles must be adequate for the range of the data being plotted. For example, if the data spans the frequency range from 25 Hz to 40 kHz, we will need 4 cycles to plot frequency values, corresponding to the decades 10 Hz - 100 Hz, 100 Hz - 1 kHz, 1 kHz - 10 kHz, and 10 kHz - 100 kHz.

When semilog graph paper is used to plot a frequency response, the horizontal (log) axis is used for frequency and the vertical (linear) axis is used for gain. Gain in a frequency response is plotted in dB, so calculated or observed values of gain must be converted to dB before plotting. On the other hand, since decibel voltage gain is a logarithmic function (db = 20 $\log_{10}$ G), gain values can be plotted on log-log paper without having to convert to decibels. The plotting of values and the interpretation of graphs on log-log or semilog paper requires practice and is best learned by doing.

Let f_1(MAX) = the larger of $f_1(C_1)$ and $f_1(C_2)$, where $f_1(C_1)$ and $f_1(C_2)$ are the lower cutoff frequencies of an ac amplifier due to coupling capacitors C_1 and C_2, respectively (see Figure 4.2). Let f_1(MIN) = the smaller of $f_1(C_1)$ and $f_1(C_2)$. If f_1(MAX) is much greater than f_1(MIN), then the frequency response of the amplifier when plotted on log-log paper between f_1(MAX) and f_1(MIN) is asymptotic to a straight line with slope 1. This line corresponds to a reduction in gain at the rate of 20 dB per decade (a tenfold decrease in gain for each tenfold reduction in frequency) or, equivalently, 6 dB per octave (a twofold decrease in gain for each twofold decrease in frequency). At frequencies below f_1 (MIN), the gain falls off at the rate of 40 dB/decade, or 12 dB/octave. Figure 4.3 illustrates this response characteristic for the case f_1(MAX) = 200 Hz and f_1(MIN) = 10 Hz.

If $f_1(C_1) = f_1(C_2)$, then, as noted earlier, the actual lower cutoff frequency is

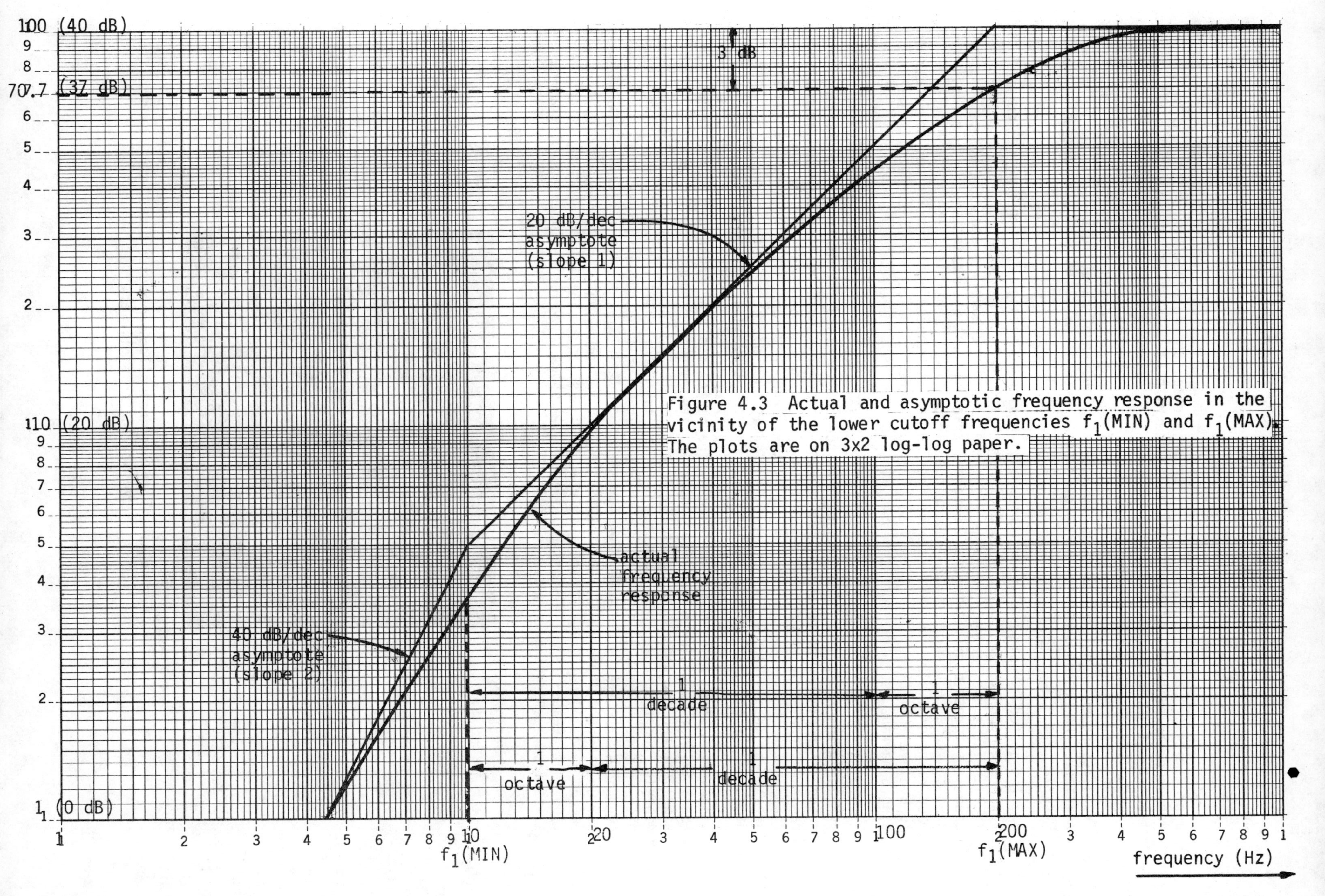

Figure 4.3 Actual and asymptotic frequency response in the vicinity of the lower cutoff frequencies f_1(MIN) and f_1(MAX). The plots are on 3x2 log-log paper.

approximately 1.5 $f_1(C_1) = 1.5f_1(C_2)$. In this case, the response characteristic on log-log paper below cutoff is asymptotic to a line with slope 2, that is, gain falls off at the rate of 40 dB/decade.

The rate at which the gain falls off as frequency increases beyond the upper cutoff frequency depends on the number and locations of the shunt capacitors connected to the amplifier. The response will be asymptotic to an integer multiple of -20 dB/decade or -6 dB/octave.

Integrated circuit amplifiers generally require connection of some external components for proper operation. Coupling capacitors are the most common example. Other examples include input and feedback resistors for gain control, decoupling and roll-off capacitors, used to suppress oscillations, and external transistor(s) with associated bias circuitry, used to increase power output. Integrated circuit audio amplifiers are available in a wide range of types and power ratings from 2W to 10W, though heat sinking may be required to achieve full rated power. Many such circuits are constructed with two amplifiers in a single package, suitable for stereo applications.

Important specifications related to the performance of an audio amplifier include frequency response, input and output impedance, distortion, and maximum output voltage swing. For minimum distortion, the frequency response should be flat through the midband range. Since audio signals may contain many frequency components simultaneously, lack of flatness in the frequency response causes some frequency components to be amplified to a greater or lesser degree than others, thus creating a distorted output. The input impedance of the amplifier should be large, to prevent loading of the audio signal source and the consequent reduction in gain. If the amplifier is of the operational type, the input impedance will be essentially the same as the series resistor connected to the input. If the amplifier is designed to drive a speaker, its output impedance should be quite low, since speaker impedances are generally 16 ohms or less. The maximum output voltage swing depends upon the power supply voltage. In the case of the 377 dual audio amplifier, which will be used in this experiment, the maximum output voltage swing is (V_{PS}-6) volts, peak-to-peak, where V_{PS} is the power supply voltage. Overdriving the amplifier by connecting a signal source whose amplitude is so large that the rated output voltage swing is exceeded will cause the output to be clipped (chopped off at the top and bottom), a condition that creates severe distortion. Figure 4.4 shows a typical clipped output that results from overdriving an audio amplifier.

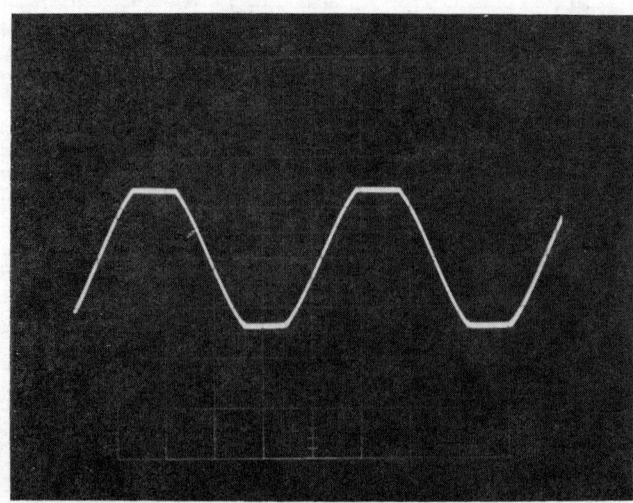

Figure 4.4 Output signal clipping caused by overdriving an amplifier.

An important specification for stereo amplifiers is the separation between channels. Ideally, any signal amplified in one channel should have zero component appearing in the output of the other channel. This separation is difficult to achieve, particularly when the amplifiers in each channel have a common power supply, which is the typical case for dual integrated circuit amplifiers. The power supply used in such applications should have a very low output impedance, to prevent signal components in one channel from being coupled through the supply into the other channel. Channel separation is usually specified in dB. A 70 dB separation means that a signal in one channel will appear in the other channel (with the latter's input grounded) at a level 70 dB below its level in the first channel.

EXERCISES

1. In Figure 4.2, $R_S = 0$, $C_1 = 0.1\ \mu F$, $R_{in} = 10K$, $R_o = 10\ \Omega$, $C_2 = 500\ \mu F$, and $R_L = 16\ \Omega$. Find $f_1(C_1)$ and $f_1(C_2)$. What is the lower cutoff frequency of the amplifier? What is the upper cutoff frequency if the bandwidth is 15 kHz?

2. In Exercise 1, to what value should the input coupling capacitor C_1 be changed in order that $f_1(C_1) = f_1(C_2)$? What would be the approximate lower cutoff frequency in that case?

3. To determine the frequency response of a certain amplifier, the output voltage was measured at 20 different frequencies between 15 Hz and 40 kHz. The minimum voltage measured was 580 mV and the maximum was 18.4 V. What type of log-log paper (how many cycles) should be used to plot this data?

4. Using the values of exercise 1, determine the approximate number of dB of attenuation below the output level at midband when the amplifier is driven at 80 Hz.

5. Using the value of C_1 calculated in exercise 2 and the other component values given in exercise 1, determine the approximate number of dB of attenuation below the output level at midband when the amplifier is driven at 9 Hz. Repeat, when the amplifier is driven at 1.8 Hz.

6. The output voltage of an amplifier is 16.3 V pk-pk at 2 kHz and 2.45 V pk-pk at 50 Hz. In each case, the input voltage is 0.25 V pk-pk. What is the change in output voltage expressed in dB? What is the change in gain expressed in db?

7. The input to one channel of a stereo amplifier is grounded. The output of the other channel is 5 V RMS. The output of the grounded channel is a 1.6 mV RMS component of the output in the other channel. What is the channel separation in dB?

8. A stereo amplifier has a channel separation of 80 dB. What voltage will appear at the output of a channel whose input is grounded when the output of the other channel is 10 V pk?

PROCEDURE

1. Connect the 377 dual audio amplifier for single channel operation, as shown in the circuit of Figure 4.5. For the first part of this experiment, we will be investigating the effect of C_1 alone on the lower cutoff frequency. For this reason, our output voltage measurements will be made directly at pin 2 of the amplifier, and the value of C_2 will not matter. Use $C_1 = .01\ \mu F$ and any value of C_2 between 10 µF and 100 µF.

2. To convince yourself of the necessity for output coupling capacitor when a speaker or some other load that is sensitive to dc is connected to the output, measure and record the dc voltage at pin 2.

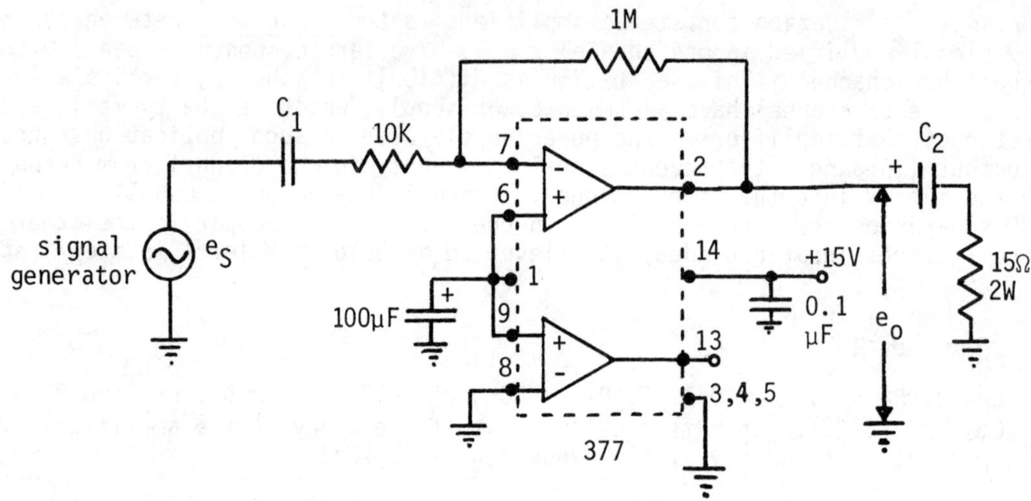

Figure 4.5

3. Connect a dual trace oscilloscope for simultaneous viewing of e_S and e_O. e_O should be observed with the oscilloscope input set for "AC," since this voltage has a large dc component. With C_1 = .01 μF, set the signal generator frequency to 10 kHz. Increase the amplitude of the signal generator output e_S until the amplifier output e_O as observed on the oscilloscope just begins to clip. Measure and record the peak-to peak clipping level.

4. Reduce the signal generator gain until the amplifier output is 2 V peak. To obtain data for a frequency response plot, measure and record the peak values of e_S and e_O at various frequencies in the midband range, below the lower cutoff frequency, and above the upper cutoff frequency. Begin by experimentally determining the gain $G = e_O/e_S$ at 10 kHz. Then determine experimentally the lower cutoff frequency $f_1(C_1)$, where the gain is .707 times its value at 10 kHz. Determine gain values at several frequencies below cutoff, going down to 50 Hz. Find the upper cutoff frequency and measure gain at several frequencies up to at least one decade beyond cutoff.

5. Connect C_1 = 0.1 μF in place of the .01 μF capacitor used in the previous step and experimentally determine the new value of the lower cutoff frequency $f_1(C_1)$. Repeat with C_1 = 1.0 μF. It is not necessary to obtain complete frequency response data through the entire frequency range in these cases.

6. To investigate the effect of C_2 alone on the lower cutoff frequency, use C_1 = 1 μF and make all input voltage (e_S) measurements at the junction of C_1 and the 10K resistor. In this way the value of C_1 has no effect. We are now prepared to find the effect of different values of C_2 on the lower cutoff frequency $f_1(C_2)$. Connect a 10 μF capacitor for C_2 (see Figure 4.5) and experimentally determine the lower cutoff frequency. Be sure to observe the polarity connections of the capacitor, as shown in the figure. Measure e_O across the 15Ω load. Repeat for C_2 = 100 μF and C_2 = 500 μF.

7. Now connect C_1 = .01 μF, C_2 = 10 μF, and replace the 15 ohm load resistor with a 10 ohm, 2W resistor. Note that $R_1C_1 = R_LC_2$, so $f_1(C_1) = f_1(C_2)$ (assuming the source resistance R_S of the signal generator and output resistance R_O of the amplifier are both small). Experimentally determine the lower cutoff frequency. Note that e_S must now be measured directly at the source (signal generator) and e_O directly across the load, since we are investigating the effect of both C_1 and C_2. Then measure the gain at several frequencies below cutoff, down to 50 Hz.

8. To determine the output resistance of the amplifier, we will measure e_0 with and without the load resistor connected. Let e_{OL} = the output voltage across the 15 ohm load and e_{ON} = the output voltage with no load. The equivalent circuit for the output of the amplifier is then as shown in Figure 4.6.

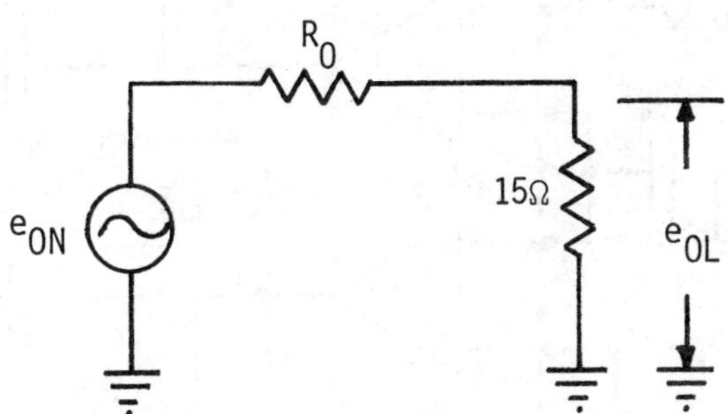

Figure 4.6 Equivalent circuit of the output of an amplifier. e_{ON} = no-load output; R_0 = output resistance; e_{OL} = output across load.

Using Figure 4.6 and the voltage divider rule, we find

$$e_{OL} = \left(\frac{15}{R_0 + 15}\right) e_{ON}$$

from which,

$$R_0 = \frac{15(e_{ON} - e_{OL})}{e_{OL}} \qquad (4)$$

With C_1 = 0.1 μF, C_2 = 100 μF (see Figure 4.5), and the frequency of the signal generator set to 10 kHz, set the peak output voltage e_{OL} to 5 V with the 15 ohm load in place. Then measure and record the peak value of e_{ON} with the load removed. (If the difference between e_{ON} and e_{OL} is very small, 1-20 mV, use a smaller value of load resistance, say 10 Ω, 2W, for improved accuracy. Modify equation (4) accordingly.)

9. To demonstrate the ability of the amplifier to drive a speaker, turn the signal generator output amplitude down to zero and connect an 8 or 16 Ω speaker in place of 15 Ω load resistor. With C_1 = 0.1 μF and C_2 = 100 μF, gradually increase the output amplitude of the signal generator until an audible tone is heard in the speaker. Vary the frequency of the signal generator between 20 Hz and 20 kHz, the nominal audio range, and note changes in the tone and volume of the audible signal, especially below the lower cutoff frequency (which should be in the neighborhood of 200 Hz).

10. With the same connections as in step 8, turn the power supply voltage down to + 8 V DC. To learn how clipping affects the tonal quality of an audio signal, set the signal generator frequency to 400 Hz and increase its amplitude until clipping is observed on the waveform at the speaker terminal. As you overdrive the amplifier to produce severe clipping, notice the distorted "tinny" sound from the speaker, due to the high frequency harmonics present in the clipped waveform.

11. Now connect the stereo amplifier circuit shown in Figure 4.7.

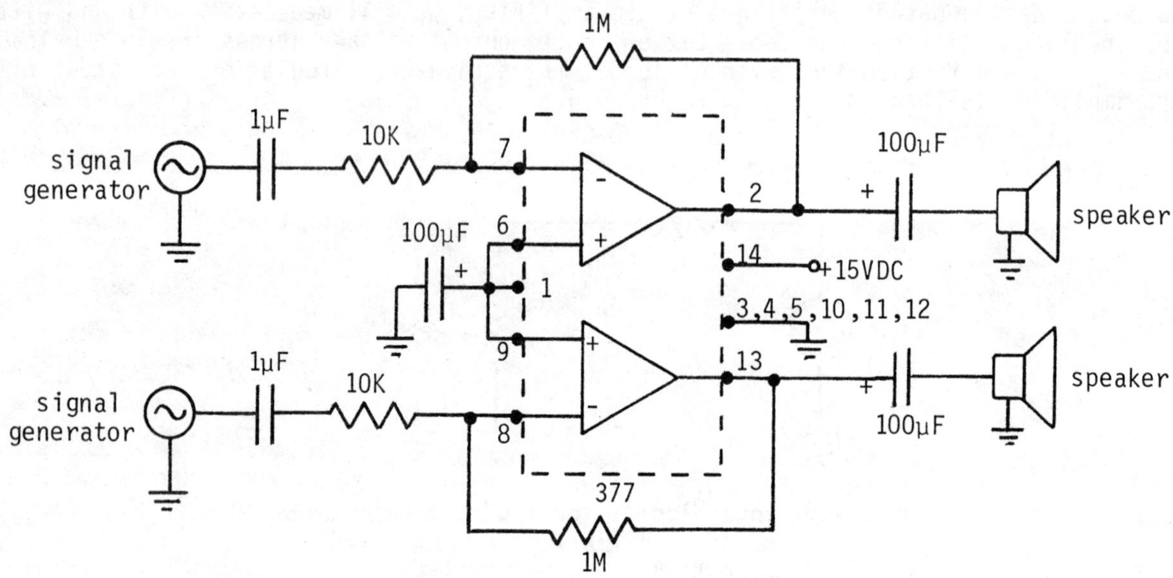

Figure 4.7

12. To verify the operation of the circuit in Figure 4.7 as a stereo amplifier, set the two signal generators to different frequencies, say 400 Hz and 2 kHz, move the two speakers a reasonable distance apart, and adjust the signal amplitudes so that the two tones can be heard at the different locations.

13. Now replace the speakers in Figure 4.7 with 15 ohm, 2W load resistors. Remove the signal generator from one of the amplifiers and connect a ground in its place. To investigate the separation between channels, we will drive one channel with a signal and measure the signal level that appears in the output of the channel whose input is grounded. Set the signal generator to 1 kHz and adjust its amplitude to obtain a 5 volt peak output level in the channel to which it is connected. Then measure the peak output of the signal in the channel whose input is grounded.

14. To determine the effect that power supply impedance has on channel separation, connect a 100 ohm resistor between the 15 V DC power supply and pin 14. This resistor simulates a high output impedance in the power supply. Readjust the power supply voltage as necessary to restore + 15 V DC at pin 14. Then repeat step 13.

QUESTIONS

1. How is the dc level that was measured at pin 2 in step 2 of the Procedure related to the power supply voltage? Why do you think this is the case?

2. Compare the peak-to-peak clipping level measured in step 3 of the Procedure with the manufacturer's specification for maximum output voltage swing. What is the average power dissipated in the 15 ohm load resistor at the maximum output level before clipping?

3. Using the data obtained in step 4 of the Procedure, plot a frequency response characteristic on either semilog or log-log graph paper. The vertical axis should be gain in dB. Identify the lower and upper cutoff frequencies on your plot. What is the bandwidth of the amplifier?

4. Based on the component values used in Figure 4.5, what is the theoretical midband

voltage gain of the amplifier? How did your measured value compare to the theoretical value?

5. What is the theoretical lower cutoff frequency $f_1(C_1)$ of the amplifier circuit investigated in step 4 of the Procedure? How does the theoretical value compare with that you measured in step 4?

6. Using the data from step 4 of the Procedure, calculate the number of dB per decade and per octave that the gain decreased with decreasing frequency below cutoff. (The first data point in this calculation should be about an octave below cutoff.) How do your values compare with theoretical values?

7. Repeat question 5 for the cutoff frequencies determined in step 5 of the Procedure. How is the cutoff frequency affected by increasing values of C_1?

8. Repeat question 5 for the values of cutoff frequency $f_1(C_2)$ determined in step 6 of the Procedure. Explain why the experimental procedure in step 6 ensures that we are finding the lower cutoff due to C_2 alone.

9. Repeat questions 5 and 6 for the actual cutoff frequency f_1, using the data obtained in step 7 of the Procedure. Why does the fact that $R_1C_1 = R_LC_2$ imply that $f_1(C_2) \simeq f_1(C_2)$?

10. Calculate the output resistance of the amplifier using the data obtained in step 8 of the Procedure. If the load resistor in step 8 of the Procedure had been progressively reduced in value until e_{OL} was exactly one-half of e_{ON}, what would be the value of R_O?

11. Why do you think the power supply voltage was reduced to 8 V in step 10 of the Procedure, before investigating the effect of clipping on the tone heard from the speaker?

12. Based on the data obtained in step 13 of the Procedure, calculate the separation between channels in dB. How does this value compare with the manufacturer's specifications?

13. Repeat question 12 using the data obtained from step 14 of the Procedure. Explain how increased power supply impedance reduces channel separation.

5 Oscillators

OBJECTIVES

1. To learn how the oscillation criterion of positive feedback is applied to the design of sinusoidal oscillators.

2. To verify experimentally the theoretical operation of the following types of oscillators: Wien bridge, RC phase shift, two-integrator, and astable multivibrator.

3. To learn how to design and predict the frequency of oscillation of the oscillator types described in (2), and to verify experimentally the design equations.

4. To learn how the 555 and/or 556 timer circuit operates, how to apply it in practical circuits, and to gain an appreciation for the versatility of this device.

EQUIPMENT AND MATERIALS REQUIRED

1. Dual trace oscilloscope.

2. Adjustable power supplies, ± 15 V DC.

3. 741 operational amplifiers (2).

4. 555 timers (2), or 556 timer.

5. Resistors: 1K (2), 2.7K, 10K (2), 33K, 100K (2), 1M (2), 10M.

6. Potentiometers: 1K, 10K, 1M.

7. Capacitors: 100pf (3), .001 μF (2), .01 μF (3), 0.1 μF (2).

DISCUSSION

An oscillator is an electronic device that generates an ac voltage (or an ac voltage superimposed on a dc level). Every oscillator must contain an active device, that is, a transistor or other type of amplifier that is capable of supplying energy to the oscillator circuit from a dc power source. Although some amplifier circuits that have been improperly designed, or improperly applied, may oscillate when oscillations are not desired, we reserve the term "oscillator" for those circuits in which oscillations are intentional, predictable, and controllable. Circuits which do not meet these criteria are said to be unstable. Examples of oscillators include the circuitry found in a laboratory type signal generator, also called a VFO, or variable frequency oscillator, and the intermediate frequency (IF) oscillators found in superheterodyne communications receivers.

The fundamental basis for oscillation in any active circuit is positive feedback. Positive feedback occurs when a portion of the output of an amplifier is supplied to its input in such a way that the portion fed back is in phase with the input. If the voltage gain of the positive feedback signal, going from input to output and back to input, is one or greater, then oscillations will be induced and sustained. Since the gain and phase shift of signals passing through an amplifier and its associated circuitry depend on the frequency of that signal, oscillation will occur only if there is some frequency at which the criterion of positive feedback at unity or greater gain is satisfied. Note that it is not necessary to supply the amplifier with a signal whose frequency satisfies the criterion; if there exists some frequency at which positive feedback with unity or greater gain occurs, then oscillations at that frequency will result. Oscillators are designed by connecting selected frequency-sensitive components (capacitors and/or inductors) in the feedback path of an amplifier so that there will exist a frequency, the one we desire, at which the oscillation criterion is satisfied.

Let $A\underline{/\phi}$ represent the (phasor form of the) gain of an amplifier, where A is the amplification factor and ϕ is the phase shift that the signal undergoes in passing from input to output. Let $\beta\underline{/\Theta}$ represent the feedback gain, where β is the proportion of the output fed back to the input and Θ is the phase shift the feedback signal undergoes in passing from output back to input. The criterion for oscillation is that the total gain of a signal passing from input to output and back to input again be $1\ \underline{/0^{\circ}}$, i.e., that

$$A\underline{/\phi}\ \beta\ \underline{/\Theta} = 1\ \underline{/0^{\circ}} \tag{1}$$

Equation (1) requires that

$$|A||B| = 1 \quad \text{and} \quad \phi + \Theta = 0^{\circ}$$

The operational amplifier is widely used as the active device in an oscillator because it has many desirable characteristics that simplify oscillator design. Its high gain makes it relatively easy to obtain the minimum (unity) gain in the path from input to output and back to input that is required for oscillation. Its high input impedance ensures that the network used in the feedback path will not be loaded, that is, will not have its gain versus frequency characteristics changed when it is connected to the amplifier. Also, the operational amplifier has a non-inverting input which is used in many designs to ensure positive feedback, that is, to maintain zero degrees of phase shift between input and output. A typical example is the Wien bridge oscillator, shown in Figure 5.1.

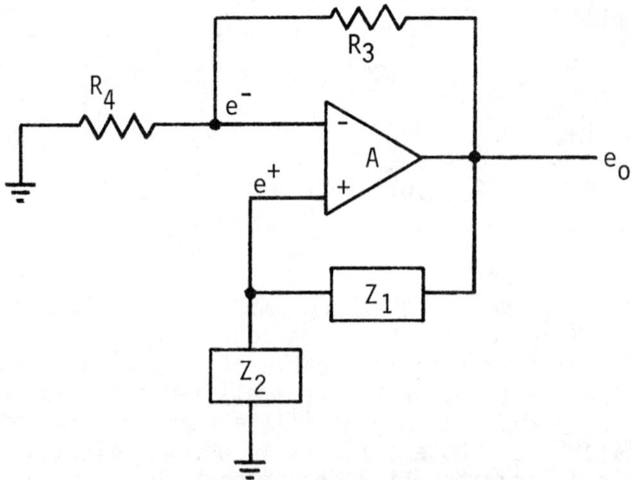

Figure 5.1 A Wien bridge oscillator.

In Figure 5.1, Z_1 and Z_2 are each a combination of resistance and reactance and together form a reactive voltage divider that delivers a portion of the output voltage back to the non-inverting input of the amplifier. The resistor-reactor combinations are designed so that they produce zero phase shift in the feedback signal, e^+, at the one frequency at which oscillation is desired. Frequencies below or above that one design frequency are phase-shifted, so positive feedback does not occur at any frequency other than the one desired.

The R_3-R_4 resistor combination provides negative feedback by delivering a portion of the output signal to the inverting input of the amplifier. These resistors are chosen so that the overall gain of the amplifier is reduced to near unity. Without this negative feedback, the high gain of the operational amplifier would result in oscillation amplitudes that exceed the limits of the output swing, that is, the output would be severely clipped.

To express the condition for oscillation mathematically, let the open loop gain of the amplifier be $A\,\underline{/0^\circ}$. The feedback signal may be expressed as $e^+ - e^-$, since this is the net voltage at the amplifier input. By voltage divider action (see Figure 5.1), we may write

$$e^+ = \left(\frac{Z_2}{Z_1 + Z_2}\right) e_o \text{ and } e^- = \left(\frac{R_4}{R_3 + R_4}\right) e_o$$

Therefore,

$$e^+ - e^- = \left(\frac{Z_2}{Z_1 + Z_2} - \frac{R_4}{R_3 + R_4}\right) e_o$$

from which we conclude that the feedback factor β is

$$\beta\,\underline{/\Theta} = \left(\frac{Z_2}{Z_1 + Z_2} - \frac{R_4}{R_3 + R_4}\right) \tag{2}$$

By equation (1), we must have

$$(A\,\underline{/0^\circ})(\beta\underline{/\Theta}) = 1\,\underline{/0^\circ}$$

which means that $A|\beta|$ must eoual 1 and that $\underline{/\beta}$ must equal zero degrees:

$$A|\beta| = 1 \qquad \underline{/\beta} = 0^\circ \tag{3}$$

Since the gain A of the operational amplifier is very large, and since from equation (3) we have

$$|\beta| = 1/A$$

it is clear that $|\beta|$ must be very nearly zero. The conditions for oscillation expressed by (3) will therefore be satisfied if

$$\frac{Z_2}{Z_1 + Z_2} = \frac{R_4}{R_3 + R_4}$$

We will use equality of these terms as our design criterion:

$$\frac{Z_2}{Z_1 + Z_2} = \frac{R_4}{R_3 + R_4} \tag{4}$$

Note that if (4) is true then $|\beta| = 0$ and furthermore $\underline{/\beta} = 0^\circ$. This latter assertion is true because $R_4/(R_3 + R_4)$ is purely resistive and therefore has zero phase angle, which means that $Z_2/(Z_1 + Z_2)$ must also have zero phase angle, which in turn means that

$$\beta = \frac{Z_2}{Z_1 + Z_2} - \frac{R_4}{R_3 + R_4}$$

has zero phase angle.

We can now focus our attention on the networks represented by Z_1 and Z_2 in Figure 5.1. Figure 5.2 shows a combination of RC networks that can be used to achieve our goal of zero phase shift in the feedback signal e^+ at one and only one frequency.

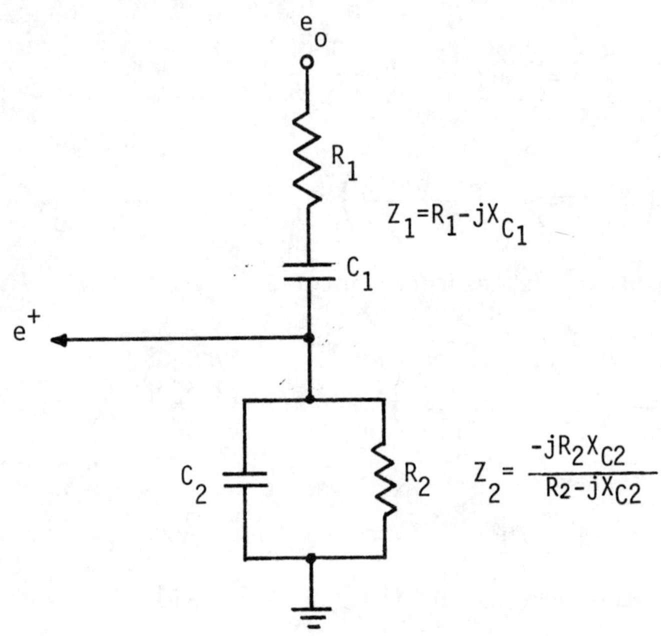

Figure 5.2 Reactive networks Z_1 and Z_2 that form a voltage divider for feedback in the Wien bridge oscillator.

We see that Z_1 is a series RC combination and Z_2 is a parallel RC combination. Therefore,

$$\frac{Z_2}{Z_1 + Z_2} = \frac{\dfrac{-jR_2X_{C2}}{R_2-jX_{C2}}}{R_1-jX_{C1}-\left(\dfrac{jR_2X_{C2}}{R_2-jX_{C_2}}\right)}$$

$$= \frac{-jR_2X_{C2}}{(R_1-jX_{C_1})\ (R_2-jX_{C_2})\ \ - jR_2X_{C_2}}$$

Multiplying the denominator out and combining real and imaginary terms, we find

$$\frac{Z_2}{Z_1 + Z_2} = \frac{-jR_2\ X_{C2}}{(R_1R_2-X_{C_1}X_{C_2})\ \ - j(R_1X_{C_2} + R_2X_{C_1}+R_2X_{C_2})}$$

This expression may be rationalized by multiplying numerator and denominator by the complex conjugate of the denominator. When this is done, we find that the imaginary part of the numerator is $R_2X_{C_2}(R_1R_2 - X_{C_1}X_{C_2})$. In order that $Z_2/(Z_1 + Z_2)$ have zero phase angle, the imaginary part of the numerator must equal zero (the denominator is real, after rationalizing). Therefore, we have

$$R_2X_{C_2}(R_1R_2 - X_{C_1}X_{C_2}) = 0$$

$$R_1R_2 = X_{C_1}X_{C_2}$$

$$R_1R_2 = \frac{1}{\omega^2 C_1 C_2}$$

Solving the latter equation for ω, we find the frequency that satisfies the oscillation criterion of positive feedback:

$$\omega = \frac{1}{\sqrt{R_1R_2C_1C_2}} \quad \text{rad/sec} \tag{5}$$

If $R_1 = R_2 = R$, and $C_1 = C_2 = C$, we find the frequency of oscillation (in Hz) is

$$f = \frac{1}{2\pi RC} \tag{6}$$

At the oscillation frequency, $X_{C_1} = R_1$ and $X_{C_2} = R_2$. Using this fact, $Z_2/(Z_1 + Z_2)$ can be shown to equal 1/3. Therefore, from equation (4), we must have

$$\frac{R_4}{R_3 + R_4} \simeq 1/3 \text{ (actually, slightly less than 1/3)}$$

From this result it is easy to show that we must choose R_3 and R_4 so that $R_3 > 2R_4$. (Derive this, as an exercise. Start with $R_4/(R_3 + R_4) < 1/3$.) Since R_3 and R_4 control the amount of negative feedback, R_3 is usually made adjustable so the user can control the distortion in the sinusoidal oscillation (prevent the signal from reaching the voltage limits of the amplifier).

Another oscillator circuit that is based on the use of operational amplifiers employs two amplifiers connected as integrators. Integrators will be discussed in detail in the next experiment, and for our purposes now we need only know the following facts:

1. An operational amplifier is connected as an integrator when the negative feedback is through a capacitor and the input to the inverting terminal is through a resistor. See Figure 5.3.

2. The magnitude of the gain of an integrator when the input is a sinusoidal wave is

$$|e_o/e_{in}| = 1/\omega RC \tag{7}$$

   where ω is the frequency of the sine wave in rad/sec.

3. When a sine wave is applied to an integrator, the output is shifted in phase by -90° with respect to the input. If this phase is combined with the 180° phase shift caused by an inverting amplifier, the net phase shift will be -270°, or $+90^\circ$. That is, the output will lead the input of an inverting integrator by 90°.

From these facts we see that if two integrators are connected in series, one as

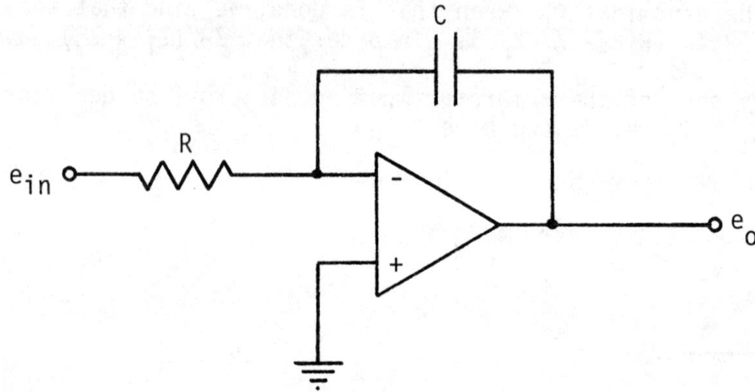

Figure 5.3 An operational amplifier connected as an integrator.

an inverter and one non-inverting, as shown in Figure 5.4, then the total phase shift in a signal passing through both of them will be $-90^0 - 270^0 = -360^0 = 0^0$ (or, equivalently, $-90^0 + 90^0 = 0^0$). We therefore have positive feedback when the output of the second integrator is connected back to the input of the first. Since the gain of the integrators depends on frequency (equation 7), there will be some frequency at which the feedback signal returns with unity gain. Oscillation will therefore occur at that frequency.

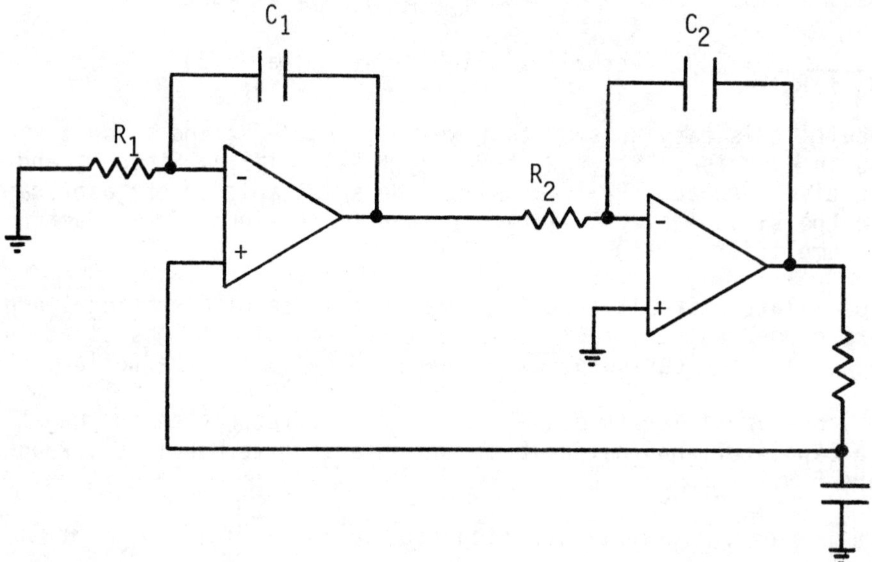

Figure 5.4 A two-integrator oscillator.

The total gain of the series-connected integrators is

$$\left(\frac{1}{\omega R_1 C_1}\right)\left(\frac{1}{\omega R_2 C_2}\right)$$

When this gain is unity,

$$\frac{1}{\omega^2 R_1 C_1 R_2 C_2} = 1$$

and we find

$$\omega = \frac{1}{\sqrt{R_1 C_1 R_2 C_2}}$$

Thus the frequency of oscillation is

$$f = \frac{1}{2\pi\sqrt{R_1 C_1 R_2 C_2}} \text{ Hz} \tag{8}$$

Resistor R_1 is adjusted very slightly to control distortion that may be caused by excessive feedback gain.

Still another type of oscillator that can be constructed using an operational amplifier is the RC phase shift oscillator, shown in Figure 5.5.

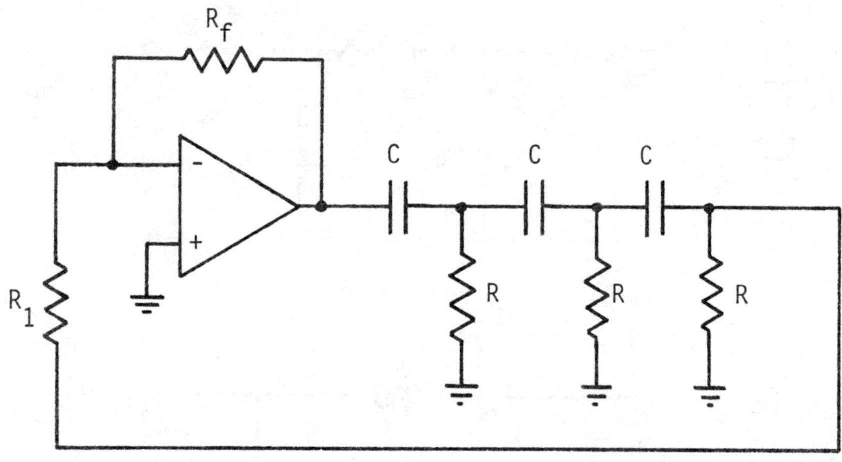

Figure 5.5 An RC phase shift oscillator.

The purpose of the cascaded RC sections is to produce 180 degrees of phase shift at some frequency. If each RC section were isolated and driven by a sinusoidal signal, we would find that its output (the voltage across its resistor) leads its input by a phase angle depending on the frequency. When all three sections are combined, there is some frequency at which the cumulative phase shift is 180°, that is, some frequency at which the output of the third (rightmost) section leads the input of the first section by 180°. Since the amplifier is used as an inverter, it causes 180° of phase shift in a signal passing through it. Thus, when the output of the RC networks is fed back to the input of the amplifier, there is a frequency at which the total phase shift from amplifier input to amplifier output and through the network back to amplifier input is 180° + 180° = 360°, or 0°. Provided the total gain through that loop is unity or greater, oscillation will result. The negative feedback supplied through R_f is used to prevent signal clipping by reducing the overall gain.

If all three RC sections are identical, then the frequency of oscillation is approximately

$$f = \frac{1}{2\pi RC\sqrt{6}} \tag{9}$$

Equation (9) is exact if resistor R_1 is large enough to neglect its loading effect on the RC phase shifting network. (Note that R_1 is in parallel with the rightmost resistor R.)

Another type of oscillator is the astable or free-running multivibrator. This device operates on a different principle from those we have discussed previously and produces a square wave output rather than a sine wave. It can be shown that a square wave of frequency f consists of an infinite number of sine waves whose frequencies are odd multiples of f: 3f, 5f, 7f, etc., called harmonics. In addition, there is a sine wave component, called the fundamental, that has the same frequency f as the square wave itself. The fundamental component is larger than any of the harmonics, so if a sine wave signal of frequency f is required, it is relatively easy to filter the fundamental out of a square wave having that frequency. Filters will be discussed in detail in a future experiment. For our purposes now, we need only realize that the circuits shown in Figure 5.6 have the property of passing low frequency signals and attenuating higher frequency signals.

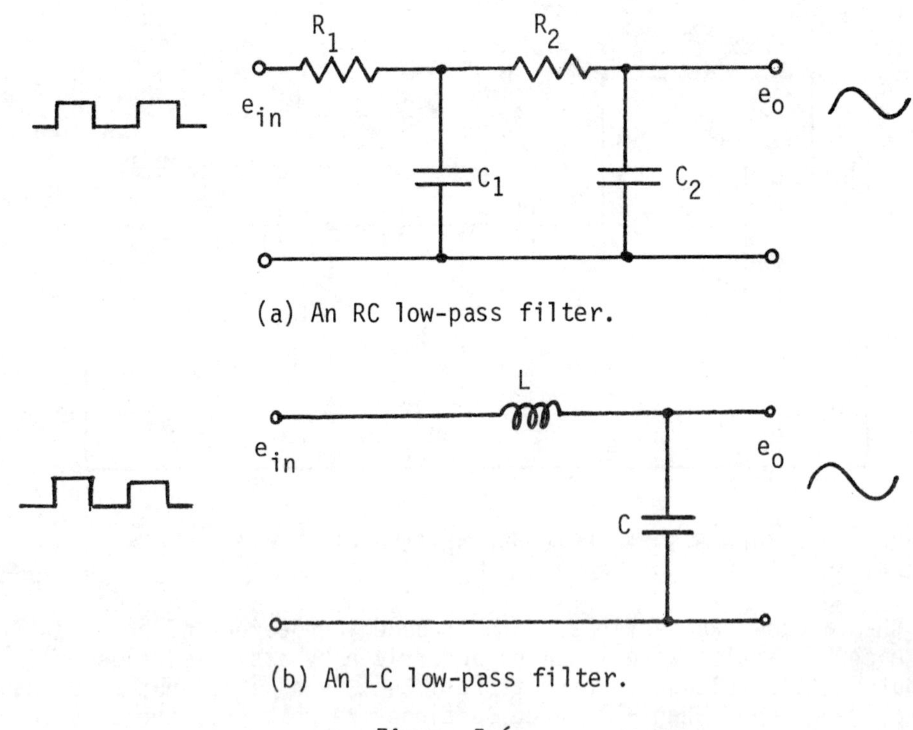

(a) An RC low-pass filter.

(b) An LC low-pass filter.

Figure 5.6

In each of Figures 5.6(a) and 5.6(b), the capacitors present a low impedance to high frequencies ($X_C = 1/2\pi fC$) and therefore shunt those components to ground. At the same time, they present a high impedance to low frequencies, so the low frequency components are passed through to the output. In 5.6(b), the inductor L presents a high impedance to high frequencies ($X_L = 2\pi fL$) and since L is in series with the signal path, it further aids in blocking high frequencies from the output. These filters are called low-pass filters. By proper choice of the RC and RL values, they can be made to attenuate all frequencies higher than the fundamental of a square wave and can therefore be used to obtain a (reasonably) sinusoidal output with a relatively low amplitude.

The 555 timer is an example of an integrated circuit that can be connected to produce a square wave oscillation. This popular and versatile device is reliable, easy to use, and requires few external components. It can also be connected to produce a single pulse of any desired width in response to an externally supplied trigger (this is called monostable operation), or to produce a square wave with duty cycle selected by the user. Duty cycle is the ratio of the time that the output is high to the total time for one cycle, i.e., to the period, often expressed in percent.

Figure 5.7 shows a block diagram of the 555 timer with the external connections necessary to create oscillation. External components R_1, R_2, and C determine the frequency of oscillation and the duty cycle.

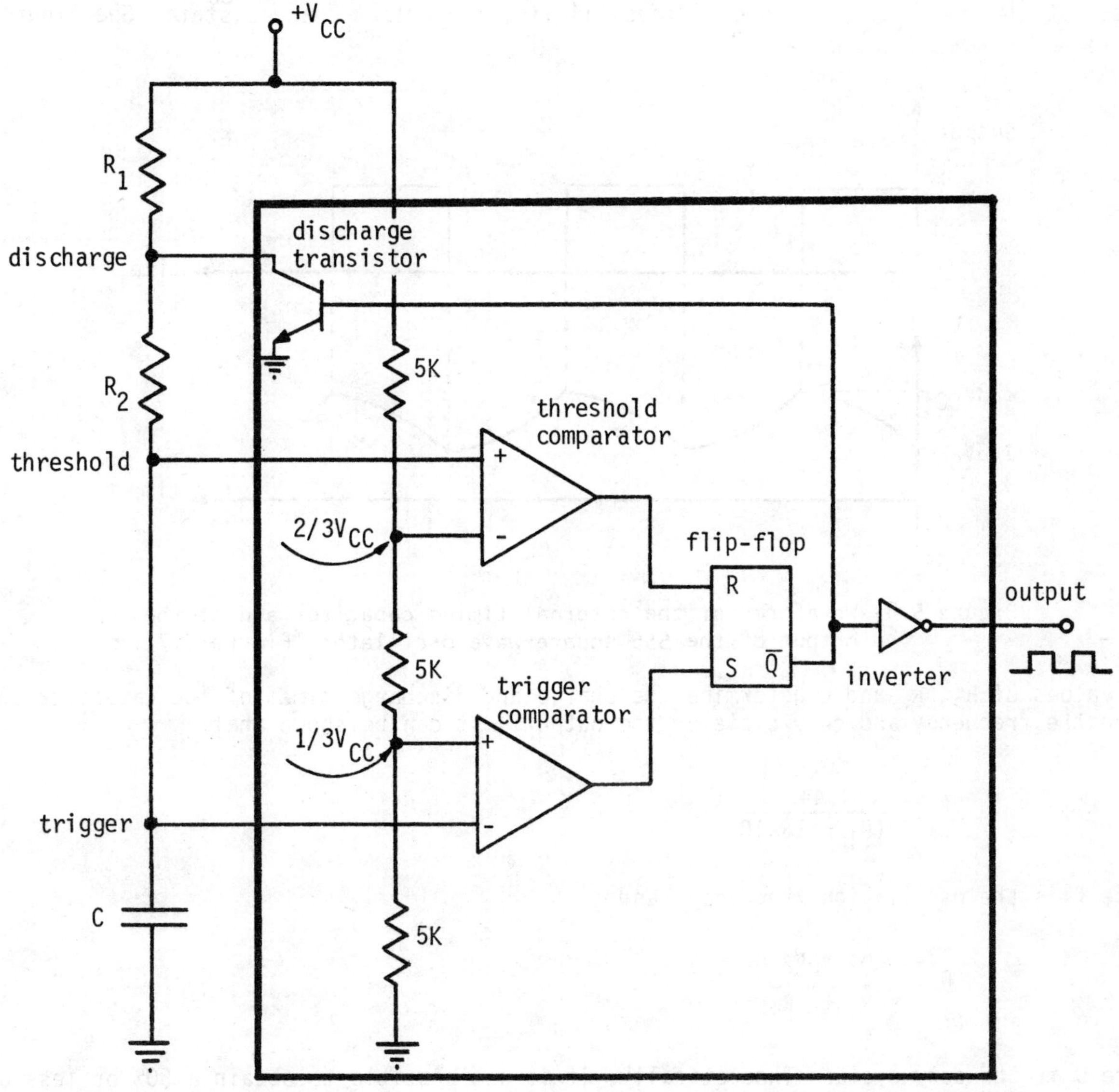

Figure 5.7 Block diagram of a 555 timer connected to produce a square-wave oscillation.

Note that the three 5K internal resistors form a voltage divider network that places 2/3 V_{CC} at one voltage comparator and 1/3 V_{CC} at the other. (A voltage comparator generates an output when its two inputs are equal; these will be covered in a later experiment.) Assuming the flip-flop is initially set, $\bar{Q}$ is low and therefore the discharge transistor is off. Consequently capacitor C charges through R_1 and R_2. When the voltage on the capacitor, V_C, reaches 2/3 V_{CC}, the threshold comparator generates a pulse that resets the flip-flop. When the flip-flop changes state, $\bar{Q}$ goes high (causing the output to go from high to low), and the discharge transistor turns on. Capacitor C then discharges into the transistor through R_2 only. When the capacitor voltage has decayed to 1/3 V_{CC}, the trigger comparator sets the flip-flop and the cycle starts over.

We see that capacitor C discharges and charges between V_C = 1/3 V_{CC} and V_C = 2/3 V_{CC} volts. Each time V_C reaches one of these limits, the output changes state. See Figure 5.8.

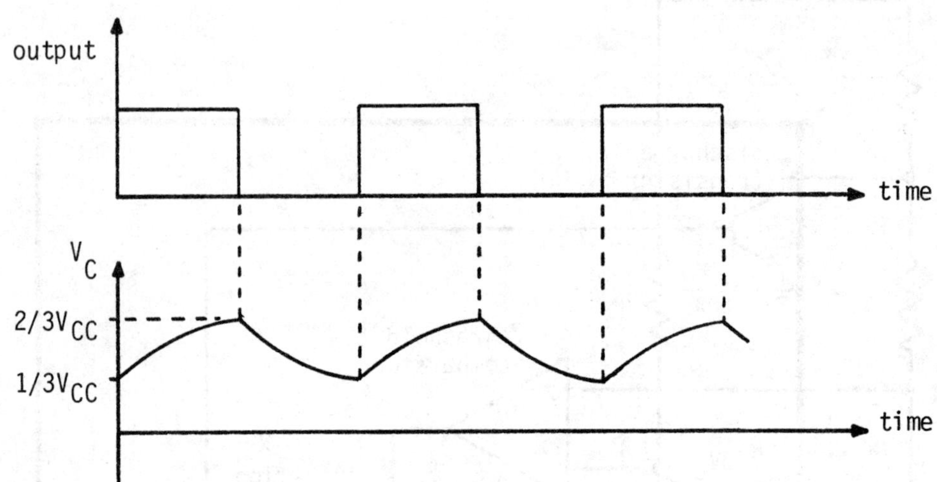

Figure 5.8 Waveforms at the external timing capacitor and at the output of the 555 square-wave oscillator (Figure 5.7).

The values of R_1, R_2 and C determine the charge and discharge times of the capacitor and hence the frequency and duty cycle of the output. It can be shown that

$$f = \frac{1.44}{(R_1 + 2R_2)C} \tag{10}$$

where f is the oscillation frequency, and

$$D = \frac{R_1 + R_2}{R_1 + 2R_2} \tag{11}$$

where D is the duty cycle. Theoretically, it is not possible to obtain a 50% or less duty cycle (why? try D = .5 in equation 11) but by selecting R_2 much larger than R_1 we can approach this limit. By connecting a diode in parallel with R_2 (with the anode connected to the discharge terminal), duty cycles less than, equal to, or greater than .5 can be obtained. The diode shorts R_2 when the capacitor is charging, so the capacitor charges through R_1 and discharges through R_2. In this case, the duty cycle is

$$D = \frac{R_1}{R_1 + R_2}$$

As noted previously, the 555 timer can also be connected to produce a single pulse of specified width when it is triggered. In this monostable or "one-shot" mode of operation, the device is connected as shown in Figure 5.9.

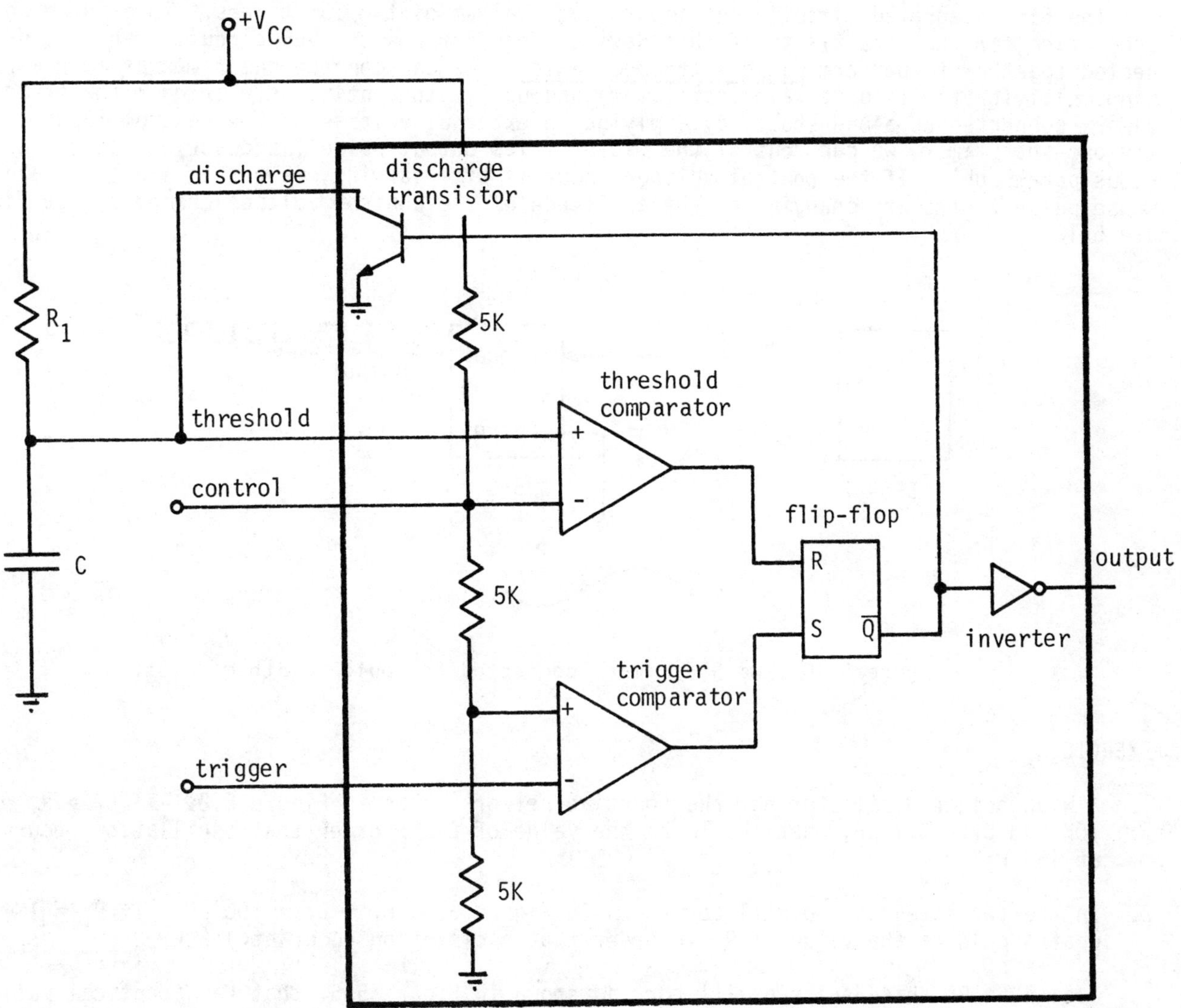

Figure 5.9 Block diagram of a 555 timer connected for monostable (one-shot) operation.

In the one-shot mode, the flip-flop is normally reset, the discharge transistor is on, and so the capacitor is fully discharged. A negative-going pulse applied to the trigger input sets the flip-flop, which turns off the discharge transistor and allows C to begin charging through R_1. When the capacitor voltage reaches 2/3 V_{CC}, the control comparator resets the flip-flop. The total time the flip-flop remained set corresponds to the time that the output remained high and is therefore equal to the width of the single output pulse.

Since this total time depends on the R_1C time constant, pulse width can be controlled by choice of the size of R_1 and C. The output pulse width is given by

$$t = 1.1R_1C \tag{12}$$

Note that one variation on this use allows an external voltage to be applied to the control input (on the threshold comparator). This changes the voltage to which the capacitor must charge before resetting the flip-flop and therefore controls the output pulse width.

The 556 integrated circuit contains the equivalent of two 555 timers. To gain an appreciation for the versatility of this device, consider how two 555 circuits can be connected together to perform pulse width modulation. We can connect one timer as a free-running multivibrator (square wave oscillator) and use it to continuously trigger the other timer, connected as a one-shot. By applying an external voltage to the control input of the one-shot timer, we can control the width of its output pulse (as described in the previous paragraph). If the control voltage input is time-varying, we obtain a pulse train whose pulse widths are changing as the amplitude of the control voltage changes. See Figure 5.10.

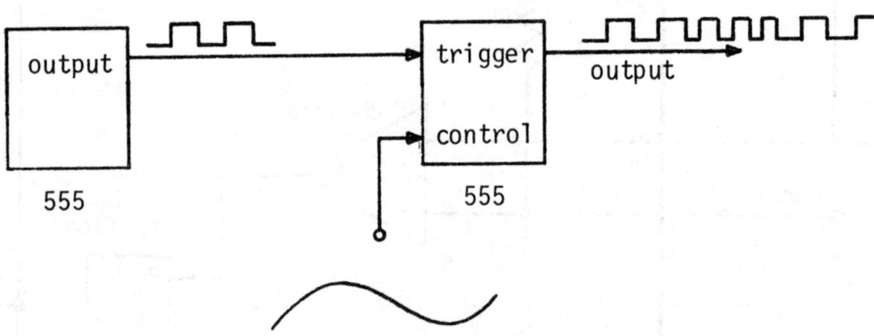

Figure 5.10 Two 555 timers connected for pulse width modulation.

EXERCISES

1. A Wien bridge oscillator has the reactive network shown in Figure 5.2. If $R_1 = R_2 =$ 10K and $C_1 = 0.1\ \mu F$, what should be the value of C_2 in order that oscillation occur at 159 Hz?

2. In the two-integrator oscillator shown in Figure 5.4, $C_1 = C_2 = 100$ pF. If $R_1 = 1M$, what should be the value of R_2 in order that oscillation occur at 1.59kHz?

3. Design an RC phase shift oscillator, as shown in Figure 5.5, that has identical values for the capacitors C and resistors R and that oscillates at 6.5 kHz.

4. What are the fundamental and fifth harmonic frequencies of a square wave whose period is 0.4 ms?

5. A 555 timer is connected as shown in Figure 5.7. If $R_1 = 1K$ and $C = .01\ \mu F$, what should be the value of R_2 in order to produce a 720 Hz square wave oscillation? What would be the duty cycle of the square wave when this value of R_2 is used?

PROCEDURE

1. Connect the Wien bridge oscillator shown in Figure 5.11. (Initial component values

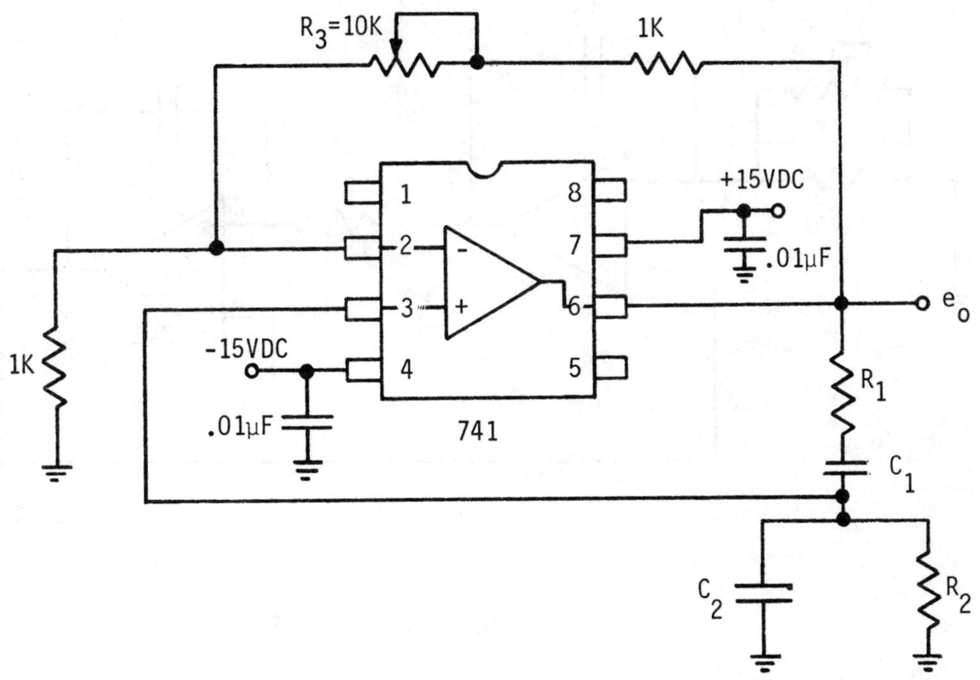

Figure 5.11

are given in step 2.) Connect an oscilloscope for observation and measurement of e_0.

2. With $R_1 = R_2 = 10K$, and $C_1 = C_2 = 0.1\ \mu F$, adjust R_3 until an undistorted sine wave appears at the output. Measure and record its frequency.

3. Repeat step 2 for $R_1 = R_2 = 10K$ and $C_1 = C_2 = .001\ \mu F$. Repeat again for $R_1 = R_2 = 10K$ and $C_1 = C_2 = 100$ pF.

4. Repeat step 2 for $R_1 = R_2 = 100K$ and $C_1 = C_2 = .001\ \mu F$. Repeat again for $R_1 = R_2 = 100K$ and $C_1 = C_2 = 100$ pF.

5. Design and connect a Wien bridge oscillator that produces an undistorted signal with frequency 25 kHz ± 5%. Use only standard value components (except a potentiometer may be used for R_3 to reduce distortion). Record the component values used in your design and the signal frequency it produced.

6. Connect the two-integrator oscillator shown in Figure 5.12. Monitor e_1 and e_2 on a dual trace oscilloscope. (See Figure 5.11 for power supply connections.)

7. With $C_1 = C_2 = C_3 = 100$ pF, adjust the 1M potentiometer until undistorted sine waves (e_1 and e_2) are observed on the oscilloscope. Measure and record their frequency and the phase shift between e_1 and e_2. (If the 1M potentiometer does not provide enough

8. Repeat step 7 with $C_1 = C_2 = C_3 = .01\ \mu F$.

9. Connect the RC phase shift oscillator shown in Figure 5.13. Monitor e_0 with an oscilloscope. (See Figure 5.11 for power supply connections.)

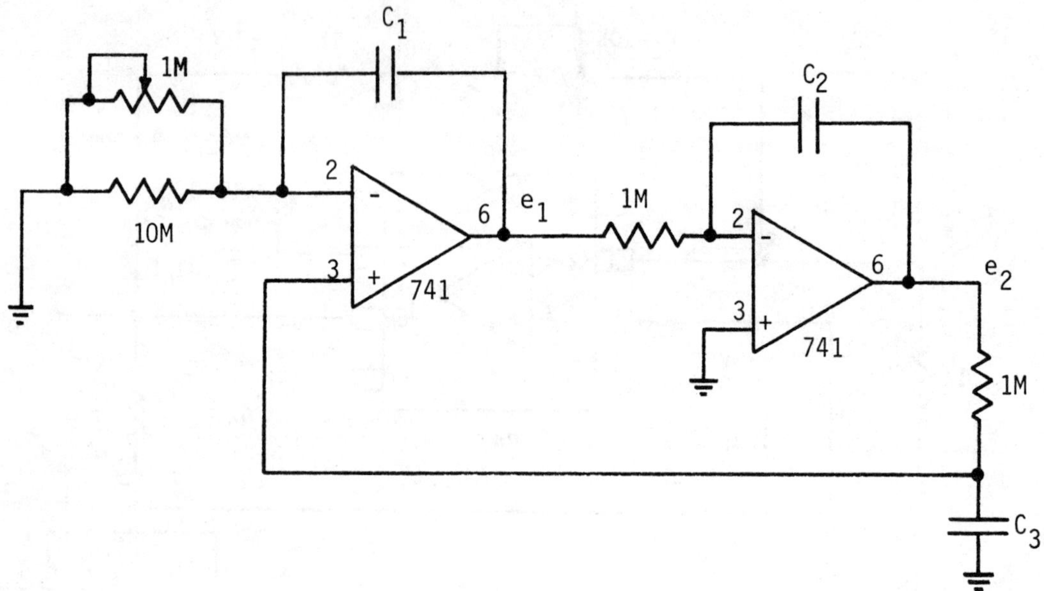

Figure 5.12

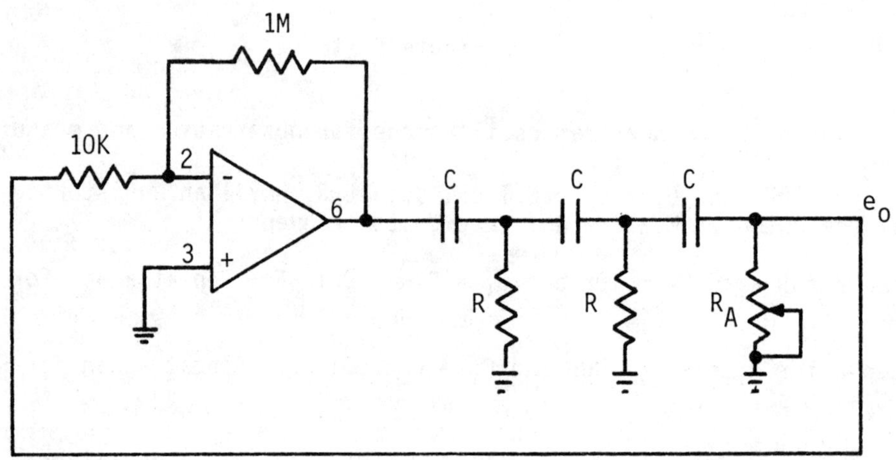

Figure 5.13

10. With each eapacitor equal to .01 μF and each resistor R equal to 1K, including the potentiometer R_A, adjust R_A to obtain an undistorted sine wave output e_o. Measure and record the frequency.

11. Disconnect the cascaded RC sections from the amplifier and connect a low impedance signal generator to the leftmost section, as shown in Figure 5.14. Be careful not to disturb the setting of the 1K potentiometer. Monitor e_s and e_o simultaneously on a dual trace oscilloscope. Set the signal generator frequency to the oscillation frequency you measured in step 10. Then measure and record the phase shift between e_s and e_o.

12. Change all resistors R in Figure 5.13, including R_A, to 10K and repeat step 10.

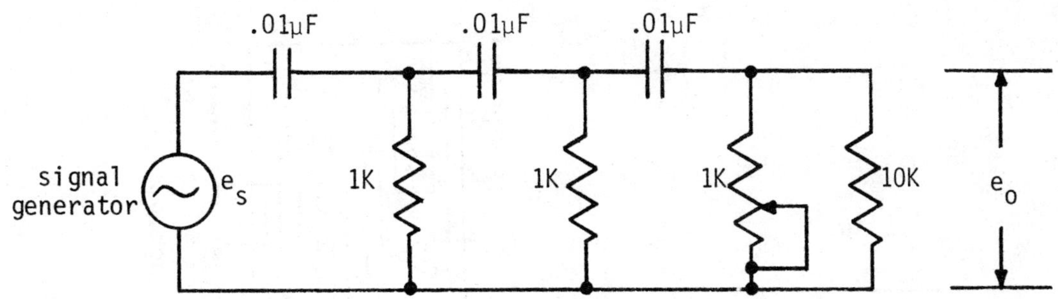

Figure 5.14

13. Using either a 555 or one timer in a 556, connect the astable (free-running multivibrator shown in Figure 5.15. Pin diagrams for both devices are shown in the figure. Monitor the output with an oscilloscope.

14. With R_2 = 1M, measure and record the frequency of the output and its duty cycle.

15. Repeat step 14 for R_2 = 100K and again for R_2 = 10K.

16. With R_2 = 82K, connect the low pass filter circuit shown in Figure 5.16 to the output of the timer. Monitor the timer output and the filter output simultaneously on a dual trace oscilloscope.

17. Adjust the 10K potentiometer in Figure 5.16 until a reasonably low distortion sine wave is observed at the output of the filter. Record the frequency of the filter input and filter output and the amplitude of the sine wave output.

18. Using the 555 or 556 timer, design a square wave oscillator that meets the following specifications:

    (a) frequency 1 kHz ± 5%

    (b) duty cycle 0.6 ± 10%

    Use only standard value components (no potentiometers). Connect your circuit and measure the frequency and duty cycle of the wave it produces. Investigate the effect of power supply changes on the performance of your circuit by reducing the power supply voltage to 13 and to 10 volts and repeating measurements. Record the component values used in your design and all measurements made.

19. Using two 555 timers or a 556, connect the pulse width modulator shown in Figure 5.17. Refer to the pin diagrams in Figure 5.15. The timer on the left is connected for astable operation. Note that the 0.1 μF capacitor discharges through zero resistance (R_2 = 0 in Figure 5.15), so the output of this timer is a sequence of very narrow pulses or "spikes." These pulses are used to trigger the timer on the right, which is connected for monostable, or one-shot, operation. Before connecting the signal generator to the control input on the one-shot, adjust its output to a 2 V peak sinusoidal frequency at 1 Hz.

20. Observe the output of the one-shot on a oscilloscope. Investigate the effects of small changes in the frequency and amplitude of the signal generator output on the appearance of the oscilloscope display. (The signal generator output should not be

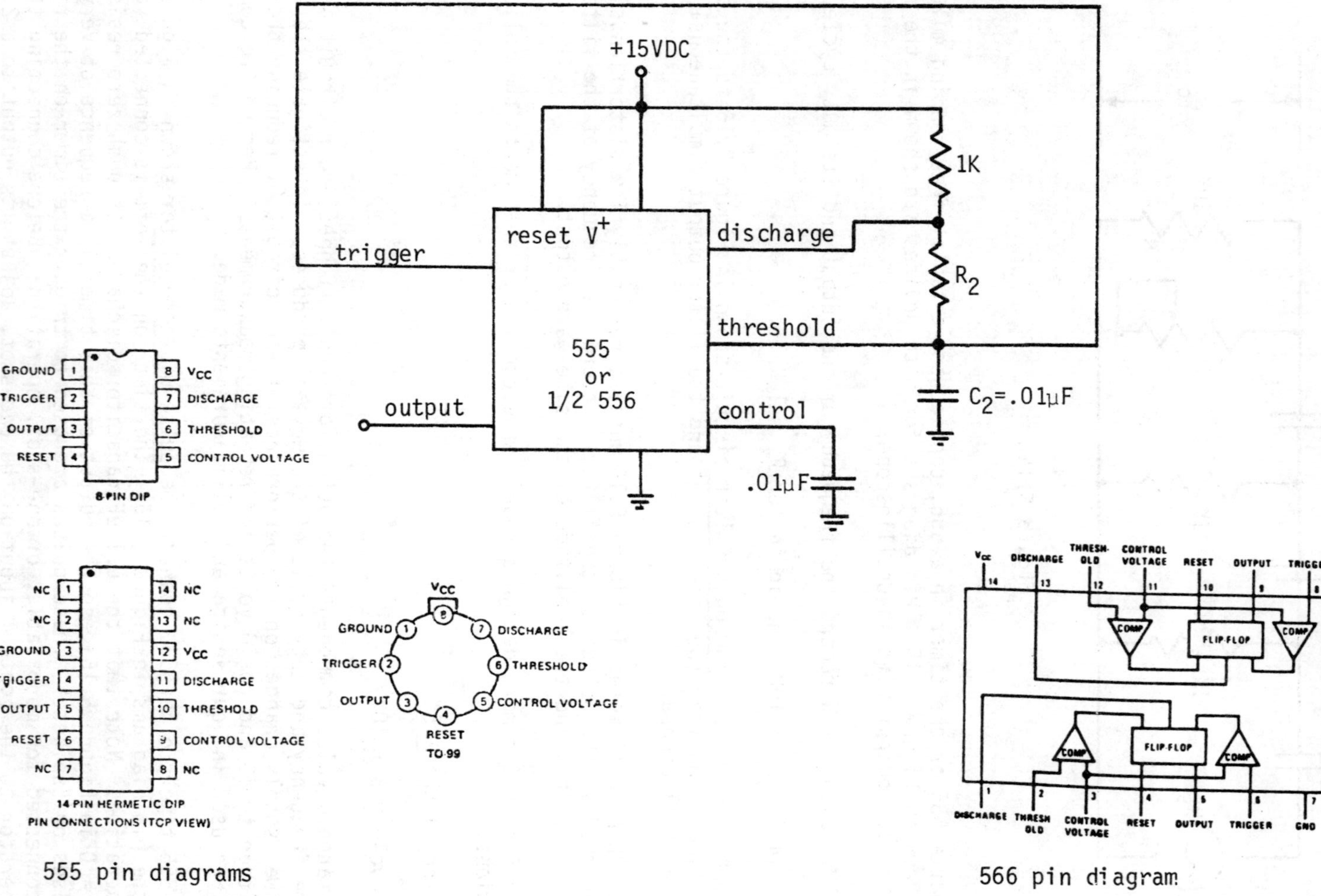

+15VDC
1K
reset V+
discharge
trigger
R2
threshold
555
or
1/2 556
output
control
C2=.01μF
.01μF
GROUND 1
TRIGGER 2
OUTPUT 3
RESET 4
8 VCC
7 DISCHARGE
6 THRESHOLD
5 CONTROL VOLTAGE
8 PIN DIP
NC 1
NC 2
GROUND 3
TRIGGER 4
OUTPUT 5
RESET 6
NC 7
14 NC
13 NC
12 VCC
11 DISCHARGE
10 THRESHOLD
9 CONTROL VOLTAGE
8 NC
14 PIN HERMETIC DIP
PIN CONNECTIONS (TOP VIEW)
555 pin diagrams
VCC
GROUND
DISCHARGE
TRIGGER
THRESHOLD
OUTPUT
CONTROL VOLTAGE
RESET
TO 99
VCC
DISCHARGE
THRESH-OLD
CONTROL VOLTAGE
RESET
OUTPUT
TRIGGER
COMP
FLIP FLOP
COMP
COMP
FLIP FLOP
COMP
DISCHARGE
THRESH OLD
CONTROL VOLTAGE
RESET
OUTPUT
TRIGGER
GND
566 pin diagram

Figure 5.15

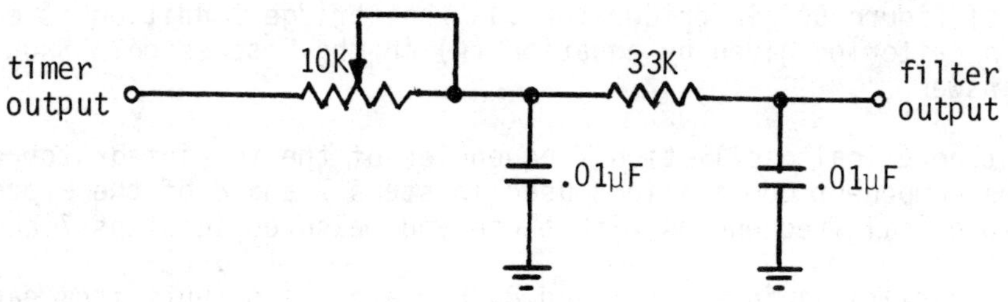

Figure 5.16

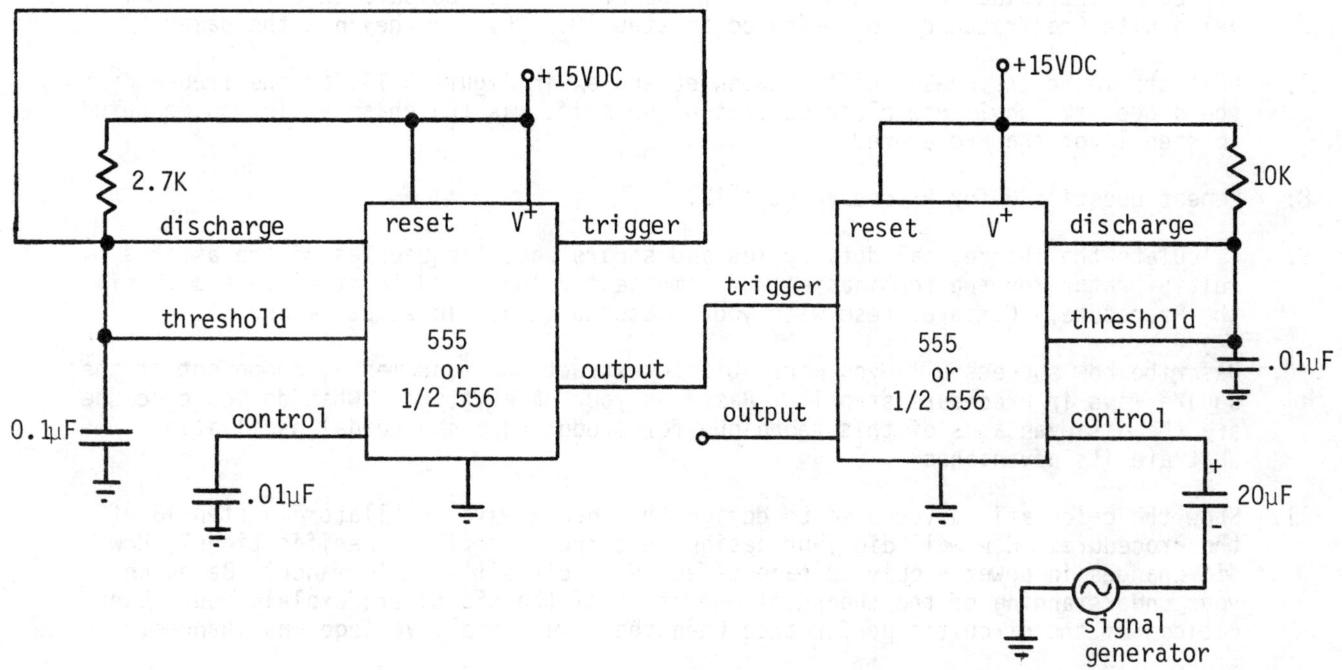

Figure 5.17

permitted to exceed 5 V p-p.) Monitor the signal generator output and the one-shot output simultaneously and form a conclusion on the relation between the two.

QUESTIONS

1. Calculate the theoretical oscillation frequency of the Wien bridge oscillator for all of the component combinations used in steps 2, 3, and 4 of the Procedure. Make a table showing the theoretical frequencies, the actual frequencies that you measured, and the percent difference. What is your conclusion, as concerns the predictability of the frequency of a Wien bridge oscillator?

2. How well did your Wien bridge oscillator design meet the specification given in step 5 of the Procedure? Draw the schematic diagram of your design showing component values used and show your design equations.

3. Why is the oscillator that you used in steps 1 through 5 called a <u>bridge</u>? (Redraw

the schematic of Figure 5.1 in bridge form.) What bridge condition is equivalent to the oscillation criterion given by equation (4) in the Discussion? Show work to justify your answer.

4. Calculate the theoretical oscillation frequencies of the two-integrator oscillator for each of the component combinations used in steps 7 and 8 of the Procedure. Compare these theoretical frequencies with those you measured in steps 7 and 8.

5. Based on your observations in steps 7 and 8, how are the outputs from each of the integrators related to each other?

6. Calculate the theoretical oscillation frequency of the RC phase shift oscillator for the component values used in step 10 of the Procedure. Compare this theoretical value with the frequency you measured in step 10. Why are they not the same?

7. What should be the phase shift between e_s and e_o in Figure 5.14, at the frequency you drove it? Why? How close to that phase shift was the phase angle you measured in step 11 of the Procedure?

8. Repeat question 6 for Procedure step 12.

9. Calculate the theoretical duty cycles and square wave frequencies of the astable multivibrator for the combinations of component values used in steps 14 and 15 of the Procedure. Compare these with your measured values in steps 14 and 15.

10. Describe how successfully you were able to extract the fundamental component of the square wave in Procedure step 17. Based on your observations, what do you conclude are the disadvantages of this technique for producing a sinusoidal oscillation? What are its advantages?

11. Show the calculations you used to design the square wave oscillator in step 18 of the Procedure. How well did your design meet the prescribed specifications? How did changes in power supply voltage affect your circuit's performance? Based on your understanding of the theory of operation of the 555 timer, explain your observations of the circuit's performance when the power supply voltage was changed.

12. Describe your observations of the one-shot output in step 20 of the Procedure. How did this output change with changes in frequency and amplitude of the signal generator output? Relate the signal generator output to the nature of the one-shot output. Why is this circuit called a pulse width modulator?

13. Based on your observations and experience in the conduct of this experiment, compare the various types of oscillators from the standpoints of

    (a) predictability

    (b) east of adjustment

    (c) economy

    (d) distortion

Design Project Number 1: Variable Frequency Signal Generator

Design an integrated circuit sine wave signal generator that will meet the following specifications:

1. Frequency continuously variable (adjustable) over the range from 1 kHz to 10 kHz.

2. Output amplitude continuously adjustable from zero V to 5 V peak, with no dc component, across a 100 ohm load resistor. (It is not necessary that the output amplitude remain constant as the frequency is varied; however, the output amplitude must be capable of being set from 0 to 5 V peak without distortion at all frequencies.)

Your design must be based on the use of $\pm$ 25 V fixed power supply voltages. Use one or more voltage regulators as necessary to provide the voltages required by your design. Do not use higher supply voltages than are absolutely necessary to meet the design specifications. Besides voltage regulators, your design should include an oscillator and an audio amplifier.

Construct your circuit and test its performance to determine how well it meets specifications. You may modify and refine your circuit after testing as necessary to improve its performance. Use potentiometers connected as variable resistors wherever you believe precision-valued components are necessary.

Write a report that contains

1. Your design equations and calculations. Justify and explain all decisions that you make relative to component selection, design tradeoffs, and design changes made after circuit tests. In particular, justify any use of non-standard or precision-valued components.

2. A schematic diagram of your final circuit.

3. Test results that show how well your circuit performed and the degree to which it met the specifications. By inserting dc ammeters in series with each of the fixed power supplies you used, determine and report the total power (VxI) that your circuit consumed when the output was set to its maximum 5 V peak value.

Design Project Number 2: 1 kHz Tone Generator

Design an integrated circuit tone generator that will meet the following specifications:

1. Generate a sine wave signal with frequency 1 kHz $\pm$ 1%.
2. Drive a speaker (6Ω to 16Ω impedance) with sufficient power to produce an audible tone. (The tone must be audible to listeners located anywhere in an average-sized room.) The sound level produced must be continuously adjustable down to an inaudible level.

Your design must be based on the use of a square wave oscillator whose output is filtered using a passive filter network. You may use any type of passive filter network you wish, provided you have access to the components necessary to construct it.

Also, you must use a fixed + 25 V DC supply for your power source. Use one or more voltage regulators as necessary to provide the voltages required by your design. Do not use higher supply voltages than are absolutely necessary to meet the design specifications.

Construct your circuit and test its performance to determine how well it meets specifications. You may modify and refine your circuit after testing as necessary to improve its performance. Use potentiometers connected as variable resistors wherever you believe that precision-valued components are necessary.

Write a report that contains

1. Your design equations and calculations. Justify and explain all decisions that you make relative to component selection, design tradeoffs, and design changes made after circuit tests. In particular, justify any use of non-standard or precision-valued components.
2. A schematic diagram of your final circuit.
3. Test results that show how well your circuit performed and the degree to which it met the specifications. By inserting a dc ammeter in series with the fixed power supply, determine and report the total power (VxI) consumed by your circuit when maximum output power is produced.

6 Voltage Summation, Integration, and Differentiation Using Op-Amps

OBJECTIVES

1. To learn how to construct operational amplifier circuits that are capable of producing outputs proportional to the sum or difference of several input voltages.

2. To verify experimentally the theoretical output of an amplifier connected to generate a scaled sum of voltages.

3. To gain experience designing operational amplifier circuits that produce a specified sum and difference of voltages.

4. To verify experimentally the theoretical output of amplifiers connected to perform mathematical integration and differentiation of sinusoidal and non-sinusoidal waveforms.

5. To learn how integration and differentiation of sinusoidal waveforms affect their amplitudes and phase angles.

6. To learn how integration and differentiation of waveforms with respect to time can be used in waveshaping applications.

EQUIPMENT AND MATERIALS REQUIRED

1. Dual trace oscilloscope.

2. $\pm$ 15 V DC power supplies. (A third, adjustable power supply may also be used; see step 1 of the Procedure.)

3. Function generator (sine, square, triangular), adjustable, 15 Hz - 1 kHz, 0 - $\pm$ 10 V peak.

4. 741 operational amplifier.

5. Resistors: 10K, 15K, 47K, 100K(2), 1M.

6. Capacitors: .001 μF, .022 μF, 0.1 μF.

DISCUSSION

Operational amplifiers are widely used to perform mathematical operations on voltages, including scaling, summation, subtraction, integration, and differentiation. Applications include analog computers, waveshaping, instrumentation systems, analog-digital conversion, and signal processing.

In Experiment 1, we investigated voltage scaling by constructing operational amplifier circuits whose gains were determined by the ratio R_f/R_1. Figure 6.1 shows an extension of this technique, which can be used for simultaneous scaling and summation of several voltage inputs.

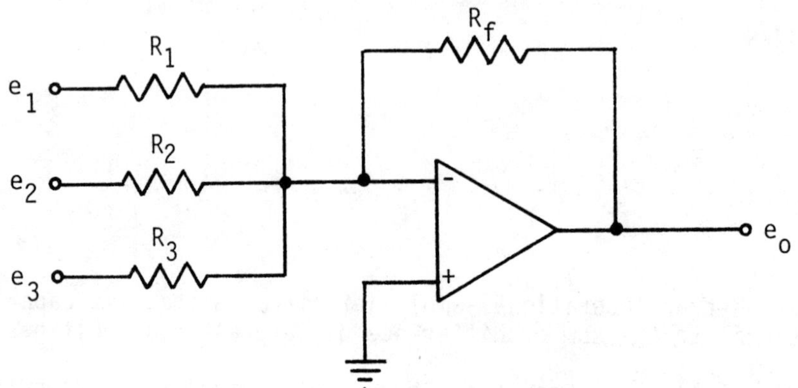

Figure 6.1 An operational amplifier connected for voltage summation.

Using a derivation very similar to that given in the Discussion section of Experiment 1, we can show, provided the input impedance to the amplifier and its open loop gain are quite large, that

$$e_o = -\left(\frac{R_f}{R_1}e_1 + \frac{R_f}{R_2}e_2 + \frac{R_f}{R_3}e_3\right) \tag{1}$$

This result can of course be extended to similar circuits containing more or less than three inputs. Equation (1) shows that the output is proportional to the (inverted) sum of the scaled inputs. Note for the special case where $R_1 = R_2 = R_3 = R$, equation (1) reduces to

$$e_o = -\frac{R_f}{R}(e_1 + e_2 + e_3) \tag{2}$$

When a dc voltage E is added to a sinusoidal wave $A\sin\omega t$, the result is $E + A\sin\omega t$, an ac signal with a dc component, which has maximum value E + A and minimum value E-A.

In some applications, it is desirable to produce a voltage which is proportional to the sum of several voltages and the difference of others. For example, suppose we want a voltage e_o equal to $10e_1 + 2e_2 - 4e_3$. Figure 6.2 shows one way e_o could be obtained, by using two operational amplifiers and taking advantage of the phase inversion that each amplifier causes. In Figure 6.2, as in all operational amplifier circuits that produce mathematical functions of input voltages, the resistors used should be precision resistors if maximum accuracy is desired, and, of course, dc offsets should be eliminated, as discussed in Experiment 2.

If the feedback resistor in an operational amplifier circuit is replaced by a capacitor, then the output voltage is proportional to the mathematical integral of the input. When a voltage is integrated (with respect to time) the output is a voltage which represents how the total accumulated area under the input waveform changes with the passage of time. Mathematical integration is represented by the symbol $\int$. To indicate that one voltage e_o is the integral of another, e_1, with respect to time t, we write $e_o = \int e_1 dt$.

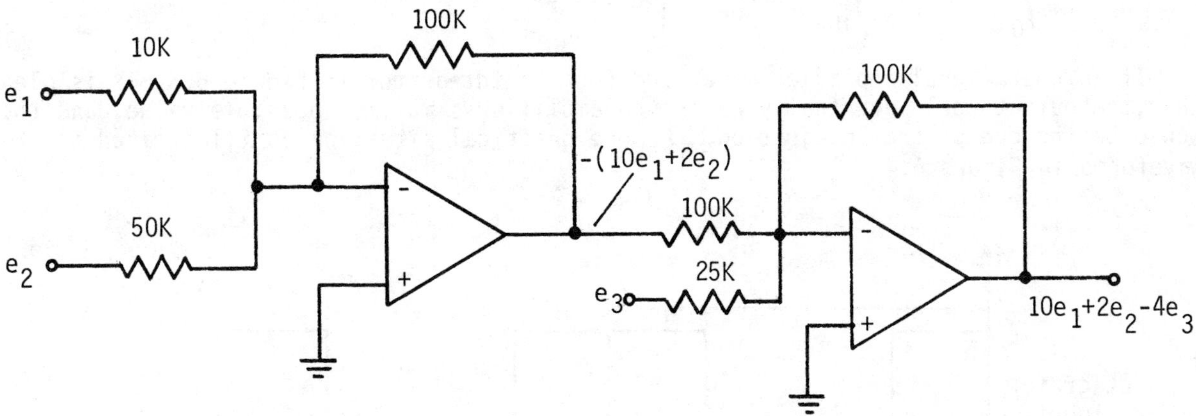

Figure 6.2 Two operational amplifiers connected to generate the output $10e_1 + 2e_2 - 4_3$.

To understand the concept of voltage integration with respect to time, study Figure 6.3.

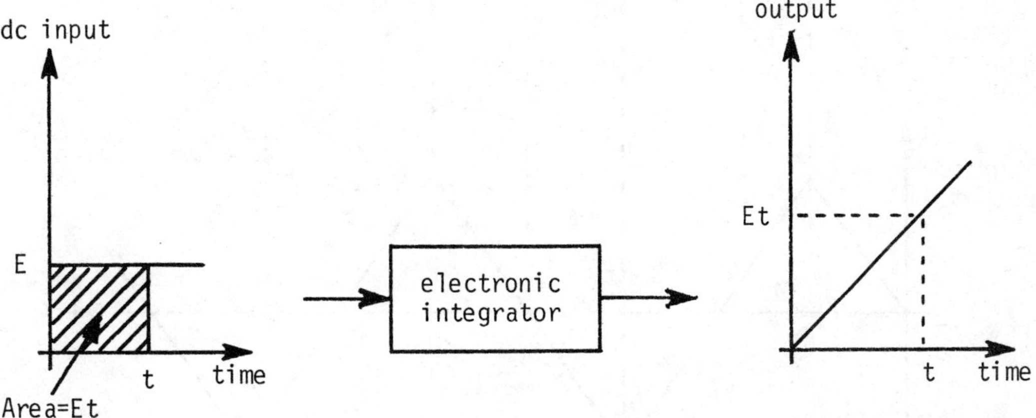

Figure 6.3 The principle of electronic integration illustrated for the case of a constant (dc) input. The output is proportional to the area under the input.

In Figure 6.3, the input to an electronic integrator is a dc voltage of E volts, beginning at t = 0. After any time t has passed, the total area under the dc level is equal to height x base = Et (shaded area shown). The output of the integrator at that time is therefore Et volts. Clearly, then, as time passes, i.e., as t increases, the output of the integrator increases linearly, as shown in the figure. The output is called a <u>ramp voltage</u>, with slope + E. (We assume this integrator does not invert.) Mathematically,

$$\int_0^t E dt = E \int_0^t dt = Et \Big]_0^t = Et$$

If an operational amplifier were used for the integrator in Figure 6.3, it is clear that the output would eventually reach the amplifier's maximum possible value, and that would be the end of the integration. A more practical situation is illustrated by the waveforms in Figure 6.4

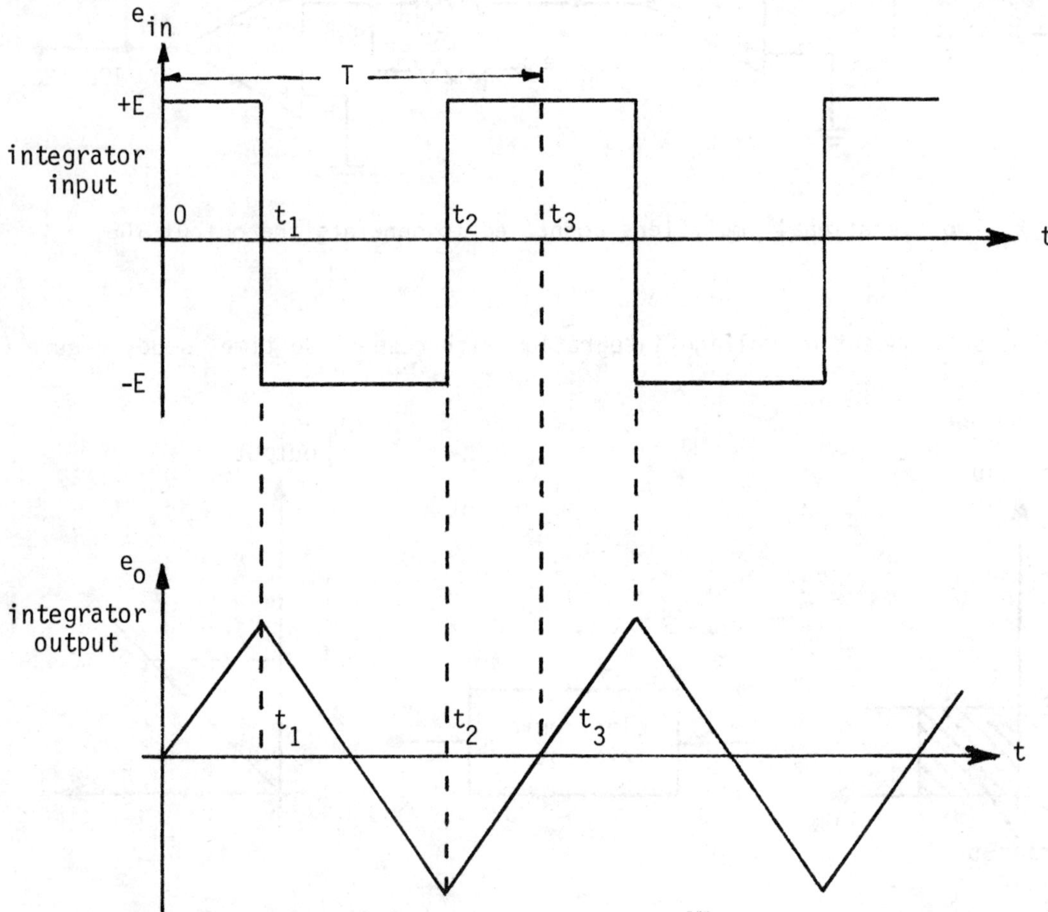

Figure 6.4 Electronic integration of a square wave produces a triangular wave.

The input waveform in Figure 6.4 can be thought of as the alternate application of positive and negative dc voltages to the input of the integrator. A negative dc voltage results in a negative area, which the integrator therefore subtracts from any positive area accumulated up to that point. In other words, the output of the integrator at any time t is the net area under the input up to that point. Note in Figure 6.4 that positive area accumulates under the input waveform until $t = t_1$. Since area is increasing during this time, the slope of the output is positive up to $t = t_1$. The input then goes negative. From $t = t_1$ to $t = t_2$ the area under the square wave is negative, and it therefore subtracts from the positive area accumulated thus far. Consequently, the net area decreases after t_1, as reflected by the negative slope of the triangular wave. At $t = t_2$ the input goes positive again, thus adding area and causing the output to increase once more. When $t = t_3$, which is the period T of the input, the total positive area accumulated equals the total negative area accumulated and the net area at that time is therefore zero. Consequently the triangular output equals zero at $t = t_3$. The entire process then repeats itself. This is an example of waveshaping: using the knowledge of how integration affects a function represented by one waveform to create another. In this case, we created a triangular wave from a square wave.

As indicated previously, an operational amplifier with a capacitor in the feedback path will perform electronic integration. See Figure 6.5

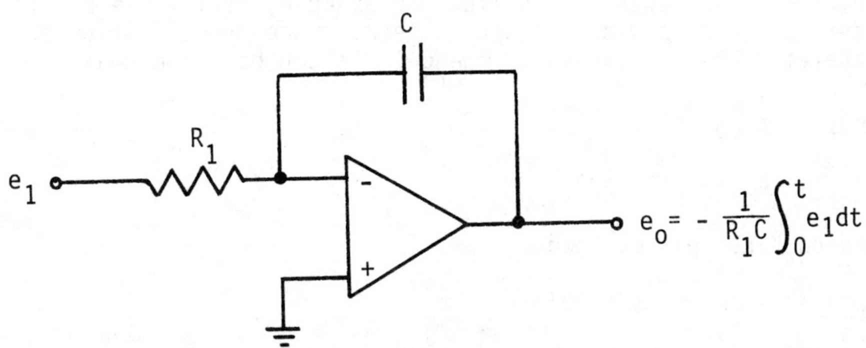

Figure 6.5 An operational amplifier connected as an integrator (with phase inversion).

Note in Figure 6.5 that the output is proportional to the integral of the input scaled by a factor equal to $1/R_1C$ and, as usual, inverted by the amplifier. Practical capacitor values used in this application are usually 1 μF or less. (Electrolytic capacitors should not be used; since e_0 goes both positive and negative, it would not be possible to maintain the correct polarity across the polarized capacitor terminals.) If the square wave shown in Figure 6.4 were applied to the integrator in Figure 6.5, the triangular output would be inverted (180^0 out of phase with that shown in Figure 6.4) and the slopes would be $\pm E/R_1C$ volts/sec, instead of $\pm E$ volts/sec as shown in Figure 6.4.

If a sine wave $e_1 = E\sin\omega t$ is applied to the integrator shown in Figure 6.5, it can be shown that

$$e_0 = -\frac{1}{RC}\int_0^t E\sin\omega t\,dt = \left(-\frac{E}{\omega RC}\right)(-\cos\omega t)$$

$$= \frac{E}{\omega RC}\sin(\omega t + 90^0) \qquad (3)$$

The output of the integrator in this case is a sine wave which leads the input by 90^0 and whose peak value $E/\omega RC$ is seen to be inversely proportional to frequency ω. As an example, suppose in Figure 6.5 that $e_1 = 10\sin 50t$, $R_1 = 100K$, and $C = 1$ μF. Then

$$e_0 = \frac{10}{(50)(10^5)(10^{-6})}\sin(50t + 90^0)$$

$$= 2\sin(50t + 90^0)$$

$$= 2\cos 50t$$

Note that the value of R_1C can be chosen to produce gain in the integration ($1/R_1C = 10$ in this example).

An electronic differentiator produces an output voltage that is proportional to the rate of change of the input voltage at any instant of time t. Differentiation produces the derivative with respect to time of a voltage waveform and is the "opposite" of integration in many senses. In fact, if e_0 is the integral of e_1, then it is the same to say that e_1 is the derivative of e_0. Study Figure 6.4 again and verify that e_1 represents the rate of change of e_0. Remember when the slope of a waveform is constant, E for example, that means that its rate of change is the constant value E volts/sec. Differentiation with respect to time is expressed by the symbol d()/dt, where the function being differentiated is written inside the parentheses. We write, for example, $d(Et)/dt = E$. It should be noted that the derivative of any dc voltage is zero, since by definition a dc voltage does not change and therefore has zero rate of change. It can be shown that

$$\frac{d(A\sin\omega t)}{dt} = A\omega\cos\omega t$$

$$= A\omega\sin(\omega t + 90^\circ) \qquad (4)$$

If a phase inversion takes place, then

$$\frac{d(A\sin\omega t)}{dt} = -A\omega\sin(\omega t + 90^\circ) \qquad (5)$$

$$= A\omega\sin(\omega t - 90^\circ)$$

True electronic differentiators are not widely used because, as shown by equations (4) and (5), differentiation produces an output which is directly proportional to frequency (again, the opposite of integration). As a result of this fact, high-frequency, low amplitude noise in the input is effectively amplified in the output, an undesirable situation. However, it is possible to design an operational amplifier circuit that produces a good approximation to the derivative of its input, over a limited frequency range.

Figure 6.6 shows a practical differentiator. Theoretically, more exact differentiation would take place if the components R_1 and C_2 were omitted from this circuit. However, due to the noise problems mentioned before, and the tendency of the circuit to oscillate without these components, their presence is necessary in the practical differentiator. Resistor R_1 helps reduce high frequency noise and capacitor C_2 helps suppress oscillations.

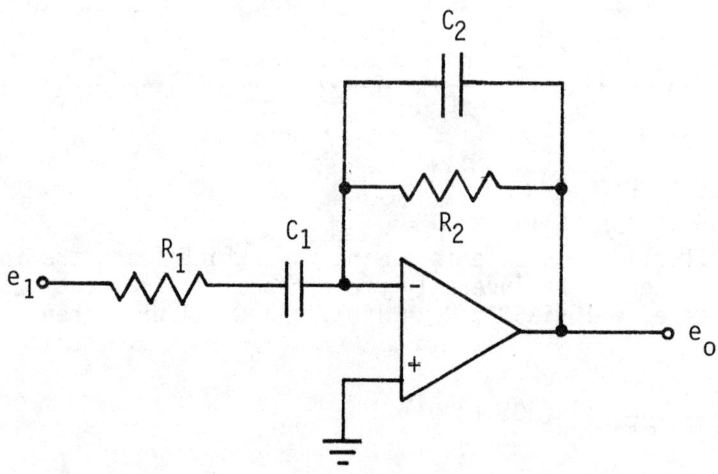

Figure 6.6 A practical electronic differentiator.

The circuit in Figure 6.6 will provide reasonably accurate differentiation up to a frequency of f_1, where

$$f_1 = \frac{1}{2\pi R_1 C_1} \text{ Hz}$$

The values of R_2 and C_2 should be chosen so that $f_2 > f_1$, where

$$f_2 = \frac{1}{2\pi R_2 C_2} \text{ Hz}$$

The gain of the differentiator in its region of differentiation is then equal to R_2C_1. If the input to the differentiator of Figure 6.6 is $A\sin\omega t$, then equation (5), modified by the gain of the differentiator, tells us that

$$e_o = A\omega R_2 C_1 \sin(\omega t - 90^\circ) \tag{6}$$

EXERCISES

1. Design an operational amplifier circuit that will produce the output $e_o = -(2e_1 + .5e_2 + 2.5e_3)$, where e_1, e_2, and e_3 are input voltages. Draw a schematic diagram of your circuit. It is not necessary to specify standard value resistors.

2. Design an operational amplifier circuit that generates the output $e_o = -(2.13e_1 + e_2)$. Use only standard 5% tolerance resistor values and draw a schematic of your design. Write the expression for e_o that results when e_1 = -9 V DC and $e_2 = \sin(2\pi \times 100t)$ V. Sketch e_o for this case.

3. In the circuit shown in Figure 6.5, R_1 = 100K and C = 0.1 μF. If e_1 = -0.5 V DC, write an equation for e_o as a function of time t. How long after the input is applied would it take the output to reach 1.0 V?

4. In the same circuit described in exercise 3, suppose $e_1 = 2\sin(2\pi \times 15t)$ V. Write an equation for e_o as a function of time t. Sketch e_1 and e_o on the same set of axes. Repeat, when $e_1 = 2\sin(2\pi \times 100t)$ V.

5. In the same circuit described in exercise 3, suppose e_1 is a 20 Hz square wave that alternates between ± 2 V. Sketch e_1 and e_o versus time on the same set of axes. Be sure to label the minimum and maximum values reached by e_o.

6. In the circuit shown in Figure 6.6, R_1 = 10K, C_1 = .022 μF, R_2 = 15K and C_2 = .001 μF. If $e_1 = 5\sin(2\pi \times 100t)$, write an equation for e_o as a function of time t. Sketch e_1 and e_o on the same set of axes.

7. A triangular wave has peak values ± 10 V and a frequency of 50 Hz. What is the slope (rate-of-change) of the voltage as it varies from -10 V to + 10 V?

PROCEDURE

1. To verify the operational amplifier's ability to sum voltages, connect the circuit shown in Figure 6.7. If the dc power supply connected to the 100K resistor is not available, e_2 may be derived through a voltage divider (potentiometer) connected across one of the ± 15 V DC power supplies. In this step e_2 will be set to a negative voltage. Set e_1 to 1 V peak at 100 Hz. Set e_2 to -9 V DC. Using an oscilloscope measure and record the output voltage e_o. Make certain that the oscilloscope input is "DC" connected and be sure to record any dc level in the output. Sketch the output waveform observed.

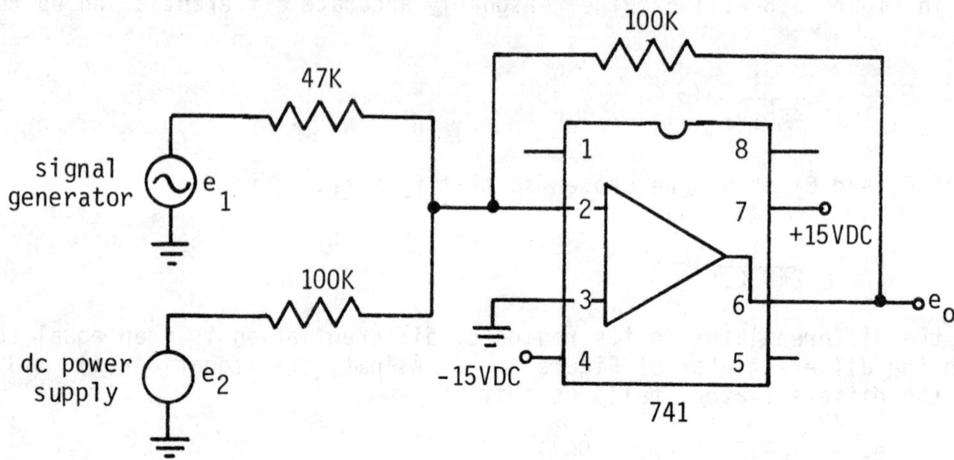

Figure 6.7

2. Slowly increase and decrease the negative dc voltage e_2 and observe the effect this has on the output waveform.

3. Reverse the polarity of the dc voltage e_2 and set it for + 9 V DC. Then repeat the measurement procedure in step 1.

4. Repeat step 2 with the positive dc voltage e_2.

5. Design a circuit using two operational amplifiers that will produce $e_o = 10e_1 - 1.5e_2 - e_3$. Connect your circuit and record e_o for e_1 = 1 V DC, e_2 = a 2 volt peak sine wave at 100 Hz, and e_3 = 5 V DC.

6. Connect the integrator circuit shown in Figure 6.8

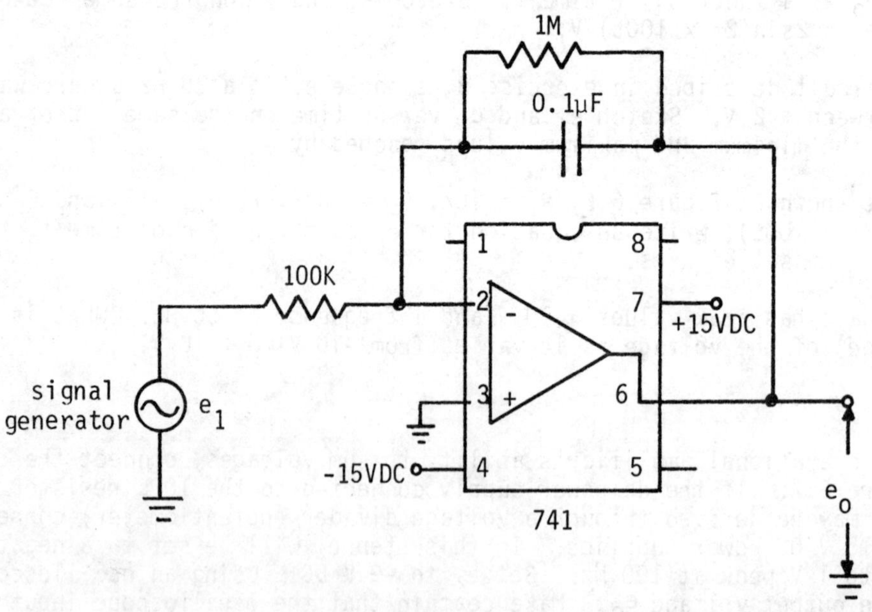

Figure 6.8

The 1M resistor connected across the 0.1 μF capacitor is used to prevent the amplifier from saturating due to the presence of any dc offset in the input. Note that the integrator would react to dc offset in the same way it would to any dc input, namely it would generate a ramp which would eventually reach the maximum output voltage of the amplifier. The 1M resistor causes the dc gain of the amplifier to be a constant 1M/100K = 10.

7. Connect e_0 and e_1 to a dual trace oscilloscope for simultaneous viewing. Set the function generator to produce a 2 volt peak sine wave at 15 Hz. Measure and record the amplitude of e_0 and its phase relation to e_1.

8. Repeat step 7 at the following frequencies: 20 Hz, 40 Hz, 80 Hz, 100 Hz, 200 Hz, 500 Hz, and 1 kHz. (Note: we are only interested in the ac characteristics of the output, i.e., peak ac value and phase, so ignore any dc offset in the output. This is best accomplished by setting the oscilloscope for ac input.)

9. Now set the function generator to produce a 20 Hz square wave that alternates between ± 2 V. Measure and record the input and output waveforms at a sufficient number of points to allow you to reproduce them in accurate sketches. Be sure to record the time points at which output voltages are measured, so that the waveform slopes can be calculated.

10. Design an operational amplifier integrator that will generate a .6 volt peak sinusoidal output when the input is a 2 volt peak sine wave at 25 Hz. Connect your circuit and verify its operation. (You may need to use a variable resistor for the input resistance, in order to obtain the value you calculated in your design.)

11. Connect the differentiator circuit shown in Figure 6.9. To aid in suppressing oscillations, connect a .01 μF capacitor between the plus and minus power supply voltages.

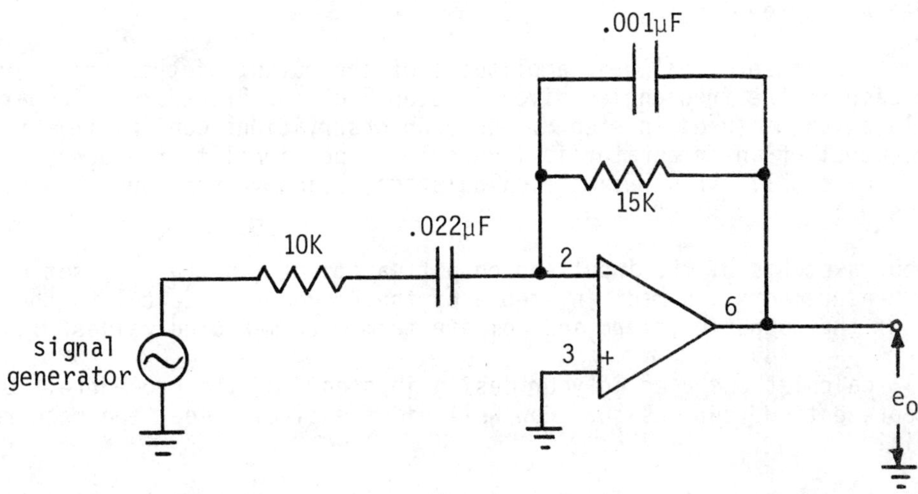

Figure 6.9

12. Connect a dual trace oscilloscope for simultaneous viewing of e_1 and e_0. Adjust the signal generator to produce a 5 volt peak sine wave at 100 Hz. Measure and record the peak value of e_0 and the phase angle of e_0 with respect to e_1. Be sure to note which one is leading.

13. Repeat step 12 at 200 Hz and again at 500 Hz.

14. Remove the 10K resistor in Figure 6.9 and connect a function generator to the .022 μF capacitor. Connect the generator to produce a triangular wave that varies between ±10 V peak values at 50 Hz. Measure the peak value of e_o and sketch its appearance. Also sketch the input triangular wave on the same set of axes.

QUESTIONS

1. Write the mathematical expression for the theoretical output of the amplifier in Figure 6.7, based on the inputs in step 1 of the Procedure. Remember to take into account the phase inversion of the amplifier. What are the theoretical maximum and minimum values? Sketch the theoretical waveform. Compare the theoretical values and the waveform with those observed in step 1 of the Procedure.

2. Describe the effect of varying the negative dc voltage e_2 on the output waveform, as observed in step 2 of the Procedure.

3. Repeat question 1 for step 3 of the Procedure.

4. Repeat question 2 for step 4 of the Procedure.

5. Show the operational amplifier circuit you designed for step 5 of the Procedure. Then repeat question 1 for the output of this circuit and the inputs used in step 5.

6. Do your experimental results generally confirm that the operational amplifier provides an accurate way to add and/or subtract voltages?

7. Write the theoretical expression for the output of the amplifier in Figure 6.8, based on the input given in step 7 of the Procedure. (Use the input as reference; i.e., assume the input has zero phase angle.) Compare your observations in step 7 with the theoretical expression.

8. Calculate the theoretical peak amplitudes of the output of the amplifier in Figure 6.8 at each of the frequencies given in step 8 of the Procedure. Compare these with the values you measured in step 8. Do your observations confirm that the amplitude of the output of an integrator is inversely proportional to frequency, when the input is sinusoidal? Show a sample calculation, using your measured values, to support your answer.

9. Show your sketches of the input and output waveforms (on the same set of axes) based on the measurements you made in step 9 of the Procedure. Calculate the theoretical slopes of the output waveform and compare them with measured values.

10. Show all calculations used in your design in step 10 of the Procedure. Draw the circuit representing your design. How well did the circuit meet the required performance criteria?

11. What is the theoretical amplitude and phase of the output of the amplifier in Figure 6.9, when the input is that given in step 12 of the Procedure? Write the theoretical expression for the output. Compare the theoretical output with that observed and measured in step 12.

12. Repeat question 11 for step 13 of the Procedure. What do your observations confirm about the relation between frequency and the amplitude of the output of a differentiator?

13. What is the maximum frequency at which the circuit in Figure 6.9 can be expected to perform differentiation?

14. Compare the theoretical output of an ideal differentiator driven by the triangular input given in step 14 of the Procedure with the output observed in step 14.

7 Active Filters

OBJECTIVES

1. To learn how operational amplifiers can be connected with resistor-capacitor networks to construct various types of filters.

2. To learn through experimental observation the distinguishing characteristics of low-pass, high-pass, and band-pass filters.

3. To learn the distinguishing characteristics of Chebyshev and Butterworth filter responses.

4. To verify experimentally the operation of VCVS, IGMF, biquad, and state-variable filters.

5. To verify experimentally certain theoretical filter characteristics, such as cutoff frequency, bandwidth, asymptotic attentuation, and ripple width.

6. To gain experience designing an active filter and testing its response.

EQUIPMENT AND MATERIALS REQUIRED

1. Dual-trace oscilloscope.

2. $\pm$ 15 V DC power supplies.

3. Sinewave signal generator, 0-100 kHz, 0-5 VRMS.

4. 741 operational amplifiers (3).

5. Resistors: 120 Ω, 180 Ω, 270 Ω, 330 Ω(4), 680 Ω, 1K, 1.5K, (2), 2.7K (2), 4.7K (2), 10K (4), 12K.

6. Potentiometer: 1K.

7. Capacitors: .047 μF (2), .005 μF, 0.47 μF (2), 0.1 μF (4).

DISCUSSION

A filter is a device that allows signals having frequencies in a certain range to pass through it while attenuating (reducing) all other signals. A low-pass filter permits all signals having frequencies below a certain cutoff frequency to pass through it and attenuates all signals with frequency greater than cutoff. Similarly, a high-pass filter passes signals having frequencies higher than a cutoff frequency and attenuates all others. A band-pass filter passes all signals in a band of frequencies between a lower cutoff and an upper cutoff frequency. A band-reject (or band-stop) filter attenuates signals having frequencies within a certain band. Figure 7.1 shows typical frequency response curves for each of these filter types.

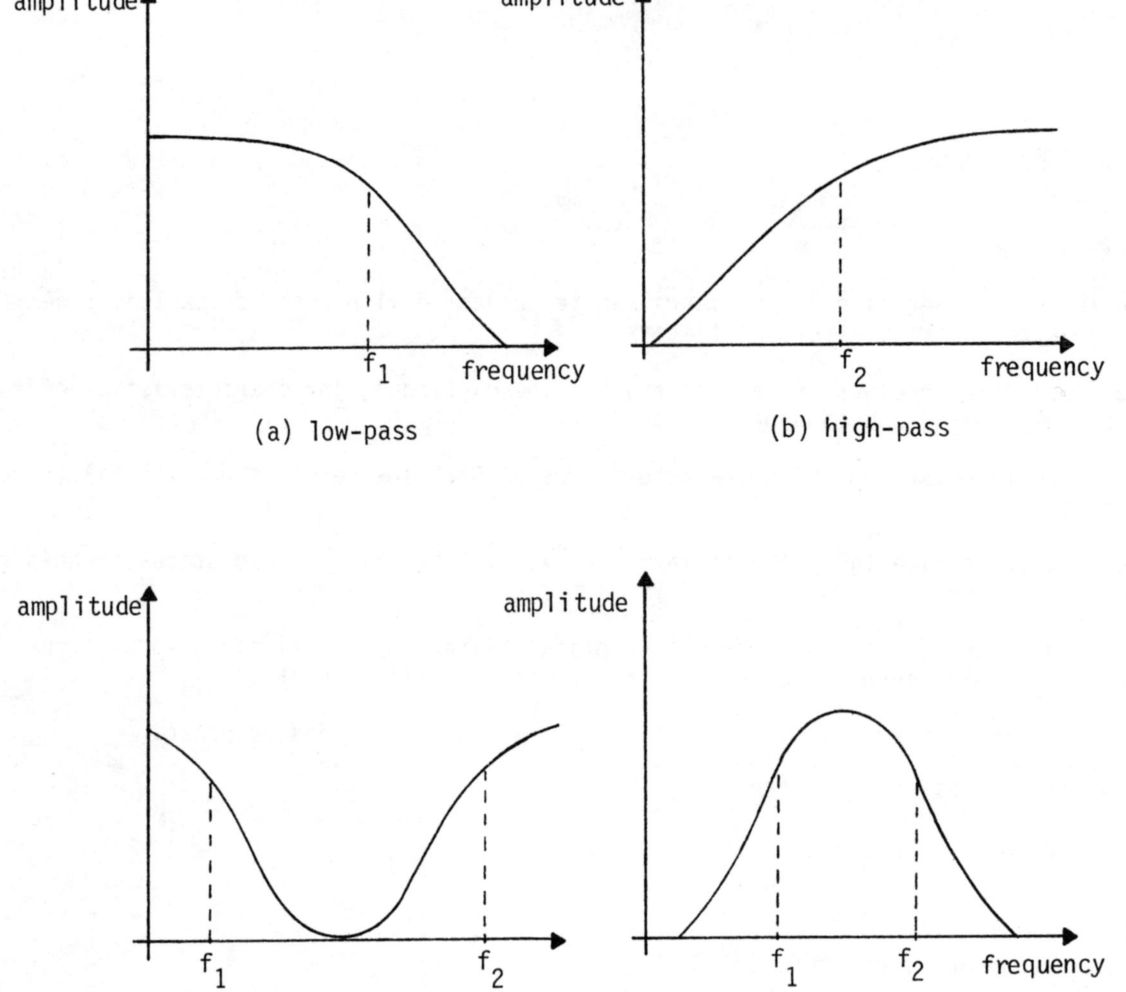

Figure 7.1 Frequency response characteristics of different types of filters.

Each frequency response characteristic in Figure 7.1 shows the magnitude of the output signal passed by the filter as the signal frequency is changed. Note in each case that signals having frequencies outside the passband are attenuated rapidly as their frequencies move further away from cutoff. The <u>bandwidth</u> of a filter is defined to be the difference between the maximum and minimum frequencies of the passband (or of the stop-band, in the case of a band-reject filter). Thus the bandwidth of 7.1(a) is f_1 and the bandwidth of 7.1(d) is f_2-f_1. Every high-pass filter has some upper cutoff frequency (not shown in Figure 7.1(b)) because no physically relizable device can have infinite bandwidth. We refer to the device as a high-pass filter when all the frequencies that are passed beyond its cutoff are frequencies that may appear at its input for a particular application. For example, if a filter has a passband between 10 kHz and 1 MHz and we are using it to filter signals whose frequency range is 5 kHz to 100 kHz, then it is for all practical purposes a high pass filter.

In practice, the kinds of signals that are most often filtered are complex waveforms consisting of many frequency components (as, for example, speech and music) rather than signals having only one frequency. The purpose of filtering, then, can be thought of as extracting only certain components from the composite waveform. As we noted in Experiment 5, a square wave is composed of a fundamental frequency and an infinite number of odd multiples of that fundamental (harmonics); by filtering the square wave, we are able to extract, for example, only the fundamental component.

An ideal filter passes, with exactly the same amplitude, all frequency components in its passband and completely "stops" (blocks) the passage of all frequency components outside its passband. Real filters do not have these properties. Real filters are classified according to certain fundamental design features that determine the degree to which frequencies in their pass-band are passed with equal amplitude and how well frequencies outside the passband are attenuated. The filter known as a Butterworth type has the characteristic that all frequencies within the passband are passed with very little variation in amplitude, although amplitudes diminish slightly as the frequency approaches cutoff. The Butterworth filter, however, does not attenuate frequency components outside its passband as well as some other types, as well as, for example, certain filters of the Chebyshev design. In order to achieve this improved attenuation, the Chebyshev design must sacrifice uniformity in its passband,; that is, its passband is not as flat as the Butterworth's. Figure 7.2 shows typical frequency response characteristics for low-pass filters of the Butterworth and Chebyshev designs (as they would appear when plotted on log-log graph paper.

Note in Figure 7.2 that the cutoff frequency f_1 of the Butterworth filter is the frequency at which the amplitude is .707 times the maximum amplitude in the passband. As can be seen in the Chebyshev characteristic, there is considerable variation in amplitudes within the passband. The cutoff frequency in this case is found at the intersection of a line drawn tangent to the lowest amplitude in the passband and the response curve outside the passband, as shown in the figure. The total amplitude variation (from minimum to maximum) within the passband is called the ripple width, RW, and is usually expressed in dB. (This should not be confused with ripple in the output of a dc power supply, which is a time-varying fluctuation.) If, for example, the maximum amplitude in the passband is 1.0 at, say 4 kHz, and the minimum amplitude in the passband is 0.707 at, say 8 kHz, then RW $= 20 \log_{10}(1/.707) = 3$ dB. The Chebyshev filter can be designed to have a small ripple width, but this is accomplished at expense of poor attenuation outside the passband.

Another way that filters are classified is by their order. The origin of this terminology and its precise meaning in filter theory is beyond the scope of this book, but we should be aware of some important practical implications of this characteristic. The order is always an integer (1, 2, etc.) and, in general, the greater the order, the more closely the filter approximates an ideal filter. Also, the greater the order, the more complex the filter, in terms of the number of components required to construct it. If a filter has order n, then its attenuation outside the passband is asymptotic to 6n dB/octave or 20n dB/decade. Figure 7.3 shows the response characteristics of second-order, low-pass Butterworth and Chebyshev filters on the same set of axes. In this figure, we compare a Chebyshev filter having an RW of 3 dB to a Butterworth filter having the same cutoff frequency.

Note in Figure 7.3 that both the Chebyshev and Butterworth frequency responses are asymptotic to 12 db/octave beyond cutoff, but that the Chebyshev has greater attenuation at any given frequency. (The attenuation is greater by 3(n-1)dB, or 3 dB in this case, since n = 2.) Note also that the second-order Chebyshev has but a single peak in the ripple in its passband. As noted previously, reduction of the ripple width in the Chebyshev passband can be achieved at the expense of attentuation outside the passband. In fact, for very small RW, the attenuation is not as good as in the Butterworth design.

A passive filter is one that is constructed from passive components only (capacitors,

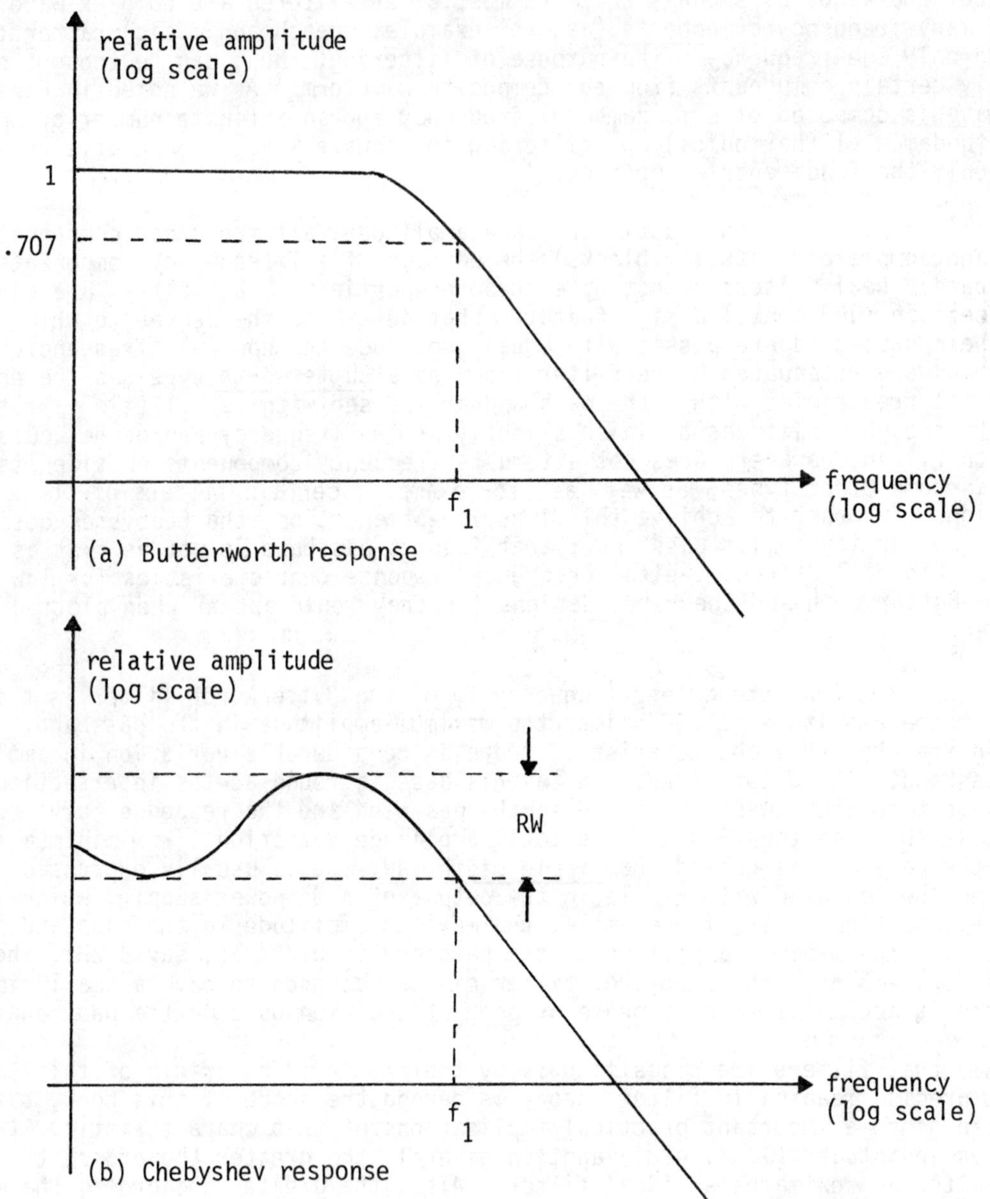

Figure 7.2 Frequency response characteristics of Butterworth and and Chebyshev low-pass filters.

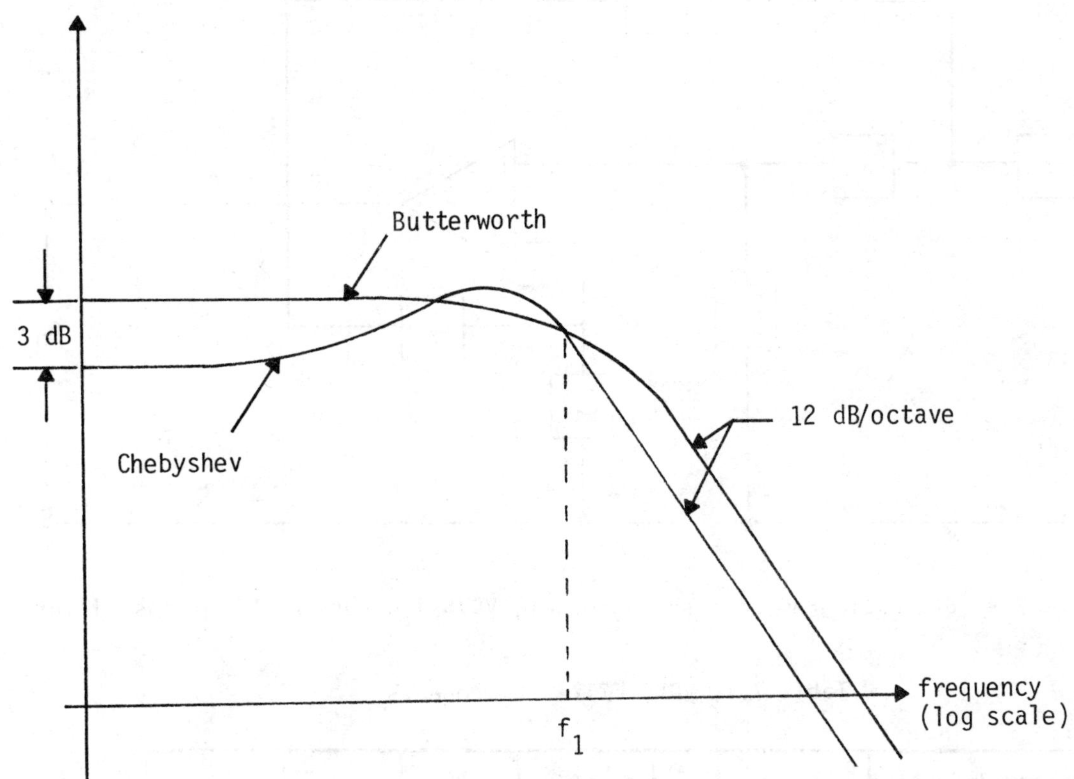

Figure 7.3 Frequency responses of second-order, low-pass Butterworth and Chebyshev filters having the same cutoff frequency.

inductors, and/or resistors). An active filter employs at least one active device, such as an amplifier, in addition to passive components. One advantage of active filters is that they can be designed to produce gain as well as filtering action; that is, signals in the passband can be amplified instead of just passed. Modern active filters are constructed from integrated circuit operational amplifiers and resistor-capacitor combinations. As we saw in Experiment 6, integrators and differentiators have outputs whose magnitudes are dependent on frequency and are therefore examples of (first-order) active filters. In this experiment we will study more complex designs. The mathematics required to analyze such filters is beyond the scope of this book, so we will content ourselves with a practical design method that is based on the use of look-up tables. This method is developed by Johnson and Hilburn in Rapid Practical Designs of Active Filters (Wiley, 1975). The material that follows is drawn from that book.

The first type of active filter we will consider is called a voltage-controlled voltage source (VCVS). By the proper choice of passive components, it can be constructed as a low-pass, high -pass, or band-pass. In its low-pass configuration, it is known as a Sallen-Key circuit. The general form of a second-order VCVS low-pass or high-pass filter is shown in Figure 7.4. (Band-pass and band-reject filters can also be constructed in the VCVS configuration, but the component arrangement is different from that shown in Figure 7.4. See Young, Linear Integrated Circuits (Wiley, 1981, p. 264))

The blocks labeled Z_A through Z_F are resistors or capacitors, depending on the type of filter desired. Table 7.1 summarizes the kind of component required in each block for each of the filter types where an R designation represents resistance and C represents capacitance.

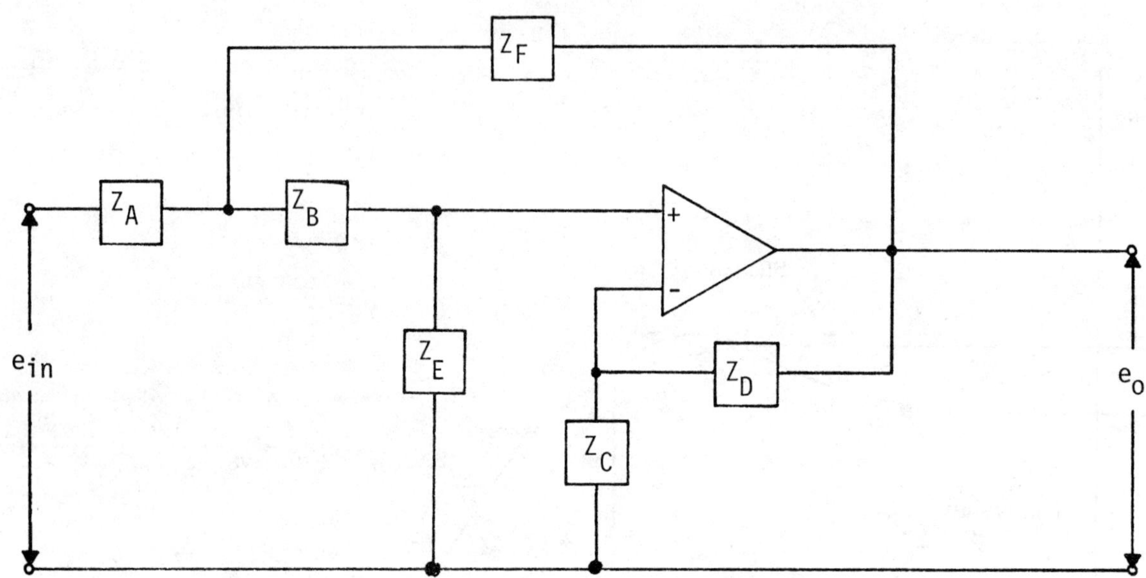

Figure 7.4 Block diagram of a second-order, VCVS low-pass or high-pass filter.

Table 7.1 VCVS Filter Components

Filter Type	Z_A	Z_B	Z_C	Z_D	Z_E	Z_F
Low-pass	R_1	R_2	R_3	R_4	C_1	C
High-pass	C	C	R_3	R_4	R_2	R_1

The design procedure for a low-pass or high-pass VCVS filter begins with a choice for the value of capacitor C. We then calculate a constant K using

$$K = \frac{10^{-4}}{fC} \tag{1}$$

where f is the cutoff frequency. The next step is to calculate values for the other components in the circuit by consulting a table of multipliers appropriate for the type of filter desired. Typical examples are shown in Table 7.2.

To illustrate the use of the tables, suppose we want to design a 4 kHz, second-order, low pass, Butterworth filter having gain 2. Choosing C = .022 μF, we find from equation (1),

$$K = \frac{10^{-4}}{(4 \times 10^3)(2.2 \times 10^{-8})} = 1.136$$

From Table 7.1, we see that $Z_A = R_1$, $Z_B = R_2$, $Z_C = R_3$, $Z_D = R_4$, $Z_E = C_1$ and $Z_F = C$. Then from Table 7.2 we find the multiplier 1.126 for R_1 in the gain 2 column, and we therefore set $R_1 = (1.126)(K) \times 10^3$ ohms, or $(1.126)(1.136) \times 10^3 = 1.28$ kΩ. Similarly, $R_2 = 2.56$ kΩ,

Table 7.2 Second-Order Filter Designs.*

Second-Order Low-Pass Butterworth VCVS Filter Designs

	Circuit Element Values[a]					
Gain	1	2	4	6	8	10
R_1	1.422	1.126	0.824	0.617	0.521	0.462
R_2	5.399	2.250	1.537	2.051	2.429	2.742
R_3	Open	6.752	3.148	3.203	3.372	3.560
R_4	0	6.752	9.444	16.012	23.602	32.038
C_1	0.33C	C	2C	2C	2C	2C

[a] Resistances in kilohms for a K parameter of 1.

Second-Order Low-Pass Chebyshev VCVS Filter Designs (2 dB)

	Circuit Element Values[a]					
Gain	1	2	4	6	8	10
R_1	2.328	1.980	1.141	0.786	0.644	0.561
R_2	13.220	1.555	1.348	1.957	2.388	2.742
R_3	Open	7.069	3.320	3.292	3.466	3.670
R_4	0	7.069	9.959	16.460	24.261	33.031
C_1	0.1C	C	2C	2C	2C	2C

[a] Resistances in kilohms for a K parameter of 1.

Second-Order High-Pass Chebyshev VCVS Filter Designs (2 dB)

	Circuit Element Values[a]					
Gain	1	2	4	6	8	10
R_1	0.640	1.390	2.117	2.625	3.040	3.399
R_2	3.259	1.500	0.985	0.794	0.686	0.613
R_3	Open	3.000	1.313	0.953	0.784	0.681
R_4	0	3.000	3.939	4.765	5.486	6.133

[a] Resistances in kilohms for a K parameter of 1.

*Reprinted from Rapid Practical Designs of Active Filters, D. Johnson and J. Hilburn.

R_3= 7.67kΩ, and R_4 = 7.67kΩ. Also, C_1=C=.022 μF. Note that the multiplier K is not used for finding the value of C_1. Extensive tables for a wide variety of VCVS filter types, including Butterworth and Chebyshev of different orders and ripple widths, can be found in the previously referenced book, Rapid Practical Designs of Active Filters. As can be seen in the previous example, values calculated from these tables are not generally standard component sizes. Precision resistors should be used if exact cutoff frequencies, gains, and attenuations are required.

Another filter type, similar in design to the VCVS, is called the infinite gain, multiple feedback (IGMF) filter. The IGMF low-pass configuration is shown in Figure 7.5.

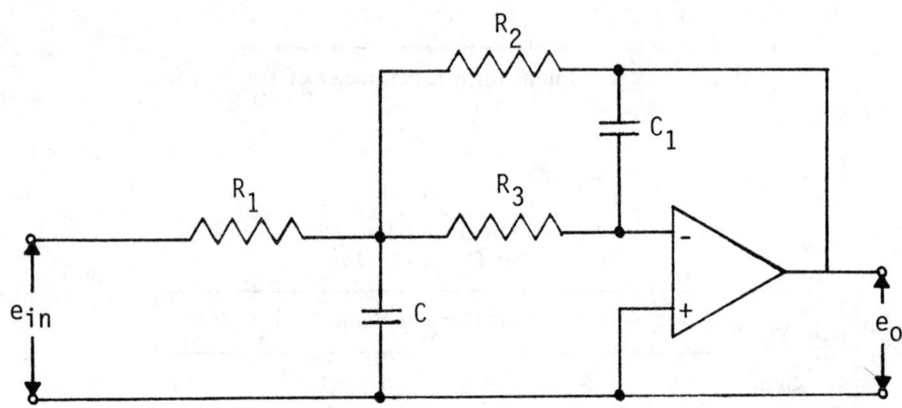

Figure 7.5 Circuit diagram of an IGMF low-pass filter. (Reprinted from Rapid Practical Designs of Active Filters, D. Johnson and J. Hilburn. Copyright © 1975, John Wiley and Sons, Inc., by permission of John Wiley and Sons, Inc.)

Note that the IGMF design requires one less component than the VCVS design. The IGMF high-pass and band-pass filters are constructed by reversing the locations of the resistors and capacitors in Figure 7.5. The design procedure for the IGMF filter is exactly the same as that for the VCVS design, except of course a different set of tables is used. As before, we first choose C and then use equation (1) to calculate the constant K. Tables for both Butterworth and Chebyshev designs, of different orders and ripple widths, can be found in the previously listed reference.

The biquad filter is a three-amplifier design that is useful in applications requiring adjustable gain and good frequency tuning. The low-pass biquad configuration is shown in Figure 7.6. The high-pass and band-pass biquad designs are similar to Figure 7.6 but have different resistor-capacitor configurations. As before, the biquad design begins with a choice of C, followed by the calculation of $K = 10^{-4}/fC$, and the use of a set of tables to determine component values. The gain of the filter shown in Figure 7.6 is R_3/R_1.

Another active filter design requiring three amplifiers is the state-variable filter. This design simultaneously provides low-pass, high-pass, and band-pass filtering, each of these functions being taken from a different point in the circuit. The general form of the state-variable filter is shown in Figure 7.7.

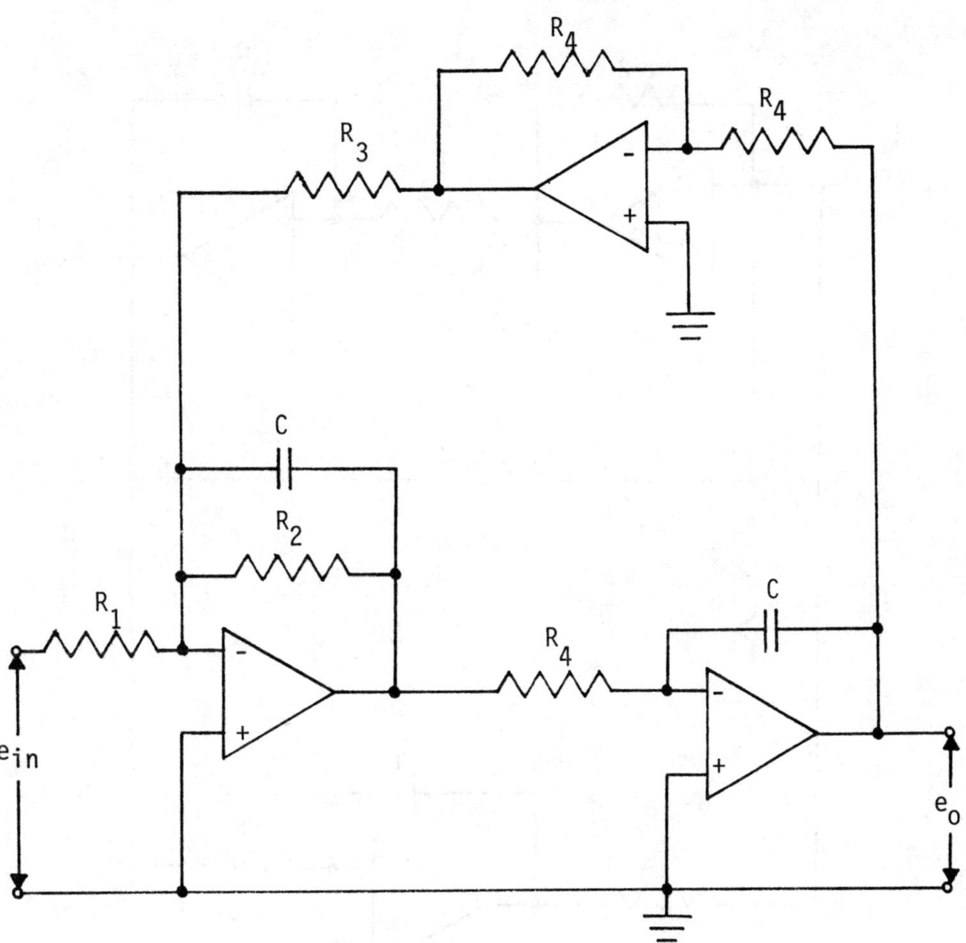

Figure 7.6 Circuit diagram of a low-pass biquad filter. (Reprinted from Rapid Practical Design of Active Filters, D. Johnson and J. Hilburn. Copyright © 1975, John Wiley and Sons, Inc., by permission of John Wiley and Sons, Inc.)

Note in Figure 7.7 that this design utilizes two integrators and one summing amplifier. The equations that specify the characteristics of the filter are amplifier. The equations that specify the characteristics of the filter are

$$\omega_o = \left(\frac{R_2}{R_4 R_a R_b C_a C_b} \right)^{1/2} \tag{2}$$

$$BW = \frac{R_o R_2}{R R_a C_a (R_o + R_3)} \tag{3}$$

$$Q = \frac{\omega_o}{BW} = \frac{R(R_o + R_3)(R_a C_a)^{1/2}}{R_o (R_2 R_4 R_b C_b)^{1/2}} \tag{4}$$

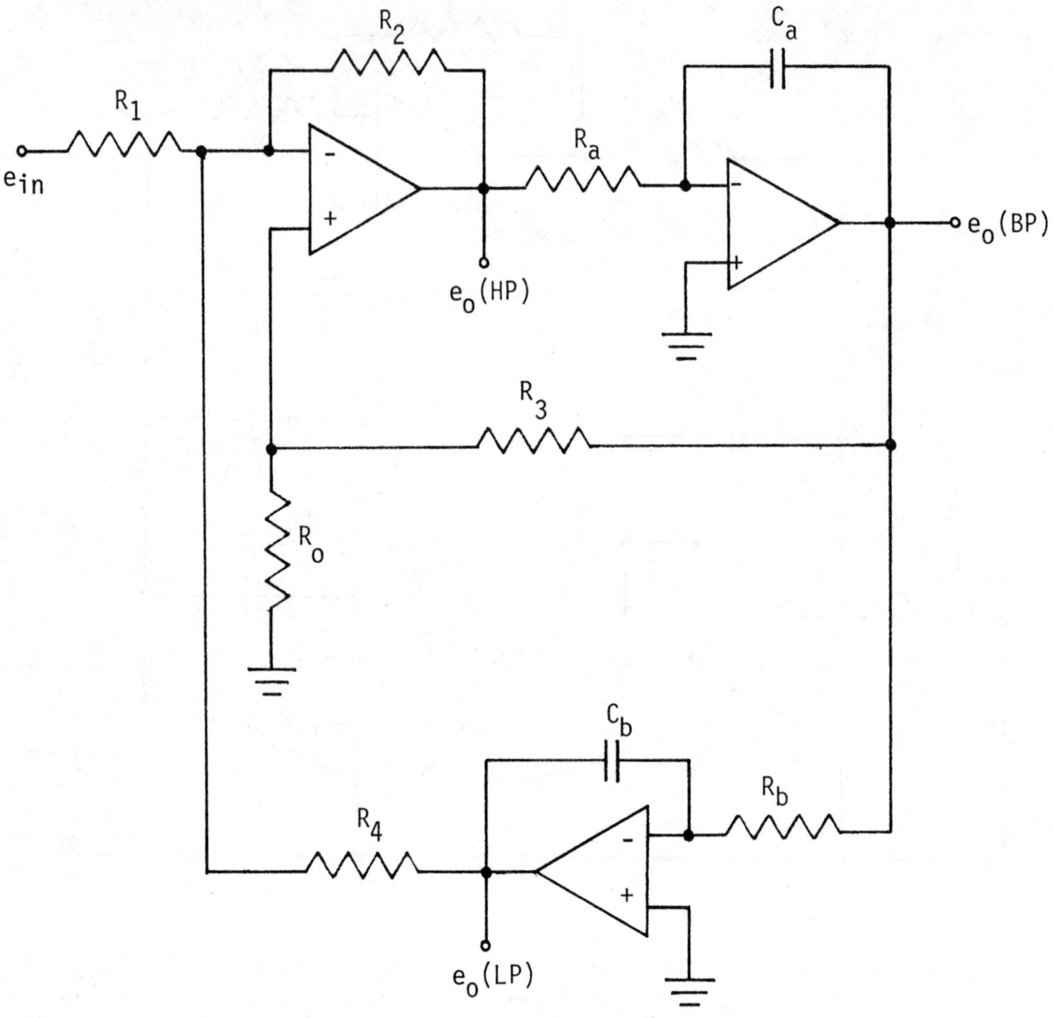

Figure 7.7 Circuit diagram of a state-variable filter showing high-pass (HP), band-pass (BP), and low-pass (LP) outputs.

where

ω_o = the center frequency of the band-pass output (rad/sec);

BW = bandwidth of the band-pass output;

and $R = R_1 || R_2 || R_4$

The Q of the circuit determines whether it has Butterworth or Chebyshev characteristics: Q = .707 for Butterworth, and Q = .885 for Chebyshev characteristics. For the Butterworth case, the cutoff frequencies for the low and high-pass outputs are each equal to ω_o, and for the Chebyshev case, the low-pass cutoff is at $0.812\omega_o$, while the high-pass cutoff is at $1.23\omega_o$. For a discussion of the state-variable design procedure, and an illustration of the Butterworth and Chebyshev frequency response of each state-variable output, see Young, Linear Integrated Circuits (Wiley, 1981).

Another type of active filter is used for delaying signals, that is, for producing an output only after a specified time has elapsed from the time of application of the input. As this behavior suggests, the filter is required to introduce a phase shift in one or more of the signals passing through it. An all-pass filter is used to introduce a specific phase shift at a specific frequency, while passing all frequencies with the same gain or attenuation. The Bessel, or all-pass, time-delay filter is used to delay all frequency components by the same amount of time, which means that each frequency component must be shifted by a different phase angle. In fact, the amount of phase shift introduced at any frequency must be directly proportional to the frequency. If, for example, the filter causes 10 degrees of phase shift in a 1 kHz signal, then it must cause 20 degrees of phase shift in a 2 kHz signal. Such a filter is said to have a linear phase characteristic. The output of an ideal all-pass, time-delay filter will be an exact replica of the input, delayed in time. Design procedures employing look-up tables similar to those we have discussed may be found for time delay filters in the previously cited (Johnson and Hilburn) reference.

EXERCISES

1. Using $Z_F = C = 0.47\ \mu F$ in Figure 7.4, design a second-order, VCVS, low-pass Butterworth filter having unity gain and a cutoff frequency of 1 kHz. Draw a schematic diagram of your circuit showing all component values. Assuming that only standard 5% tolerance resistors are available, what resistor values would you use in your design?

2. What kind of filter (low-pass, high-pass, etc.) is represented by the circuit of Figure 7.9?

3. What kind of filter (low-pass, high-pass, etc.) is represented by the circuit of Figure 7.10?

4. The output of a Chebyshev filter varies between 2.05 VRMS and 2.30 VRMS as the frequency of the input is changed through the range of the filter's passband. What is the ripple width in dB?

5. Using $Z_A = C = 0.1\ \mu F$ in Figure 7.4, design a VCVS, second-order, high-pass, Chebyshev filter having a cutoff frequency of 1 kHz, a gain of 2, and a ripple width of 2 dB. Draw a schematic diagram of your circuit showing all component values. Assuming that only standard 5% tolerance resistors are available, what resistor values would you use in your design?

6. What is the gain of the active filter shown in Figure 7.11?

7. Determine the center frequency bandwidth, and Q of the bandpass output of the state-variable filter shown in Figure 7.12.

PROCEDURE

NOTE: Connect a 0.1 μF capacitor between the plus and minus terminals of the power supplies used in all op-amp circuits below.

1. Connect the VCVS Butterworth filter shown in Figure 7.8.

2. Monitor e_{in} and e_o on a dual-trace oscilloscope. Set the signal generator output to 1 volt peak and sweep its frequency while observing the oscilloscope. In this way determine experimentally the filter type (low-pass, high-pass, band-pass, or band-reject.)

3. Experimentally determine the cutoff frequency. Measure and record e_o and e_{in} at

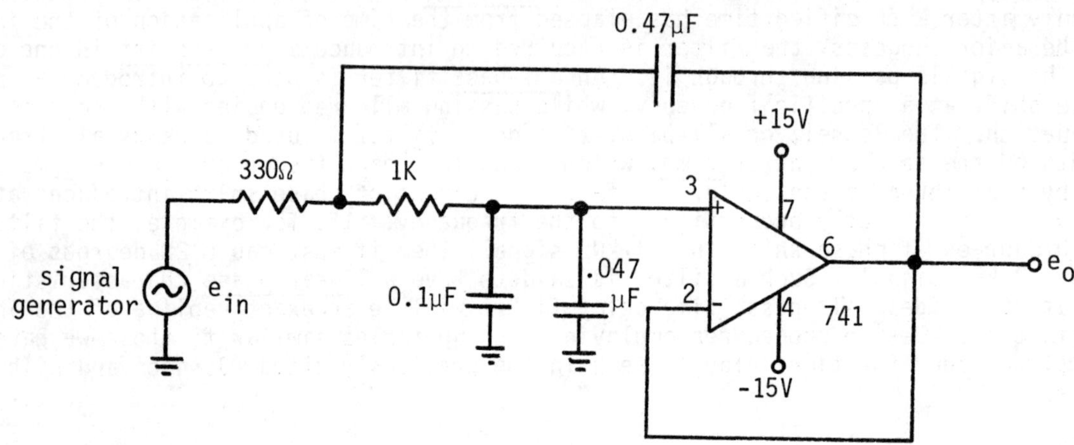

Figure 7.8

selected frequencies in the range from one decade below to one decade above cutoff. Make a sufficient number of measurements to plot a response characteristic and to determine its slope.

4. Construct the VCVS Butterworth filter shown in Figure 7.9.

5. Repeat steps 2 and 3 for the circuit of Figure 7.9.

6. Connect the IGMF Chebyshev filter shown in Figure 7.10

7. Repeat steps 2 and 3 for the circuit of Figure 7.10. Make sufficient additional measurements to determine the ripple width (RW) in the pass-band, so the cutoff frequency can be determined accurately. Also measure the phase shift of the output with respect to the input at 3 frequencies within the passband, lying in a one-octave interval of the passband.

8. Design a VCVS, second-order, high-pass, Chebyshev filter having a cutoff frequency of 1 kHz, a gain of 2, and a ripple width of 2 dB. Then repeat step 7. Use standard value, 5% resistors.

9. Construct the biquad Butterworth filter shown in Figure 7.11. Power supply and input/output pin numbers are the same as in previous circuits.

10. Repeat Procedure steps 2 and 3 for the circuit of Figure 7.11.

11. Replace R_1 with a 1K potentiometer connnected as a variable resistance and determine the effect on the filter's response characteristics of changes in this resistance.

12. Restore R_1 to its original value and replace R_2 by a 1K potentiometer connected as a variable resistance. Determine the effect on the filter's response characteristics of changes in this resistance.

13. Construct the state-variable filter shown in Figure 7.12. Power supply and input/outpin numbers are the same as in previous circuits.

14. Experimentally determine the center and cutoff frequencies of the band-pass output,

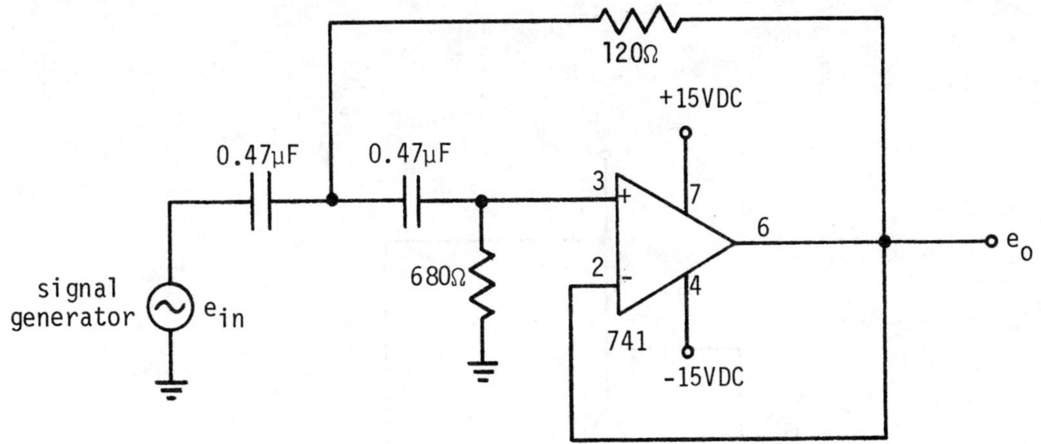

Figure 7.9

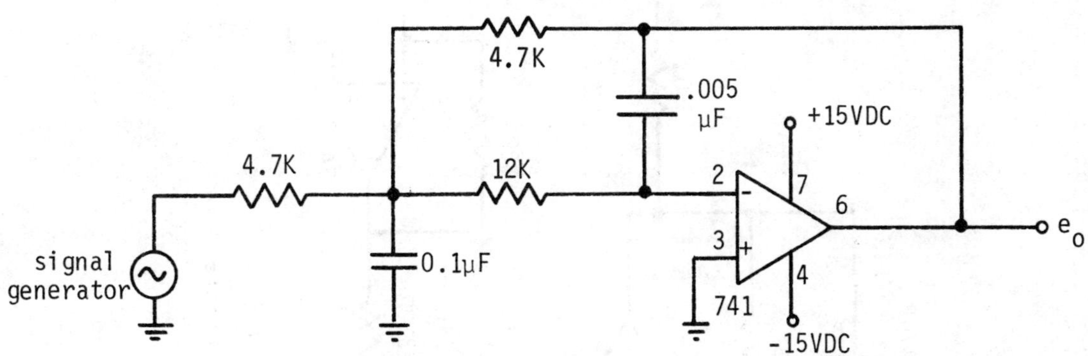

Figure 7.10

taking a sufficient number of measurements to construct an accurate plot of the response characteristic and its slopes outside the passband. Also, measure the cutoff frequencies and response characteristics of the LP and HP outputs.

15. Replace R_A and R_B with 2.7K resistors and experimentally determine the new center frequency of the BP output and the cutoff frequencies of the LP and HP outputs.

16. Restore R_A and R_B to their original values and replace C_A and C_B with .047 μF capacitors. Then repeat step 14.

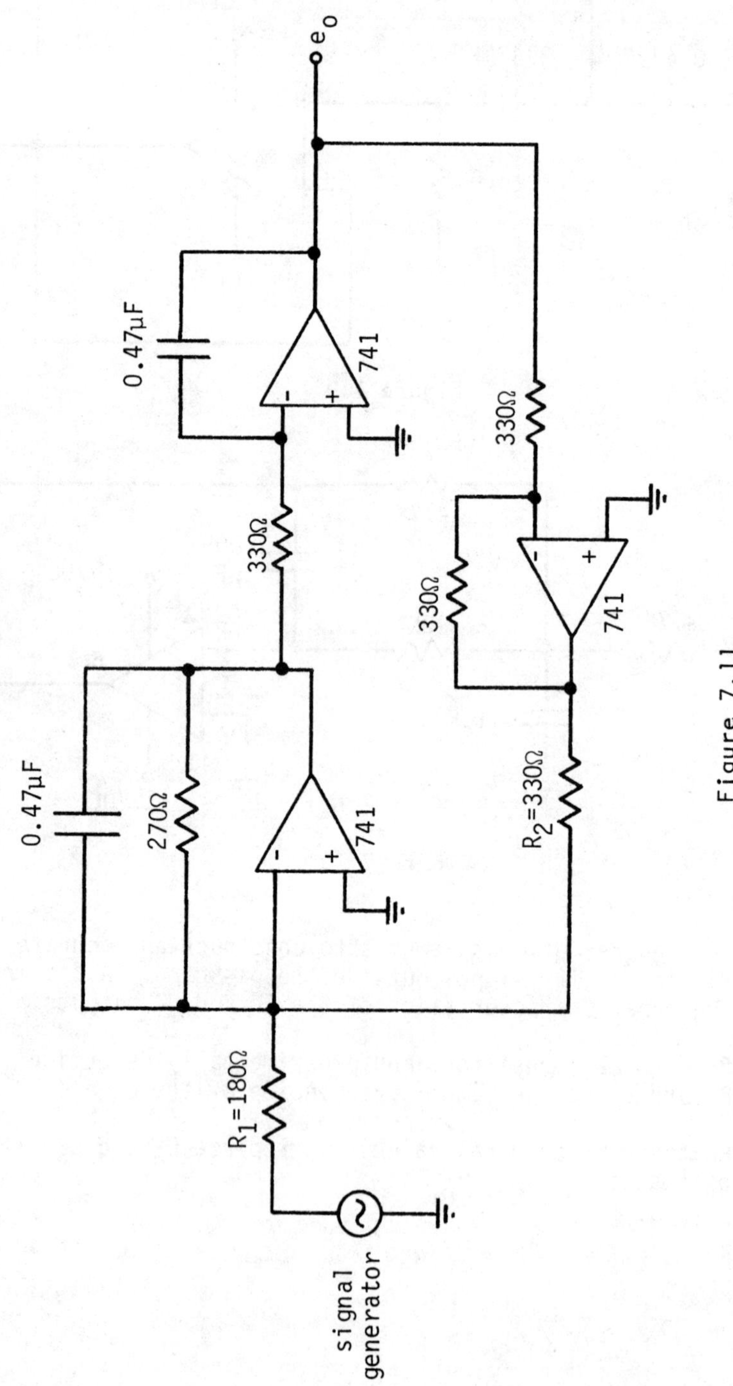

e_o
0.47μF
741
330Ω
0.47μF
270Ω
741
330Ω
330Ω
741
R_1=180Ω
R_2=330Ω
signal generator

Figure 7.11

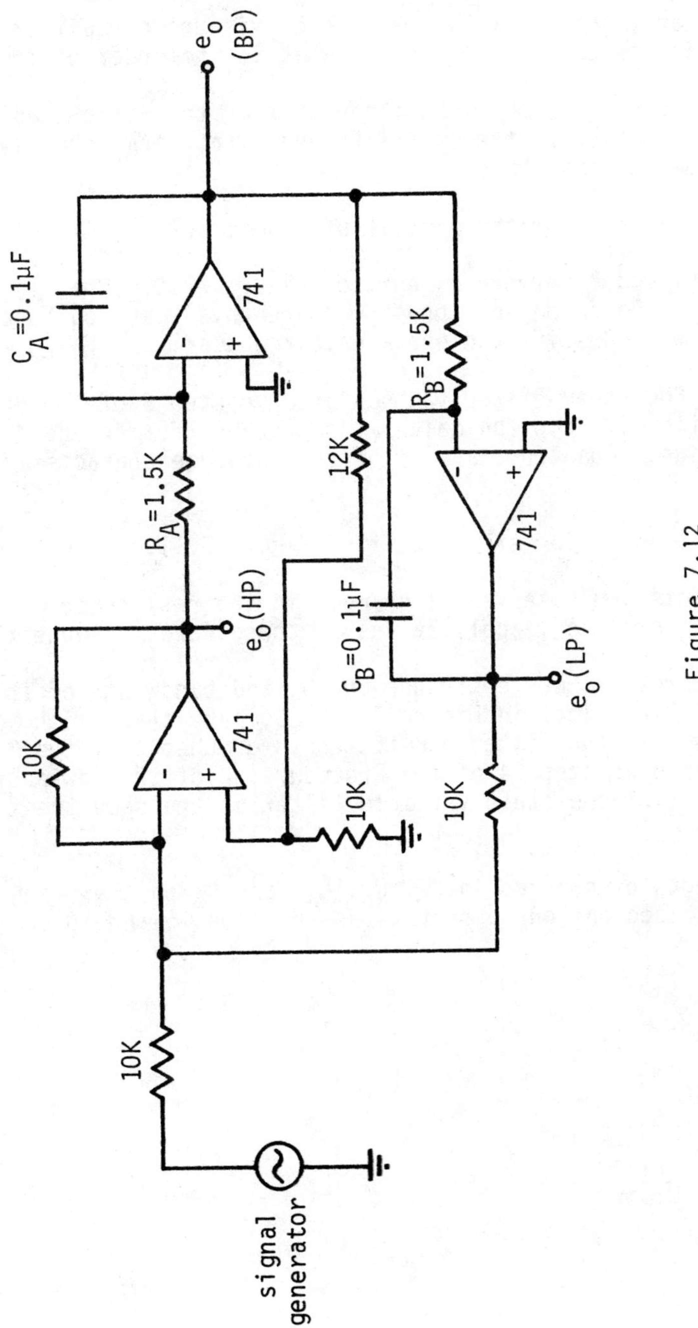
signal generator
10K
10K
741
10K
e_o(HP)
R_A=1.5K
C_A=0.1μF
741
e_o (BP)
12K
R_B=1.5K
741
C_B=0.1μF
e_o(LP)
10K

Figure 7.12

QUESTIONS

1. Based on your experimental results, answer the following questions:

   What type of filter (high-pass, low-pass, etc.) is the circuit of Figure 7.8? What is its gain? What is its cutoff frequency? What is the order of the filter?

2. Construct a plot of the response characteristic of the filter, using log-log graph paper. Show the location of the cutoff frequency and draw the asymptote of the characteristic outside the pass band.

3. Repeat questions 1 and 2 for the circuit of Figure 7.9.

4. Repeat questions 1 and 2 for the circuit of Figure 7.10. What is the ripple width of this filter in dB? Based on your phase measurements, can you conclude that this is a linear phase filter? Justify and explain your answer.

5. How well did the Chebyshev filter you designed in step 8 of the Procedure meet the required specifications? Show the calculations you performed and draw a schematic diagram of your design. Construct a plot of the response characteristic on log-log graph paper.

6. Repeat questions 1 and 2 for the circuit of Figure 7.11.

7. Describe the effects you observed of changes in the resistance values R_1 and R_2 in Figure 7.11. What practical applications do these results suggest?

8. Calculate the theoretical center frequency, Q, and bandwidth of the band-pass output of the state variable filter in Figure 7.12. Compare these with your experimentally determined values. Is the filter a Butterworth or Chebyshev type? Using the experimental data obtained in step 14 of the Procedure, plot the low-pass, bandpass, and high pass responses of the state variable filter on the same sheet of log-log graph paper.

9. Describe the effects of changes in R_A, R_B, C_A, and C_B on the state variable filter characteristics, based on your observations in steps 15 and 16 of the Procedure.

Design Project Number 3: Signal Decomposition into Audio Frequency Bands

Design an active filter circuit that will meet the following specifications:

1. Produce 3 simultaneous outputs representing the following components of an input signal: its component in the frequency band from 0 to 400 Hz, its component in the frequency band from 400 Hz to 1 kHz, and its component in the frequency band from 1 kHz to 5 kHz.

2. Each frequency component produced (filtered) with a gain of two.

3. The filter circuit driven by the output of an amplifier whose gain can be varied over a sufficient range to present a 3 VRMS signal to the filter, for audio inputs in the range from 0.3 VRMS to 3 VRMS.

You may use any two laboratory power supplies, set for whatever voltages you need. Voltage regulator(s) must be used if any additional supply voltages are required.

Construct your circuit and test its performance to determine how well it meets specifications. In particular, test it when the audio input is a single 200 Hz sine wave, a single 500 Hz sine wave, a single 2 kHz sine wave, and a 250 Hz square wave. If available, connect the speaker terminals from a radio receiver to the input of your circuit. Then observe the frequency components displayed on an oscilloscope when the radio is tuned to produce speech and music. You may modify and refine your circuit after testing as necessary to improve its performance. Use potentiometers connected as variable resistors wherever you believe precision-valued components are necessary.

Write a report that contains

1. Your design equations and calculations. Justify and explain all decisions that you make relative to component selection, design tradeoffs, and design changes made after circuit tests. In particular, justify any use of non-standard or precision-valued components.

2. A schematic diagram of your final circuit.

3. Test results that show how well your circuit performed and the degree to which it met the specifications.

Design Project Number 4: Signal Synthesizer

Design a signal synthesizer that meets the following specifications:

1. Filters a 500 Hz square or triangular wave to obtain three separate outputs simultaneously: its fundamental, its third harmonic, and its fifth harmonic.

2. Sums the fundamental component in specified ratios with either the third or fifth harmonic, as shown in the table below:

Synthesized Waveform Number	Input Waveform	Fundamental	Third Harmonic	Fifth Harmonic
1	Square	3	1	-
2	Square	1	1	-
3	Square	1	3	-
4	Triangular	3	-	1
5	Triangular	1	-	1
6	Triangular	1	-	3

As an example of how this table should be interpreted, waveform number 1 could be synthesized by summing a 3 V peak fundamental with a 1 V peak third harmonic, or by summing a 600 mV peak fundamental with a 200 mV peak third harmonic, and so forth.

It will of course be necessary to amplify the third and fifth harmonics that are recovered from the square or triangular wave input. Gain may be introduced in either the filtering or summing process, or both.

You may use an external (laboratory) signal generator to produce the square or triangular wave inputs. You may use any two fixed-voltage laboratory power supplies. Use voltage regulator(s) as necessary if additional supply voltages are required. Use potentiometers connected as variable resistors whenever you believe precision-valued components are necessary.

Construct your circuit and test its performance. You may modify and refine your circuit after testing as necessary to improve its performance.

Write a report that contains:

1. Your design equations and calculations. Justify and explain all decisions that you make relative to component selection, design tradeoffs, and design changes made after circuit tests. In particular, justify any use of non-standard or precision-valued components.

2. A schematic diagram of your final circuit.

3. Test results that show how well your circuit met the design specifications. Include accurate sketches of each of the synthesized waveforms, as observed on an oscilloscope. Be sure to show horizontal and vertical scales, and label the values at significant points (such as peaks).

8 Analog-to-Digital (A/D) and Digital-to-Analog (D/A) Converters

OBJECTIVES

1. To learn the basic properties of digital (binary) numbers and how they can be used to represent analog values.
2. To learn how the R-2R ladder network is used in digital-to-analog conversion.
3. To investigate experimentally the properties of an integrated circuit D/A converter.
4. To learn how the method of successive approximation is used to perform digital-to-analog conversion.
5. To investigate experimentally the properties of a successive approximation A/D converter.
6. To verify experimentally A/D and D/A conversion by using one device to drive the other.

EQUIPMENT AND MATERIALS REQUIRED

1. Dual trace oscilloscope.
2. dc voltmeter 0-5 V, 0-10 V.
3. Function generator (sine, square, triangular), 0.1 Hz - 10 kHz, 5 V pk.
4. Adjustable power supplies, 0-10 V DC (2); + 5 V DC supply (or 7805 voltage regulator).
5. 7805 voltage regulator (optional).
6. DAC0801 D/A converter.
7. 7493 (2) binary counters.
8. ADC0804 A/D converter.
9. Light emitting diodes (LED's) (8)
10. Ge or Si diode, 50 mA.
11. Resistors: 100 Ω, 1K (4), 3.9K, 4.7K (4), 10K (2), 100K.
12. Potentiometer: 1K.
13. Capacitors: 220 pF, .01 μF (2), 0.1 μF (2).

DISCUSSION

When the voltage or current in a circuit can vary continuously between two limits, that is, can be any value between those limits, we say it is an analog type signal. An example is the output voltage of an audio amplifier. When the voltage or current can be only one of two distinct values (except for the very brief time when it is switching from one value to the other) it is said to be a digital signal. Digital signals are used to represent binary numbers, numbers whose digits are all zeroes or ones. Each binary digit is called a bit, and each bit of a binary number can be thought of as the multiplier of a certain power of 2. As an example, the 4-bit binary number 1101 is equivalent to $1 \times 2^3 + 1 \times 2^2 + 0 \times 2^1 + 1 \times 2^0 = 8 + 4 + 0 + 1 = 13$. Note that the rightmost bit, the least significant bit (LSB), multiplies $2^0 = 1$ and that the powers of 2 increase from right to left, 2^1, 2^2, etc. The most significant bit (MSB) is the leftmost bit, which in this example is a 1 and multiplies $2^3 = 8$.

When a binary number is represented in digital form, one of the two permissible voltage levels is used for each 1 in the number and the other level is used for each 0 in the number. For example, if + 5 V represents a 1 and 0 V (ground, not open circuit) represents a 0, then the binary number 1101 would be represented by the sequence of voltages + 5 V, + 5 V, 0 V, + 5 V. If this 4-bit number is generated in parallel form, then one terminal or signal line would be required for each bit, and each terminal would be at 0 V or + 5 V depending on whether the corresponding bit is a 0 or a 1.

If a binary number has n bits, then it can have 2^n different combinations of 1's and 0's, and can therefore have 2^n different values. The largest number it can represent, all bits equal to 1, has value 2^n-1. For example, an 8-bit binary number has $2^8 = 256$ different values from 00000000 to its largest value, $11111111 = 2^8-1 = 255$. A binary counter is a device that produces a binary output representing the total number of pulses that have been applied to its input. For example, a 4-bit binary counter, starting with the count 0000, would have the output 0001 = 1 after one pulse has appeared at its input, 0010 = 2 after two pulses have appeared, and so forth. After $2^n-1 = 2^4-1 = 15$ pulses have occurred, the output of the counter would be 1111 and the next pulse would reset it to 0000. Figure 8.1 shows the waveforms that would be generated at each bit of the output of a 4-bit binary counter when the input is a continuous sequence of pulses (a square wave). We can verify that the output of the counter is the binary number corresponding to the total number of pulses that have occurred at its input by simply examining the states (1 or 0) of each bit at any point in time in Figure 8.1. For example, at a point in time between the 12th and 13th input pulse, we see in Figure 8.1 that the output states are 1100, or binary 12 (the MSB, 1, is at the bottom of Figure 8.1). We also see that after the count reaches 1111, or 15, all outputs revert to 0 on the next input pulse, so future pulses simply cause the entire sequence to be repeated. Note that the binary counter can also be regarded as a frequency divider: the waveform for each bit has 1/2 the frequency (twice the period) of the preceding less significant bit. The 7493 integrated circuit is a 4-bit binary counter that we will use in this experiment to produce waveforms similar to those shown in Figure 8.1. In fact, we will cascade two such counters to construct an 8-bit counter.

An n-bit digital-to-analog (D/A) converter produces an analog voltage at its output that is directly proportional to the value of an n-bit binary number presented in digital form to its input. For example, an 8-bit D/A converter will have 8 inputs, one for each bit of an 8-bit number, and a single analog output. Since the binary input can have $2^8 = 256$ different values, the analog output will be one of 256 different voltages. The output is therefore not truly analog, that is, it cannot be any value in its range, only one of 256 possible values.

A passive D/A converter can be constructed using a so-called R-2R ladder network. Each resistor in this network has either resistance R ohms or resistance 2R ohms. Figure 8.2 shows an example of a 4-bit R-2R D/A converter in which R = 10K and 2R = 20K.

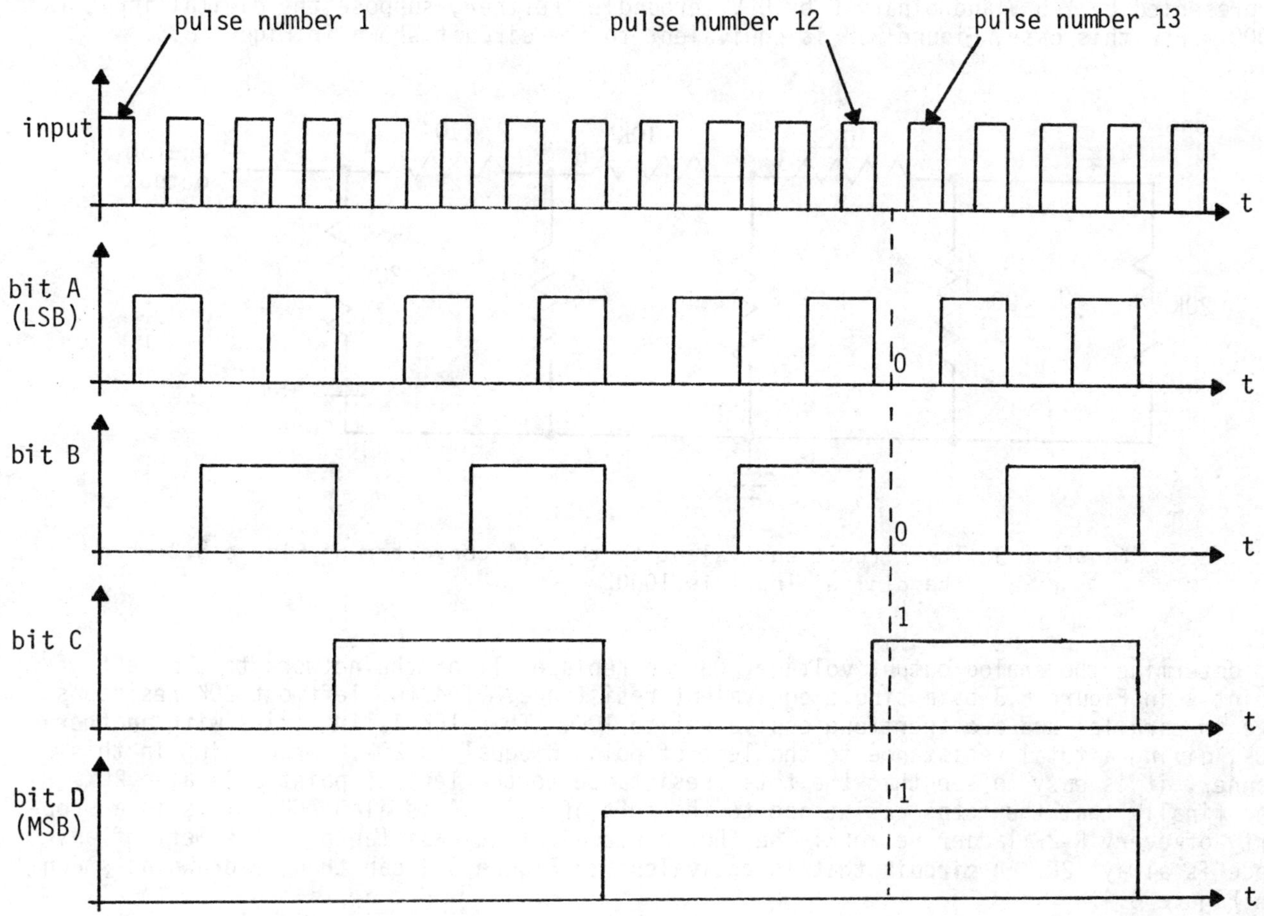

Figure 8.1 Input and outputs of a 4-bit binary counter. Note that the output is 1100 = 12 after the 12th input pulse.

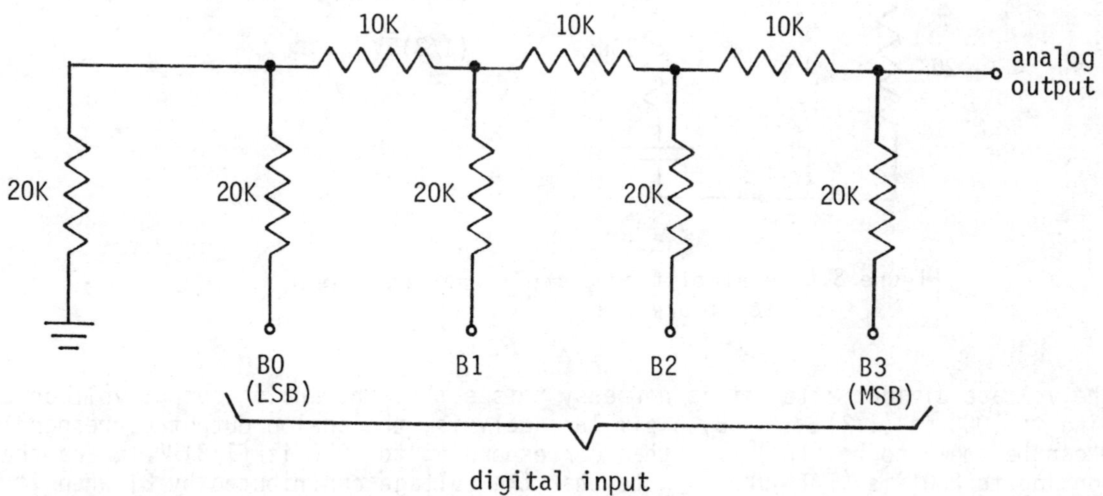

Figure 8.2 A four-bit, R-2R digital-to-analog (D/A) converter.

To illustrate the operation of the R-2R ladder network, let us suppose that binary 1 is represented by + 5 V and binary 0 by 0 V (ground). Further, suppose the digital input is 1000. For this case, Figure 8.2 is equivalent to the circuit shown in Figure 8.3.

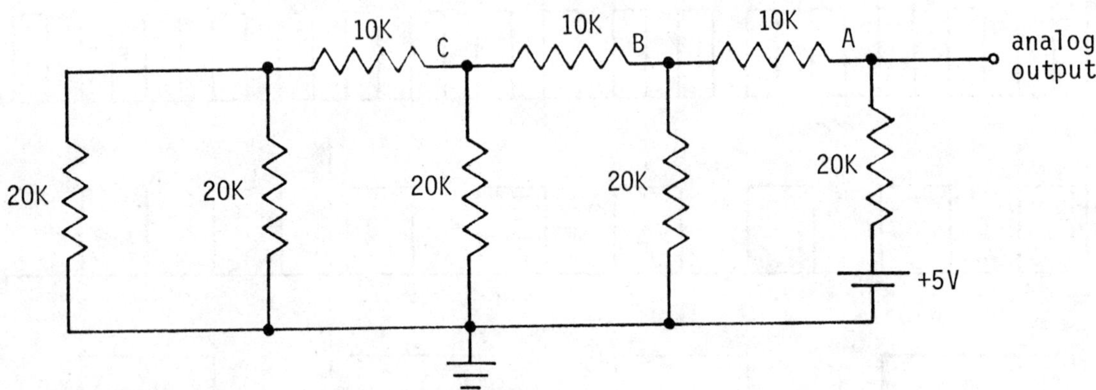

Figure 8.3 The circuit equivalent to the D/A converter of Figure 8.2 when the digital input is 1000.

To determine the analog output voltage, we can replace all of the network to the left of point A in Figure 8.3 by a single equivalent resistance. The two leftmost 20K resistors are in parallel and are therefore equivalent to 10K. This 10K is in series with another 10K, giving a total resistance to the left of point C equal to 20K. Proceeding in this manner, it is easy to see that the total resistance to the left of point B is also 20K. and finally that the total resistance to the left of point A is also 20K. This is a property of every R-2R ladder network: The Thevenin equivalent resistance to the left of any node is always 2R. A circuit that is equivalent to Figure 8.3 can then be drawn as shown in Figure 8.4.

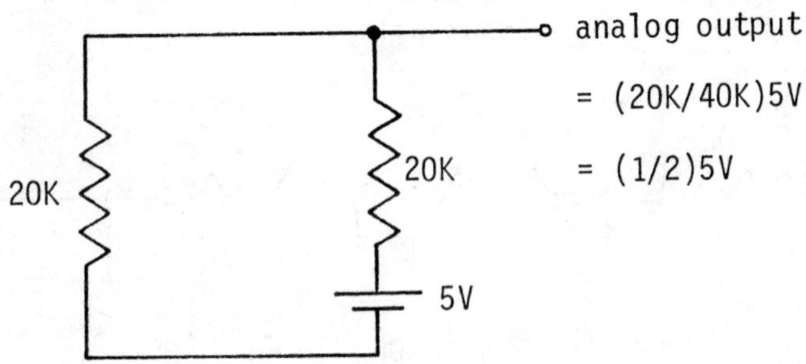

Figure 8.4 A simplified circuit and equivalent to Figure 8.3.

Using the voltage divider rule, it is now easy to see that the analog output voltage corresponding to 1000 is (1/2) 5V. By a similar analysis, the analog output corresponding to 0100 can be shown to be (1/4)5V, that corresponding to 0010 is (1/8)5V, and that corresponding to 0001 is (1/16)5V. Note that the voltage contributed by B1 when it is a 1 is 2 times the contribution of B0, B2 produces a voltage that is twice that of B1, and so forth, each bit position contributing twice the voltage of its predecessor as we move toward the MSB. Thus the contributions to the output voltage are weighted by powers of 2 in exactly the same way that the bit positions of the binary number are weighted when computing its value. Using the principle of superposition, we can find the output voltage corresponding to any binary input simply by adding the contributions of each bit position that has a 1. For example,

$$1010 = (1/8)5\text{ V} + (1/2)5\text{ V} = 3.125\text{ V}$$

and

$$1111 = (1/16)5\text{ V} + (1/8)5\text{ V} + (1/4)5\text{ V} + (1/2)5\text{ V} = 4.6875\text{ V}$$

Integrated circuit D/A converters incorporate active circuitry as well as a ladder network. Typically, each bit of the binary input controls an electronic switch that applies a fixed reference voltage or current to the ladder network. Also, the analog output is available through an amplifier that provides gain and isolates the output from the ladder network. In this experiment we will use the DAC0801 8-bit D/A converter. This device produces an output current that is proportional to the value of the binary input, namely

$$I_{OUT} = I_{REF}(D/256)$$

where I_{REF} is a reference current that is set by an external resistor and D is the value of the binary input. We will measure the voltage across an external load resistor through which output current flows and thus obtain a voltage proportional to the binary input.

An n-bit analog-to-digital (A/D) converter produces an n-bit binary output whose value is proportional to an analog input. Many A/D converters incorporate a D/A converter as an integral part of their design. An example is the A/D converter that uses the method of successive approximation. In this method, a series of trial binary numbers are developed, each value of which is compared to the analog input. Based on the result of each comparison, the trial binary number is either increased or decreased to form the next trial number. This process continues until the trial binary number has a value that equals the analog input. Figure 8.5 shows a block diagram of a 4-bit A/D converter that uses this method of conversion.

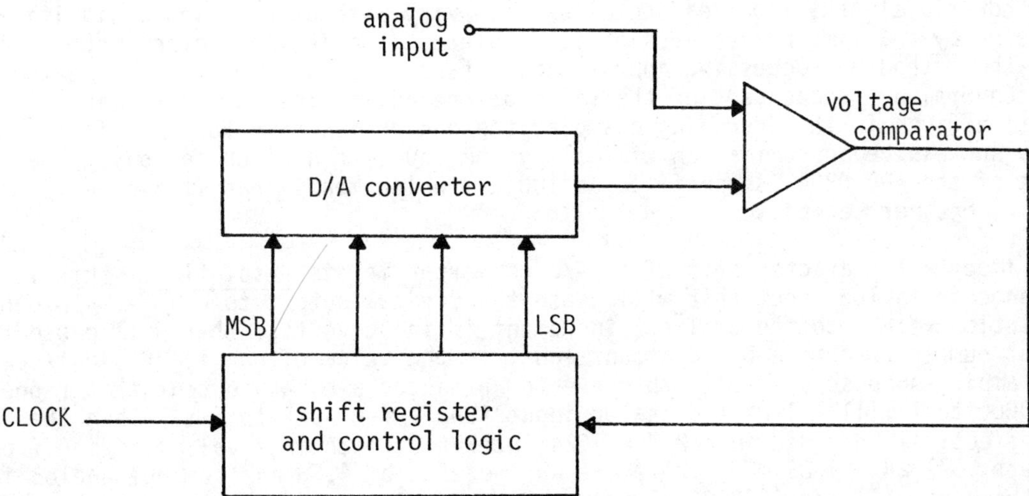

Figure 8.5 Block diagram of a four-bit analog-to-digital (A/D) converter that uses the method of successive approximation.

The method is best understood by way of an example. For simplifying purposes, suppose the analog input in Figure 8.5 ranges from 0 to 15 V and that the D/A converter produces an analog output that ranges from 0 to 15 V as its binary input ranges from 0000 to 1111. The clock input shown in Figure 8.5 is a fixed frequency square wave that is used to synchronize the sequence of operations that follow. At the time of the first clock pulse, the shift register places a 1 in the MSB position and a zero in every other position, so the input to the D/A converter is 1000, the first trial binary number. Suppose now that the analog input, which we shall refer to as the "unknown," is, say, 13 V. Since the input to the D/A converter is 1000, its output is 8 V. The purpose of the voltage comparator is to determine if the D/A output (and therefore the trial binary number) is equal to the unknown analog input. In this case the unknown input is the larger (13 V > 8 V), so we conclude that 1000 is too small a binary number to represent 13 V. The output of the voltage comparator therefore causes the logic circuitry to shift a 1 into the next position of the trial binary number, while retaining the 1 in the MSB, thus creating 1100 as the new trial number. This takes place at the time of the second clock pulse. Since 1100 has value 12, the comparator once again determines that the unknown analog input is larger (13 V > 12 V) and so causes the logic circuitry and shift register to create 1110 as the next trial number, during the third clock pulse. This time 1110 = 14 is too large (14 V > 13 V), so the 1 that was most recently inserted in the trial number is changed to a 0 and a 1 is placed in the next position, the LSB in this case. Therefore, during the fourth and last clock pulse, the trial number is 1101 and it is determined that this value, 13, is equal to the unknown analog input. The process terminates with 1101 as the final output of the A/D converter.

In this experiment we will use an integrated circuit 8-bit A/D converter, the ADC 0804, that employs the method just described. There are several other methods of A/D conversion, including several that incorporate binary counters, and complete descriptions of their theory of operation may be found in the literature. See, for example, Young, Linear Integrated Circuits (Wiley, 1981).

An important characteristic of an A/D converter is the total length of time it takes to convert an analog input to a digital output. This time is called the conversion time and its reciprocal is the number of conversions per second that it can perform. In practice, the analog input to an A/D converter is usually changing with time rather than being a static, dc level. Consequently, the conversion time of the converter is important in determining how rapidly the input can be allowed to vary without becoming a significantly different value by the time one conversion is completed. Unlike some other methods of A/D conversion, the method of successive approximation takes the same amount of time to perform every conversion, independent of the value of the analog input. As we can see from the description given in the preceding example, the conversion time for a n-bit A/D converter using successive approximation will always be equal to n clock periods. The conversion time of the ADC 0804 is specified as 100 μs, which means that it can perform 10,000 conversions per second.

Another important characteristic of an A/D converter is its resolution. This is the smallest change in analog input that will cause the digital output to change. In other words, resolution refers to the smallest increment in input voltage that will cause the binary output number to change by an amount equal to the value of one least significant bit. For example, suppose a 10-bit A/D converter produces a binary output that ranges from 0000000000 to 1111111111 as the analog input ranges from 0 V to 5 V. Then the input voltage is effectively divided into 2^{10} = 1024 intervals, each interval differing from the adjacent one by 5/1024 = 4.88 mV. Therefore an increase of 4.88 mV from one analog input voltage to the next will add 0000000001 to the binary output, and we say that the resolution is 4.88 mV. More often, the resolution is specified simply as being equal to the number of bits of the A/D converter. Given the number of bits, the user can easily calculate the resolution in terms of input voltage, as in the preceding example, depending on the total range of the analog voltage in a particular application. Also, the resolution may be stated as the percentage one LSB is of the total number of binary numbers that can be generated. In the preceding example, the resoltion could be specified either as simply 10 bits, or as

$$(1/1024) \times 100\% = .0976\%$$

EXERCISES

1. When the input to an 8-bit D/A converter is 10000000 the output is 5 V. What is the output when the input is 00000001? What is the output when the input is 11001101?

2. Describe completely the waveform you would expect to see at the output of the D/A converter in exercise 1 if its 8-bit input is driven by the 8-bit output of a binary counter. The counter is driven by a 10 kHz square wave (so the counter produces 00000000, 00000001, 00000010, ..., 11111111, 00000000, ..., and so forth).

3. When the input to an 8-bit A/D converter ranges from 0 to 5 V, the output ranges from 00000000 to 11111111. What change in input voltage causes the output to change from 00000000 to 00000001?

4. What is the resolution of the A/D converter in exercise 3 in volts? In bits? In percent?

5. The conversion rate of an A/D converter is 8000 conversions/sec. What is its conversion time?

PROCEDURE

1. Connect the D/A converter circuit shown in Figure 8.6. (The 7805 is used optionally to reduce the number of power supplies required.)

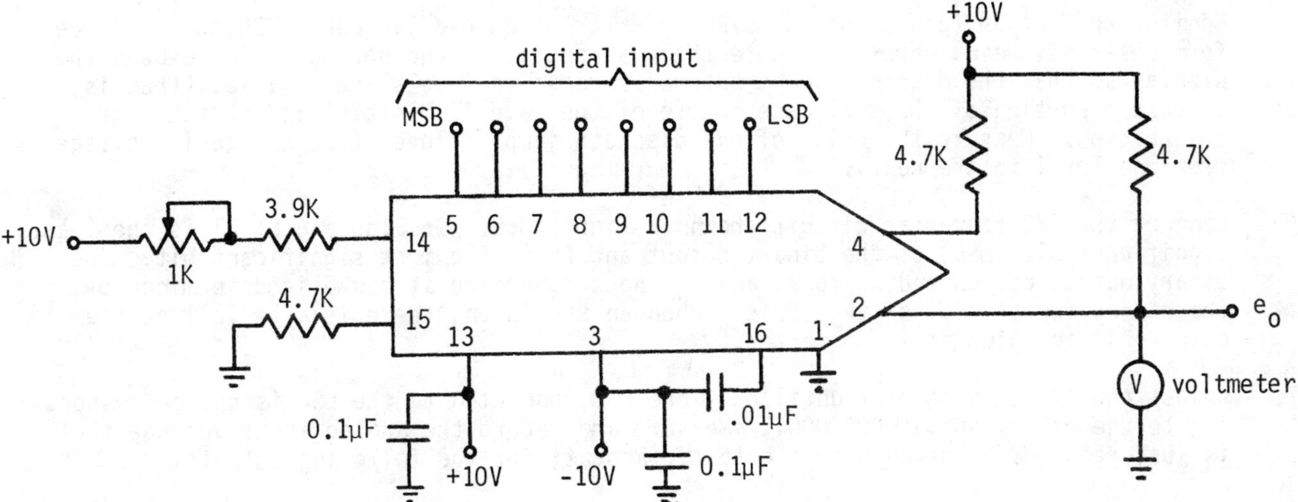

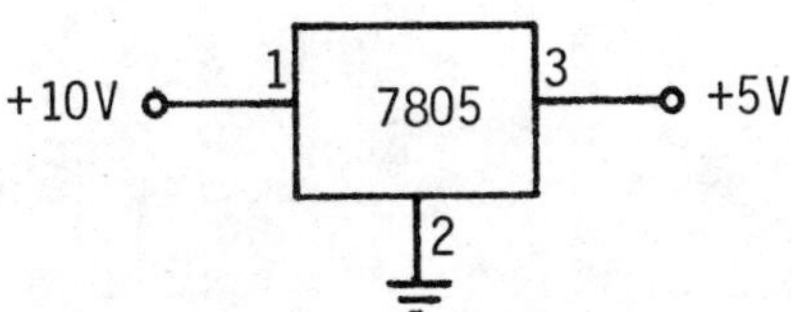

Figure 8.6

With all digital input bits connected to ground (pins 5 through 12), adjust the 1K potentiometer until the output voltage is exactly zero volts. Use a millivoltmeter or a very sensitive scale on a multimeter for this measurement.

2. Connect the least-significant bit (pin 12) of the input to + 5 V and, with all other input bits connected to ground, measure and record the output voltage. Repeat, for each of the remaining inputs; that is, measure the output when each input is connected to + 5 V and all other inputs are zero.

3. Now measure and record the output voltage for each of the following binary inputs:

   01011000

   10010000

   11000000

   00000111

   00001000

   11111111

4. In this step, we will construct an 8-bit binary counter that when driven by a clock signal (square wave), will generate the sequence of 256 binary numbers: 00000000, 00000001, . . .,11111111, and continually repeat this sequence. Two cascaded 7493 4-bit counters will be used for this purpose. The binary sequence will be applied to the D/A converter and we will examine the analog output that results. Connect the circuit shown in Figure 8.7 Before connecting the square-wave generator, adjust it to produce a 10 kHz, 5 V peak square wave.

5. Monitor the output (pin 2) of the D/A converter on an oscilloscope. Sketch the waveform observed, being careful to note the peak amplitude and period. Also expand the display so that the discrete jumps between successive levels are visible, (that is, so that a portion of the staircase nature of the wave is visible) and sketch your observation. Measure the value of one discrete jump in level (the change in voltage from one level to the next).

6. Connect the A/D converter circuit shown in Figure 8.8. In Figure 8.8, B1 is the least significant bit (LSB) of the binary output and B8 is the most significant bit. The binary output corresponding to an analog input connected at pin 6 is determined by the states (on or off) of the LED's. When an LED is on, there is a logic 1 at the output bit to which it is connected.

7. Adjust the 1K potentiometer until only the LED connected to the LSB is on, corresponding to the binary output 00000001. Measure and record the analog input voltage that is just required to produce this output. Repeat, for the following outputs:

   (b) 00000010

   (c) 00000011

   (d) 00000100

   (e) 00001000

   (f) 00010000

   (g) 10000000

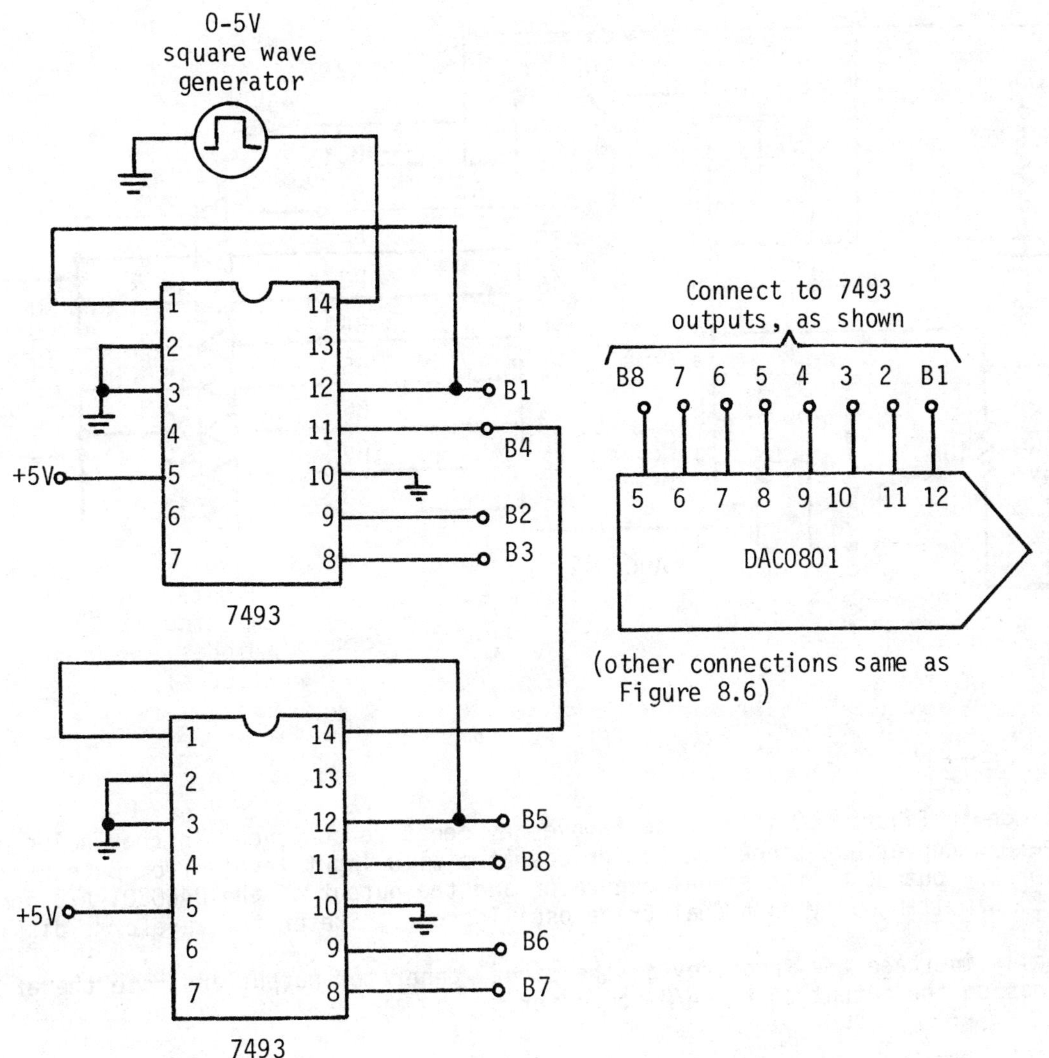

Figure 8.7

(h) 10000001

(i) 11111111

8. Replace the potentiometer in Figure 8.8 with a very low frequency triangular or sawtooth signal that varies from 0 to 5 V with a frequency of about 0.1 Hz or less. Observe the LED display that results. Repeat for a very low frequency square wave.

9. In this step we will convert an analog signal to digital form and then back to analog. We will investigate how well the A/D and D/A conversions are performed by comparing the analog output of the D/A converter with the analog input of the A/D converter. Connect the circuit shown in Figure 8.9. Before connecting the signal generator, set it to produce a 5 V peak sine wave at 50 Hz.

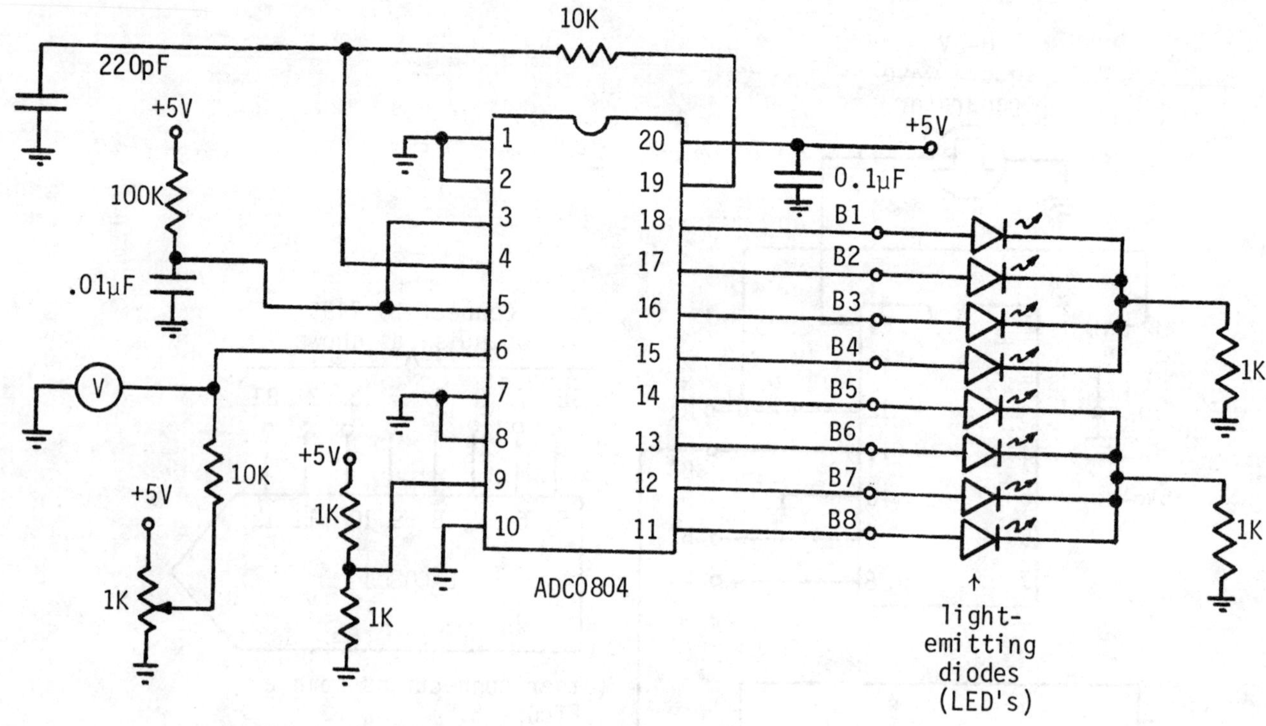

Figure 8.8

10. The diode in Figure 8.9 is used to remove any negative component in the analog input to the A/D converter, since the latter operates with input levels from 0 to + 5 V. Monitor the output of the signal generator and the output of the DAC0801 A/D converter (pin 2) simultaneously on a dual trace oscilloscope. Sketch the waveforms displayed.

11. Gradually increase the frequency of the signal generator output and note the effect this has on the output of the A/D converter.

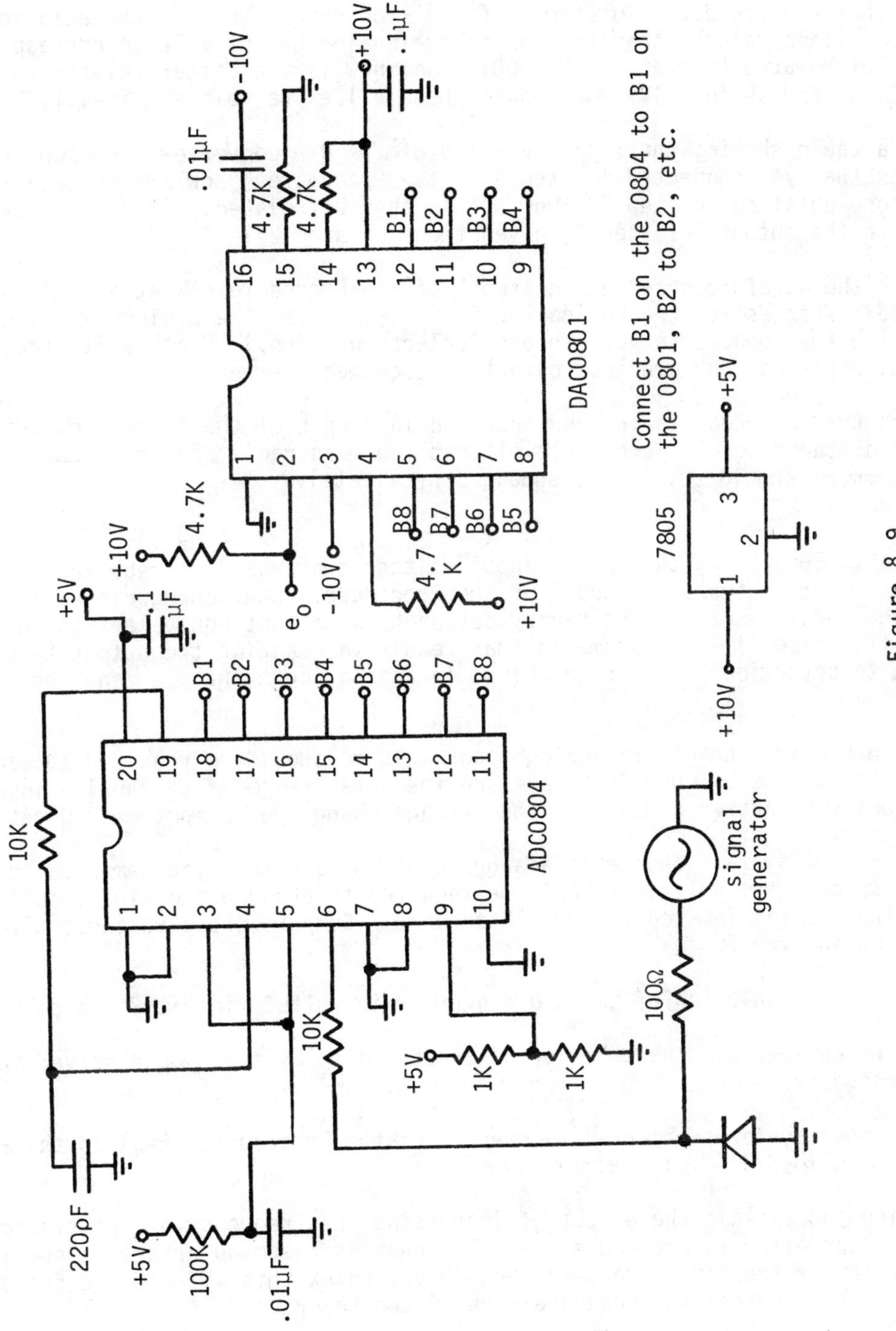
ADC0804
DAC0801
7805
signal generator
+5V
+10V
-10V
10K
100K
220pF
.01μF
.1μF
1K
4.7K
4.7 K
100Ω
e_o
B1
B2
B3
B4
B5
B6
B7
B8
Connect B1 on the 0804 to B1 on the 0801, B2 to B2, etc.

Figure 8.9

QUESTIONS

1. Make a table showing the output voltage of the D/A converter for each of the binary inputs that you connected in step 2 of the Procedure. What is the relationship between these voltage values (that is, how is the analog output voltage corresponding to 00000001 related to that for 00000010, and how is the latter related to that for 00000100, and so forth)? How should these values be related, ideally?

2. Make a table showing the output voltage of the D/A converter for each of the binary inputs that you connected in step 3 of the Procedure. How are these values related to those obtained in step 2? How should they be related? (That is, how would you compute the outputs of step 3, given those of step 2)?

3. Sketch the waveform observed in step 5 of the Procedure. What are its peak value and period? What is its theoretical period? (Hint: use the period of the square wave used to drive the counter in your theoretical computation.) What is its theoretical peak value, based on your previous output voltage measurements?

4. Sketch the expanded display you observed in step 5 of the Procedure. What was the value of the discrete jump in level that you measured? What previous output voltage measurement should this value equal, approximately?

5. Make a table showing the analog input voltage that was just required to produce each of the binary outputs in step 7 of the Procedure. What change in analog input voltage is required to increase the binary output by an amount equivalent to adding 00000001 to it? (You made five measurements that result in changing the output by 00000001. Use these to determine the average change in analog input that is required to produce this result.)

6. What theoretical change in analog input voltage should be required to cause the binary output to change by 00000001? (Assume the total range of the analog input is 0 to 5 V.) How does your answer compare to the average change you computed in Question 5?

7. Based on the average change in analog input voltage that you computed in Question 5, what analog input voltages should be required to produce the binary outputs (e), (f), (g), (h) and (i) in step 7 of the Procedure? Compare these voltages with those you measured in step 7.

8. What is the resolution of the A/D converter in volts? in bits? in percent?

9. Describe and explain the behavior of the LED display that you observed in step 8 of the Procedure.

10. Sketch the waveforms observed in step 10 of the Procedure. Explain the appearance of the output of the D/A converter.

11. Describe and explain the effect of increasing the frequency of the sinusoidal input to the A/D converter in Procedure step 11. What is the manufacturer's specified conversion time for the A/D converter? How do you think this affects the frequency at which the A/D - D/A system in Procedure step 10 can be driven?

Design Project Number 5: Function Generator

Design an integrated circuit function generator that will meet the following specifications:

1. Generate sine, square, and triangular waveforms over a continuously adjustable frequency range from 200 Hz to 1 kHz. The output waveform should be switch selectable; that is, any one of the three waveforms should be capable of being switched to a single output terminal. The triangular wave should have only positive values.

2. Output voltage continuously variable from zero volts to 5 volts peak across a 10K load. It is not necessary that the output amplitude remain constant when the frequency is changed.

3. Produce an 8-bit digital output of the triangular wave in addition to the continuous (analog) output. The digital output should be 00000000 whenever the sine or square wave output is selected. The digital output should be 11111111 whenever the triangular wave reaches + 5 V.

Your design should be based on the use of a sine wave oscillator, a voltage comparator that produces a square wave, and an integrator. You may use any two laboratory power supplies, set for whatever voltages you need. Voltage regulator(s) must be used if any additional supply voltages are required.

Construct your circuit and test its performance to determine how well it meets specifications. Verify the operation of the digital output by determining the binary numbers generated when a test input is set to various DC levels between 0 and + 5 V. You may modify and refine your circuit after testing as necessary to improve its performance.

Write a report that contains

1. Your design equations and calculations. Justify and explain all decisions that you make relative to component selection, design tradeoffs, and design changes made after circuit tests. In particular, justify any use of non-standard or precision-valued components.

2. A schematic diagram of your final circuit.

3. Test results that show how well your circuit performed and the degree to which it met the specifications.

9 Analog Switches and Multiplexers

OBJECTIVES

1. To learn how to use a digitally controlled analog switch in a circuit.
2. To learn how to interpret and verify manufacturers' specifications for an analog switch.
3. To learn how analog switches are used to construct a multiplexer and to verify multiplexer operation.
4. To learn how a multiplexer may be used as a demultiplexer.
5. To gain experience interpreting multiplexer and demultiplexer outputs in terms of the binary signals applied to their address inputs.

EQUIPMENT AND MATERIALS REQUIRED

1. Dual trace oscilloscope.
2. dc voltmeters (2), 0-5 V DC.
3. Power supplies, ± 5 V DC.
4. Function generator (sine, square), 5 V pk, 1 kHz.
5. 4052 multiplexer/demultiplexer.
6. 7493 binary counter.
7. Resistors: 1K (5).
8. Potentiometer: 1K.

DISCUSSION

An <u>analog switch</u> (or digital-analog switch) is an electronic device that can be turned <u>on</u> or <u>off</u> by a digital control signal. Typically, the application of a logic "1" level to the digital input turns the switch on and permits conduction of a continuous, analog type signal between one of its analog terminals and the other. When the digital input is switched to the logic "0" level, the switch is turned off, thus preventing further conduction between its analog terminals. In short, an analog switch behaves very much like a conventional, manually-operated, mechanical switch, where the manual operation is replaced by a digital control signal.

An analog switch is of course an active device; most integrated circuit versions are monolithic, CMOS devices using field effect transistors. Many are available with more than one switch in a single package. Since the analog switch does not truly open and close mechanical contacts like a conventional switch does, its ON resistance is likely to be higher, and its OFF resistance lower, than the corresponding quantities of a mechanical switch. One of the important electrical specifications of the analog switch is its ON resistance, R_{on}, which may be in the neighborhood of 100 Ω, but which varies significantly with supply voltage, input signal voltage, and temperature. The degree to which the analog switch is able to shut off conduction when it is in its OFF state is measured by the specification of its leakage current, typically in the pico or nano ampere range.

Other important specifications for the analog switch are related to its digital control function. The propagation delay time, for example, is the time between application of the digital switching signal and the time that a change in state is effected. This time may vary from less than 100 to several hundred nano-seconds depending upon the device and the capacitive load connected to the switch. Also important is the minimum voltage V_{IH} that the digital input will recognize as a logic "1" level, and the maximum voltage, V_{IL}, that it will recognize as a "0" level.

Other specifications that may need to be checked for a particular application include the -3 dB cutoff frequency, the total harmonic distortion introduced into the signal by the switch, the maximum rated current, power supply requirements, and power dissipation.

Multiplexing is the process of using only one signal path for the transmission of several signals. In frequency miltiplexing, for example, a number of signals with different frequencies may be transmitted simultaneously through one signal line. Time multiplexing, which we will study in this experiment, is the process whereby different signals are transmitted through the same signal path at different intervals of time. There is never more than one signal on the line at any given time. A time multiplexer is like a commutator, or rotating switch, which sequentially connects multiple input signals to a common output line. The length of time that any one signal is connected to the output of the miltiplexer depends upon the rate at which the device samples the multiple inputs. See Figure 9.1.

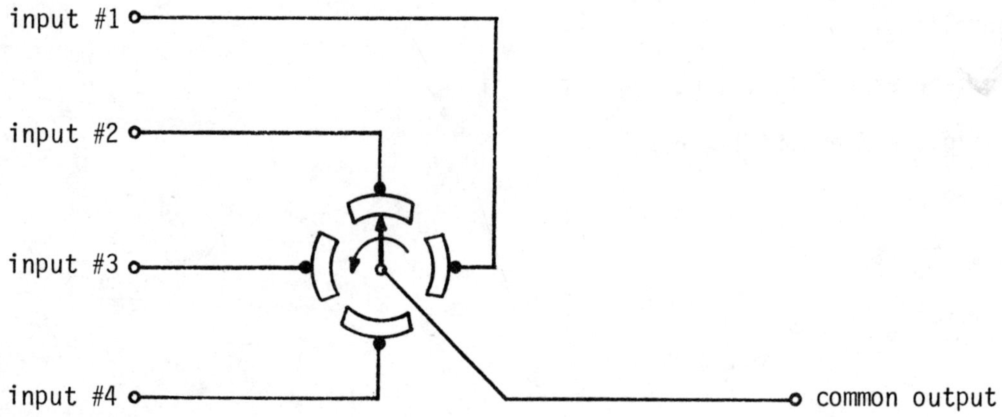

Figure 9.1 A commutator, or rotating switch, whose operation is like that of a multiplexer.

Integrated circuit multiplexers are constructed by incorporating several analog switches into a single package. Each switch has a separate digital control input and all the switches have one analog terminal connected together to form the common output. In normal operation, only one analog switch is ON at any given time, so only one of the multiple inputs is connected to the common output at any time. Typically, the digital inputs will be driven by a sequence of pulses which turn each switch in sequence to its ON state.

The 4051, 4052, and 4053 CMOS series is a typical family of integrated-circuit analog multiplexers with several additional features. These devices have built-in binary-to-decimal decoders, so a given switch can be turned on by applying the digital code corresponding to its number to the address inputs of the device. For example, if the device has 4 address inputs, then code 0000 connected there would select switch number 0, 0001 would select switch number 1, 0010 switch number 2, and so forth. This feature makes it convenient to drive the multiplexer directly from a binary counter: as the counter steps through its binary sequence, each switch is turned on in succession.

The 4051 is an 8-channel multiplexer, i.e. it can connect 8 inputs in succession to a common output. The 4052 is actually two 4-channel multiplexers in one package: it has 4 inputs (designated X-inputs) which can be connected to a single X-output, and 4 Y-inputs which can be connected to a single Y-output. The 4053 contains three 2-channel multiplexers. Figure 9.2 shows functional block diagrams of the three devices in this series. A careful study of these diagrams will aid in understanding the operations we have described. Also shown are the "truth tables" that reveal which channel (switch) is "ON" for each possible combination of address inputs. In Figure 9.2, the four X channels are designated 0X, 1X, 2X, and 3X. In this experiment, we will refer to these as X_0, X_1, X_2, and X_3 respectively. Note that address inputs have letter designations. For example, a 4052 address is the binary number BA, and from the truth table we see that B = 1, A = 0 will turn on channel 2 in both the X and Y multiplexer section. If the address pins of the 4052 are driven by the continuously repeating sequence 00, 01, 10, 11, 00, ..., then X OUT/IN will be connected to X_0, X_1, X_2, X_3, X_0,... and so forth, in succession. Another feature of the 4051/52/53 series is the incorporation of an inhibit input signal that may be used to disable or deselect the entire device. The inhibit input signal overrides address inputs, and when it is at the logic "1" level all switches are in their OFF states. As can be seen in the truth tables of Figure 9.2, when the inhibit input is "1", it makes no difference what the address inputs are, insofar as channel selection is concerned. The fact that all channels are off, regardless of the address inputs, is conveyed in the truth tables by the letter X.

Finally, each device may be used as a "demultiplexer," in which a signal originating on one line is transmitted to multiple output lines in succession. In this mode of operation the device behaves like a distributor, i.e., it distributes one input to multiple outputs. This is accomplished simply by connecting the input signal to the common terminal described before as the common output (designated COMMON OUT/IN on the diagrams) and taking the outputs from the multiple signal channels.

EXERCISES

1. A 4052 multiplexer has its "A" address input connected to 5 V and its "B" address input connected to 0 V. Between which sets of pins can conduction take place?

2. An analog switch has its analog input connected to + 5 V and its output connected through a 1K resistor to ground. When the switch is in its "ON" state, the voltage across the 1K resistor is 4.5 V. What is the "ON" resistance, R_{ON}, of the switch? (Hint: R_{ON} and the 1K resistor form a voltage divider.)

3. A 4052 multiplexer has its X-inputs connected to the following voltages:

   X_0 = 0 V, X_1 = 1 V, X_2 = 2 V, and X_3 = 3 V. The A and B address inputs are driven by the following sequence of binary numbers (A is the least significant bit):

   00, 01, 10, 11, 00, 01, ..., and so forth. Sketch the output that appears at the X OUT/IN pin. Repeat, if the binary sequence is 11, 10, 01, 00, 11, 10, ..., and so forth.

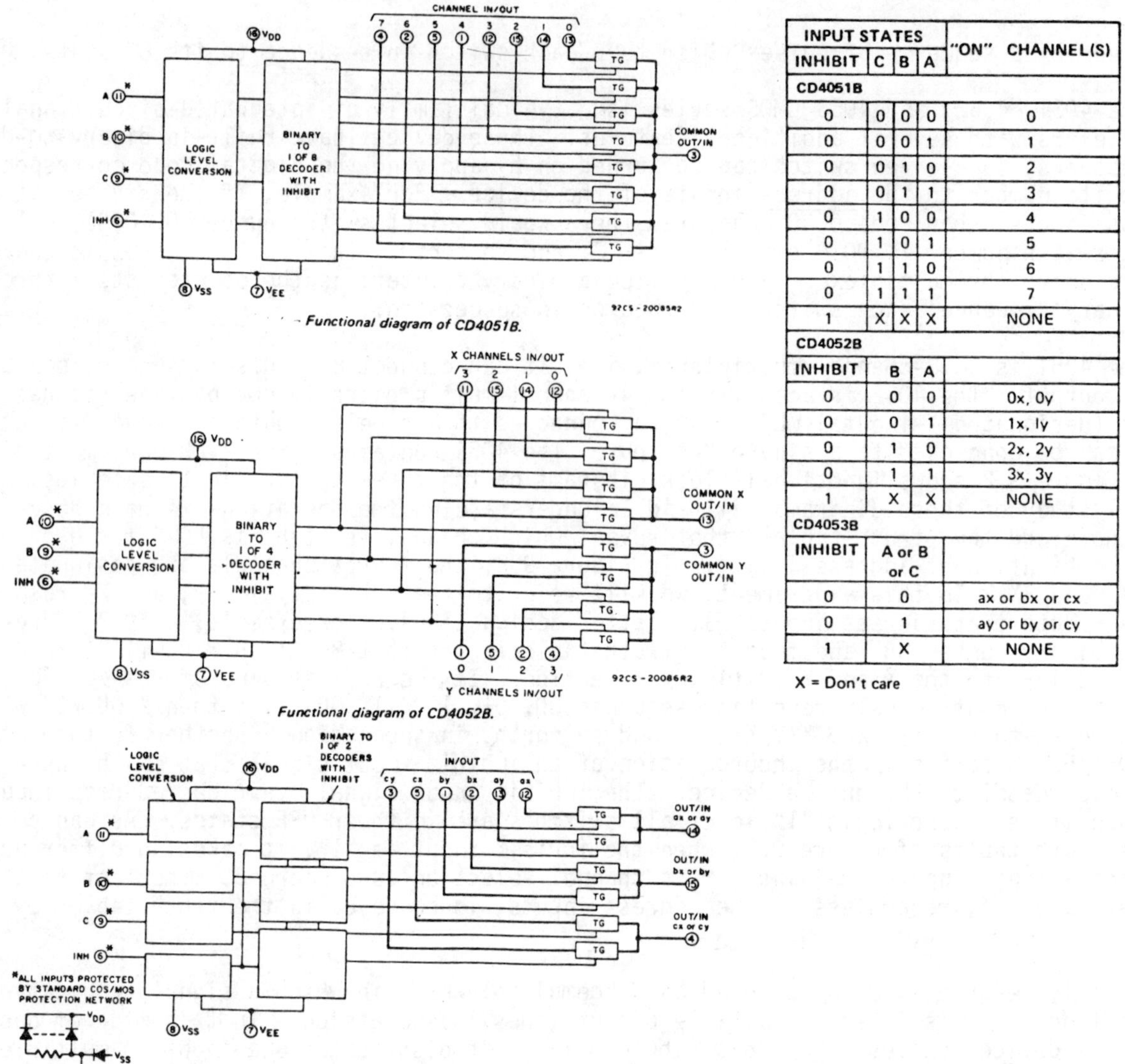

INPUT STATES				"ON" CHANNEL(S)
INHIBIT	C	B	A	
CD4051B				
0	0	0	0	0
0	0	0	1	1
0	0	1	0	2
0	0	1	1	3
0	1	0	0	4
0	1	0	1	5
0	1	1	0	6
0	1	1	1	7
1	X	X	X	NONE
CD4052B				
INHIBIT		B	A	
0		0	0	0x, 0y
0		0	1	1x, 1y
0		1	0	2x, 2y
0		1	1	3x, 3y
1		X	X	NONE
CD4053B				
INHIBIT	A or B or C			
0	0			ax or bx or cx
0	1			ay or by or cy
1	X			NONE

X = Don't care

Figure 9.2 Block diagrams and truth tables for the 4051, 4052, and 4053 multiplexers/demultiplexers.

4. A 4052 multiplexer is used as a demultiplexer by connecting an input signal to its "X OUT/IN" (pin 13). Assume this signal is a + 5 V dc level, and the address pins A and B are driven by the binary sequence 00, 01, 00, 01, ..., where A is the least significant bit. Describe the signals that would appear at X_0 and X_1 (pins 12 and 14). Make a sketch that shows A, X_0 and X_1.

5. Repeat exercise 4 if the binary sequence at pins A and B is 00, 10, 00, 10, ...

PROCEDURE

1. Connect the circuit shown in Figure 9.3. The 4052 is a CMOS device and must be handled with appropriate care to avoid damage from static charge caused by handling. Note that the A address input is connected to zero volts, while the B address is connected to a variable voltage derived from a 1K potentiometer. When the B address is zero, channel X_0 will be selected, thus enabling the 5 volt level connected there to appear at the common out/in terminal. We will investigate the voltage level required to change the state of the switch.

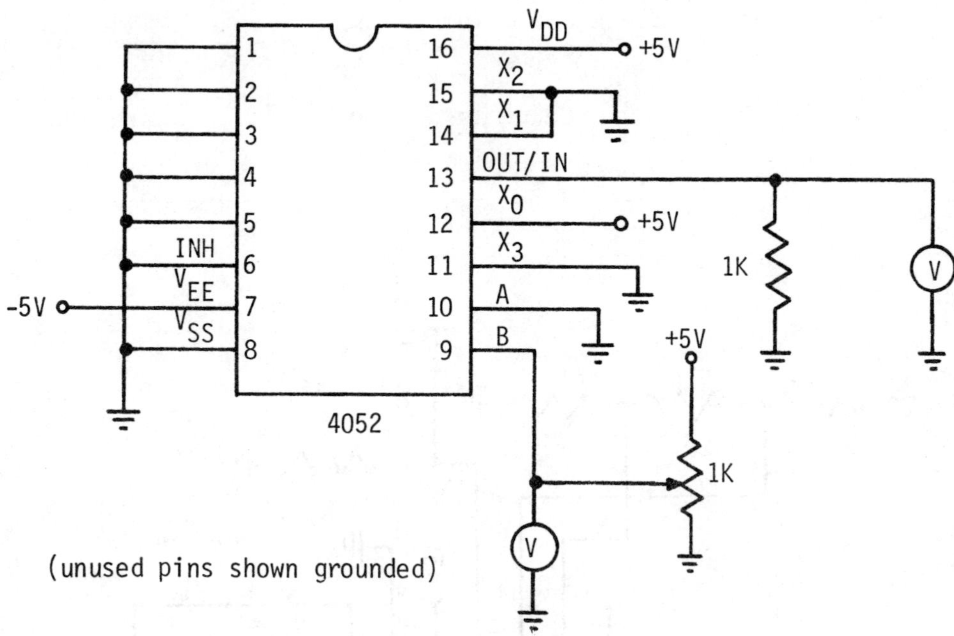

Figure 9.3

2. Monitor the voltage at the common out/in terminal (pin 13 of the 4052) and the voltage applied to the B address input (pin 9). Slowly adjust the potentiometer back and forth through its range and observe the switching action which takes place at pin 13. Record the voltage levels at pin 9 which just cause the switch to turn on and to turn off.

3. Turn off all power and connect the inhibit input (pin 6) to + 5 V. Then repeat step 2.

4. Restore pin 6 to 0 V and connect address input B (pin 9) to 0 V. To obtain data for calculating the value of R_{ON}, carefully measure and record the exact value of the voltage applied to the X_0 input (pin 12) and the exact value of the output voltage at pin 13.

5. Now connect the circuit shown in Figure 9.4. The three 1K resistors form a voltage dividing network which will apply different voltage levels to the X channel inputs. The 7493 is a binary counter whose Q_A and Q_B outputs (the least significant and the next to the least significant bits of the count) will be used to drive the A and B address pins on the 4052. Leave the connections to pins 9 and 10 on the 4052 and to pin 14 on the 7493 temporarily open.

6. Adjust the output of a square wave generator to approximately + 4.5 V peak at about 1 kHz. Then connect this output through the 1K resistor to pin 14 on the 7493 counter. Monitor pins 9 and 12 (Q_A and Q_B) of the 7493 on an oscilloscope to ensure that the counter is operating properly.

7. Once you have the counter operating properly, make the Q_A and Q_B connections shown in Figure 9.4 to pins 10 and 9 of the 4052. Monitor and record the waveform at the output of the multiplexer (pin 13) using an oscilloscope. To help you interpret your results, monitor and record Q_A simultaneously with the multiplexer output, and repeat with Q_B. Note carefully the time points where switching action takes place, in relation to changes in Q_A and Q_B. Synchronize the oscilloscope externally with the Q_B signal.
8. Now interchange the connections to pins 9 and 10 of the 4052 and repeat step 7.

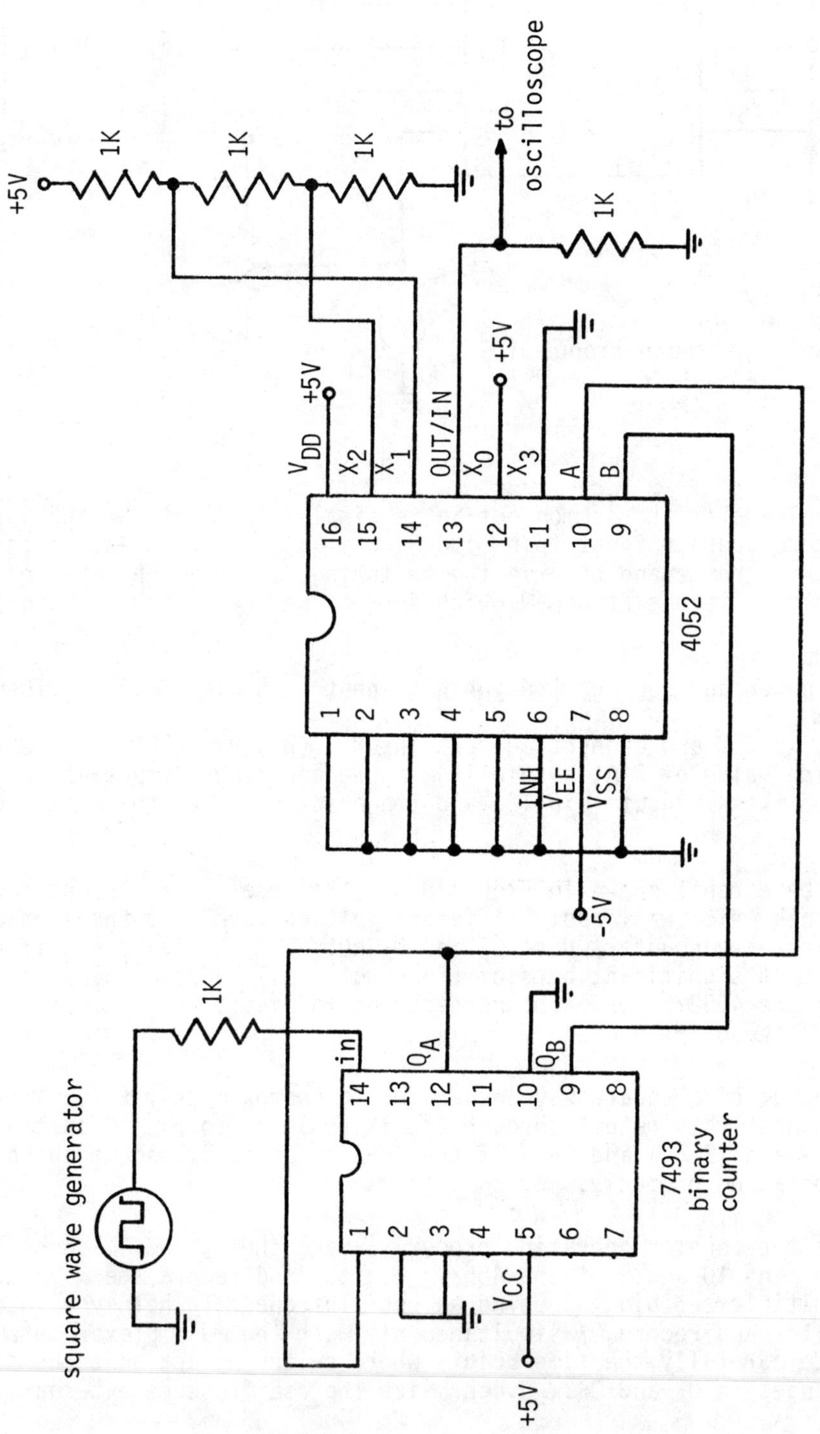

square wave generator
1K
+5V
7493
binary
counter
in
Q_A
Q_B
V_CC
4052
V_DD
X_2
X_1
OUT/IN
X_0
X_3
A
B
INH
V_EE
V_SS
-5V
+5V
1K
1K
1K
1K
to
oscilloscope

Figure 9.4

9. Connect the circuit shown in Figure 9.5. Here the 4052 is connected as a demultiplexer, with a sinusoidal input applied to the common out/in terminal (pin 13) of the 4052 and outputs taken from channels X_0 and X_1. Leave the connections to pin 14 on the 7493 and to pin 10 on the 4052 temporarily open.

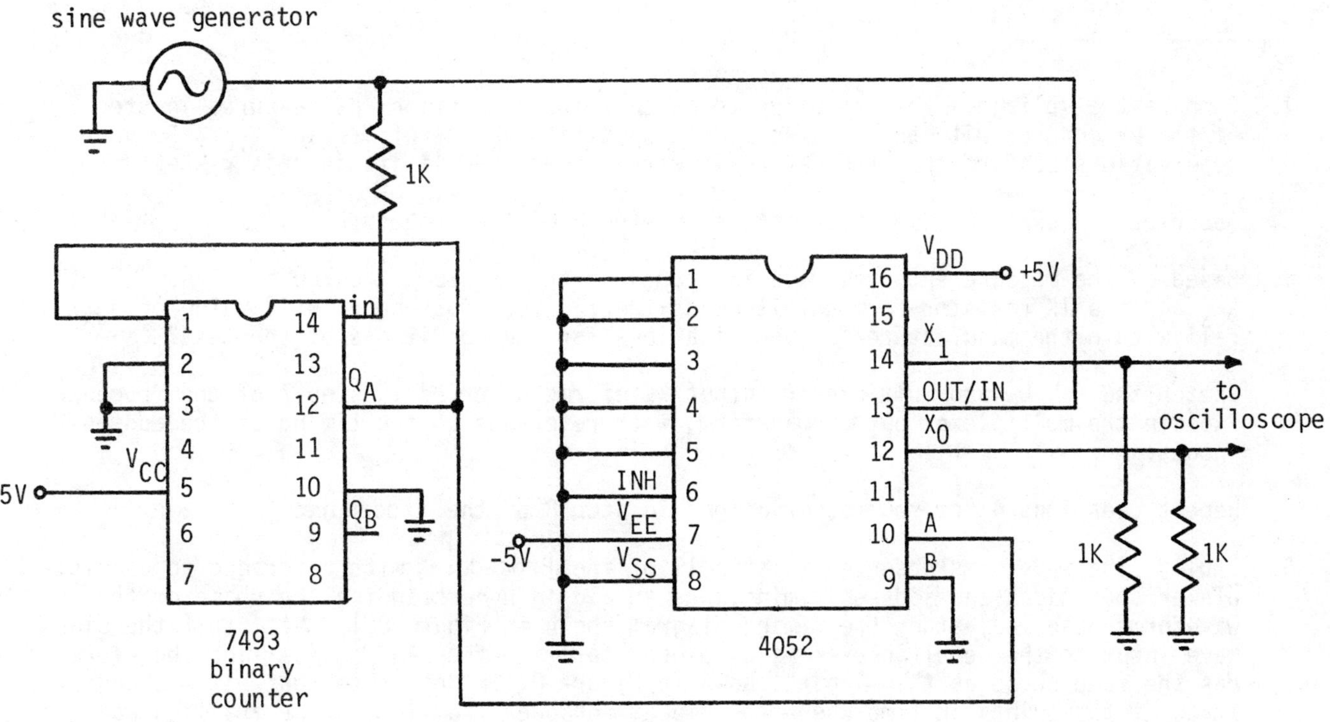

Figure 9.5

10. Adjust the output of a sine wave signal generator to approximately 4.5 volts peak at about 1 kHz. Then connect this output through the 1K resistor to pin 14 of the 7493. Monitor pin 12 on an oscilloscope to ensure that the counter is operating properly.

11. Connect pin 12 (Q_A) of the 7493 to pin 10 (A address input) of the 4052. Monitor X_0 and X_1 (pins 12 and 14) of the 4052 on a dual trace oscilloscope. Also view Q_A with X_0 and with X_1. Record the waveforms observed. Synchronize with Q_A.

12. Interchange the connections to pins 9 and 10 of the 4052 and monitor and record the waveforms observed on pins 12 and 14.

13. Now restore the ground connection to the B address input (pin 9) and connect the A address input (pin 10) of the 4052 to the Q_B output of the counter (pin 9 on the 7493). Record the X_0 and X_1 waveforms observed in this case. Synchronize with Q_B.

QUESTIONS

1. Compare the voltage levels required to cause switching action, as measured in step 2 of the Procedure, with the manufacturer's specified values of V_{IH} and V_{IL}. Do your observations confirm that the device is within specified limits in this respect?

2. Describe and explain your observations in step 3 of the Procedure.

3. Based on the measurements you made in step 4 of the Proceudre, calculate R_{ON}. (Hint: R_{ON} and the 1K resistor form a voltage divider.) Does your calculated value of R_{ON} fall within the manufacturer's specifications for the conditions of the test?

4. Sketch the Q_A, Q_B and multiplexer output waveforms observed in step 7 of the Procedure. Explain the multiplexer output waveform, with reference to the timing of the address inputs.

5. Repeat questions 4 for your observations in step 8 of the Procedure.

6. Explain the waveforms observed in step 11 of the Procedure, with reference to demultiplexer operations and address timing. As an aid in understanding the phase of the waveforms observed, study the timing diagram shown in Figure 8.1. Note that the sine wave input to the demultiplexer is also used to drive the 7493 counter and therefore has the same phase as the "input" shown in Figure 8.1. Note also that bit A changes state at the points in time where the "input" changes from 1 to 0(the trailing edge of the input). Similarly, bit B changes state on the trailing edges of bit A.

7. Repeat question 6 for the waveforms observed in steps 12 and 13 of the Procedure.

8. Suppose a demultiplexer application required that the signals appearing at X_0 and X_1 in step 11 of the Procedure each consist of 8 complete cycles of the sine wave generated by the signal generator. How could you modify the experimental procedure to accomplish this result?

10 Amplitude Modulation and Demodulation

OBJECTIVES

1. To understand the process of amplitude modulation in terms of its mathematical representation as a product of two waves.
2. To learn how to distinguish between full-carrier AM and suppressed carrier AM in terms of the appearances of their waveforms.
3. To learn the meaning of certain AM waveform characteristics, including modulation index and frequency spectrum, and how they are affected by the level and frequency of the modulating and carrier signals.
4. To learn how to measure modulation index.
5. To construct and verify the operation of an integrated-circuit amplitude modulator.
6. To construct and verify the operation of an integrated-circuit demodulator.
7. To construct and verify the operation of an envelope detector.

EQUIPMENT AND MATERIALS REQUIRED

1. Dual trace oscilloscope.
2. Power supplies: + 12 V DC, -8 V DC.
3. Variable frequency function generator (sine, square, triangular), audio frequency.
4. Sinewave generator, 100 kHz.
5. 1496/1596 balanced modulator/demodulators (2).
6. Resistors: 56 Ω (3), 100 Ω (2), 820 Ω (3), 1K (7), 3.3K (2), 4.7K (2), 6.8K, 10K (2).
7. Potentiometer: 10K.
8. Capacitors: .005 μF (3), 0.1 μF (5), 1μF

DISCUSSION

Modulation is the process of changing some characteristic of a waveform in direct proportion to the instantaneous changes in the level of another waveform. For example, frequency modulation (FM) means changing the frequency of a sine wave in accordance with changes in the magnitude of another waveform. In amplitude modulation (AM), the amplitude (peak value) of a high frequency sinewave is varied in proportion to the instantaneous changes in the level of a lower frequency waveform. The high frequency sinewave is called the carrier and the low frequency waveform is called the modulating signal.

An AM modulator is a device that accomplishes amplitude modulation as described above. Figure 10.1 illustrates the process of amplitude modulation. The figure shows that when the carrier is modulated by a small dc level, the result is a modulated signal that has relatively small amplitude. It also shows that a large dc level results in a proportionately larger modulated signal. Finally, when the modulating signal is a slowly varying sinewave, the amplitude of the modulator output is seen to vary accordingly: when the input is large, the amplitude of the AM signal is large, and conversely.

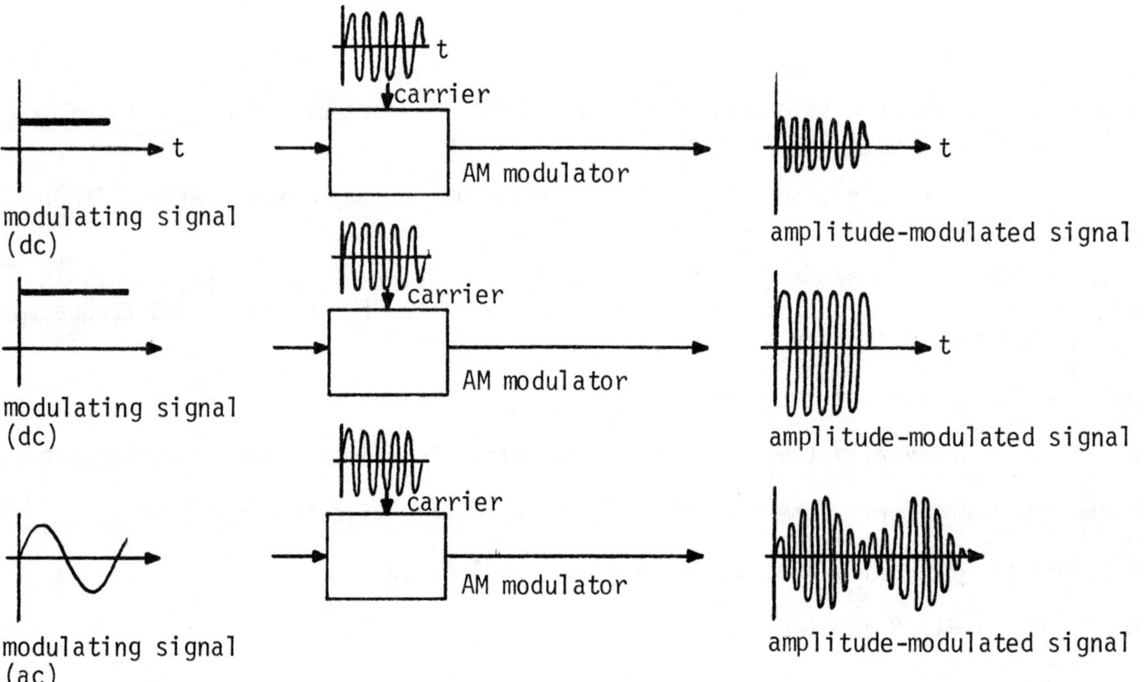

Figure 10.1 Amplitude modulated (AM) waveforms produced by dc and ac modulating signals.

Note in Figure 10.1 that the shape of the modulating signal is reproduced in both the positive and negative peaks of the carrier. The negative peaks are the mirror image of the positive peaks. This variation in the peak values is called the envelope of the carrier.

It is instructive to study a mathematical formulation of the modulation process. Towards that end, let us suppose that the modulating signal is the sinewave:

$$e_m = A_m \sin\omega_m t \qquad (1)$$

where A_m is the peak value of the modulating signal and ω_m is its angular frequency in radians/sec. Let the carrier signal be

$$e_c = A_c \sin\omega_c t \qquad (2)$$

where A_c is the peak value of the carrier and ω_c is its angular frequency in radians/sec.

The positive envelope of the modulated signal is therefore $A_c + e_m$, the peak value of the carrier added to the instantaneous values specified by e_m. For example, if the carrier has 5 V peak value and the modulating signal is 2sin100t, then this envelope is 5 + 2 sin100t, a waveform that alternates between 5 + 2 = 7 V and 5 - 2 = 3 V with a frequency of 100 rad/sec. (The mirror image of this envelope, the negative envelope, is $-(A_c + e_m)$.) The composite (AM) wave is therefore represented by expressing a sinewave (the carrier) whose peak value is the time-varying envelope:

$$e = (A_c + e_m)\sin\omega_c t$$

$$= (A_c + A_m \sin\omega_m t)\sin\omega_c t \qquad (3)$$

It is important to note that amplitude modulation, as expressed by (3), involves multiplication of the two time-varying quantities e_m and e_c. Unfortunately, the word "mixing" is used conventionally to mean both amplitude modulation and the summation of different signals. Amplitude modulation is a non-linear process requiring a device (a modulator) that is capable of performing multiplication, while summation is a linear process that can be accomplished by an operational amplifier, as we have seen.

Equation (3) can be rewritten as follows:

$$e = A_c\left(1 + \frac{A_m}{A_c}\sin\omega_m t\right)\sin\omega_c t$$

$$= A_c(1 + m\sin\omega_m t)\sin\omega_c t \qquad (4)$$

where $m = A_m/A_c$ is called the modulation index. The modulation index has value less than or equal to one and is often expressed as a percent. Thus, 50% modulation means that the peak value of the modulating signal is one-half the peak value of the carrier. Figure 10.2 shows oscilloscope photographs of 50% and 100% modulated waveforms. In each case the modulating wave is a single frequency. Notice that the envelope of the 100% modulated waveform varies all the way from zero to its peak of $A_m + A_c$. If A_m exceeds A_c ($m \geq 1$), the waveform is said to be overmodulated, and distortion results.

The modulation index can be determined experimentally by measuring the total voltage difference between the positive peak of the envelope and its mirror image, and the total difference between the negative peak and its image. Figure 10.3 illustrates the method. Note in Figure 10.3 that for 100% modulation, $E_2 = 0$ and therefore m is calculated as

$$m = \frac{E_1 - E_2}{E_1 + E_2} = \frac{E_1}{E_1} = 1$$

The frequency spectrum of a waveform can be displayed in the form of a plot that shows the relative magnitudes of the various frequency components of the waveform. The horizontal axis is frequency and the vertical axis is magnitude (often in dB). For example, the frequency spectrum of a single sine wave would have the appearance of a single vertical line drawn at the appropriate point on the frequency scale and having length proportional to its amplitude. When a single sine wave modulates a carrier, the frequency spectrum of the modulated wave consists of a component at the carrier frequency f_c, and the sum and difference frequencies $f_c + f_m$ and $f_c - f_m$, where $f_c = \omega_c/2\pi$ and $f_m = \omega_m/2\pi$. Figure 10.4 shows a typical spectrum of this type.

As shown in Figure 10.4, the sum and difference frequencies are called the upper and

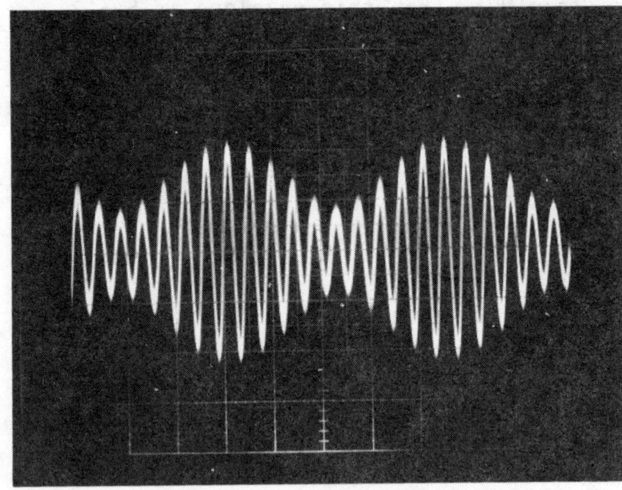

Figure 10.2 (a) $f_c = 10f_m$

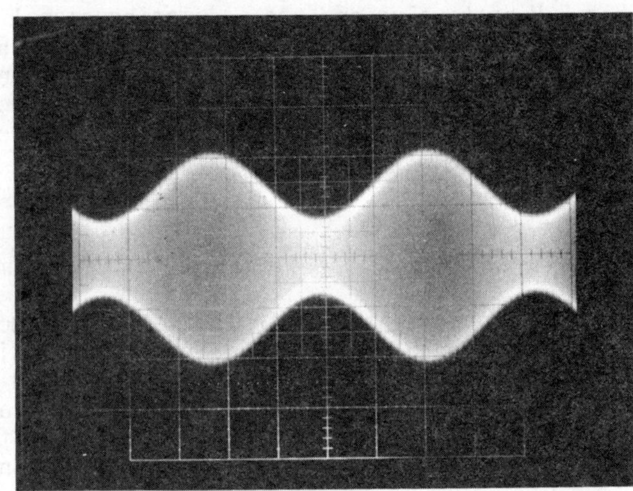

Figure 10.2 (b) $f_c = 100f_m$

50% modulated AM waveforms

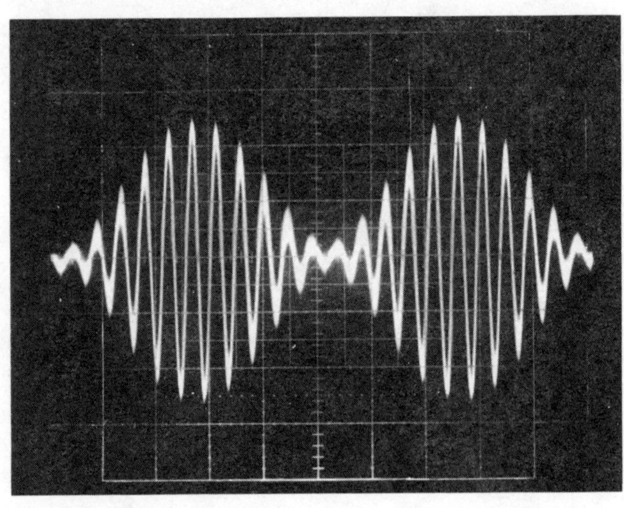

Figure 10.2 (c) $f_c = 10f_m$

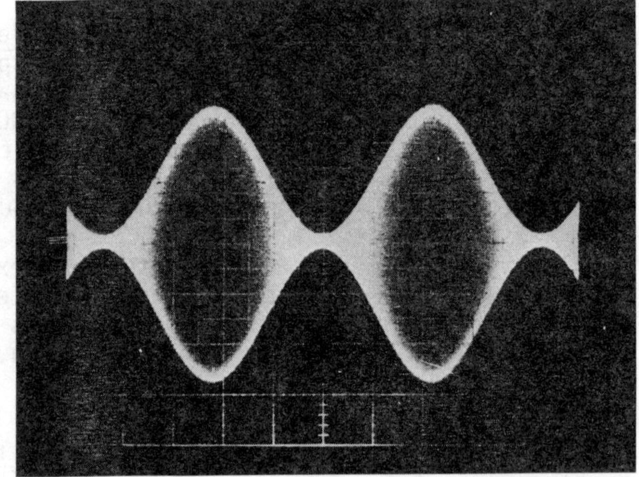

Figure 10.2 (d) $f_c = 100f_m$

100% modulated AM waveforms

Figure 10.2 Amplitude modulated (AM) waveforms. f_c = carrier frequency; f_m = modulating frequency.

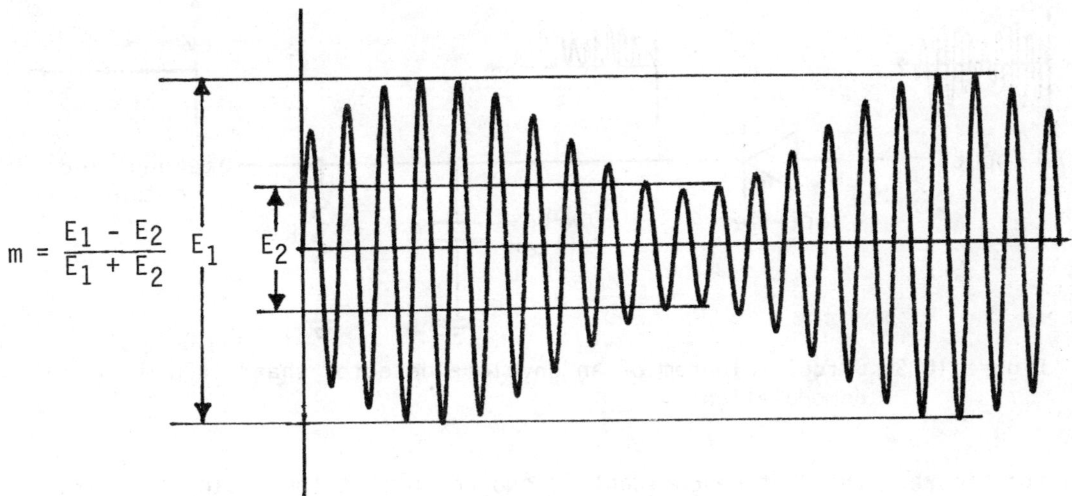

Figure 10.3 Calculation of the modulation index m from experimental measurements.

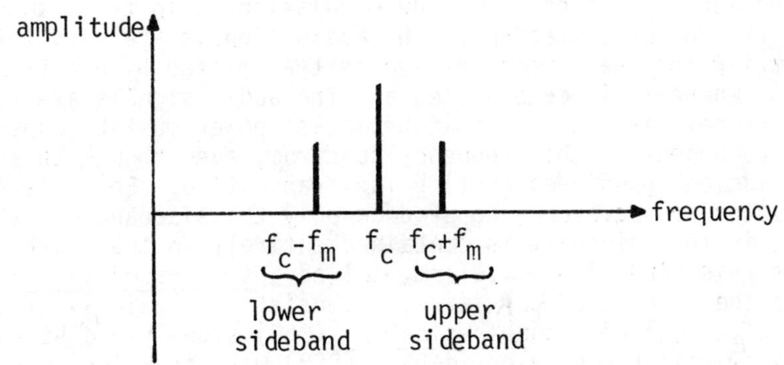

Figure 10.4 Spectrum of an AM wave showing carrier frequency and sidebands. In this example, each sideband consists of a single frequency.

lower sidebands, respectively. In practice, the modulating signal is usually a complex waveform containing many frequency components itself (such as speech, or music) and so the upper and lower sidebands of the AM spectrum consist of many sum and difference frequencies. The amplitude of a sideband frequency component is always less than or equal to one-half the magnitude of the carrier, depending on the modulation index. For 100% modulation, the sideband component reaches its maximum value of one-half the carrier level.

Demodulation is the process of recovering the modulating signal from the composite (AM) waveform. A circuit that is widely used to perform demodulation is the envelope detector, shown in Figure 10.5. The circuit consists of a diode that rectifies the AM wave, followed by a low-pass RC filter. Typical waveforms at various points in the circuit are

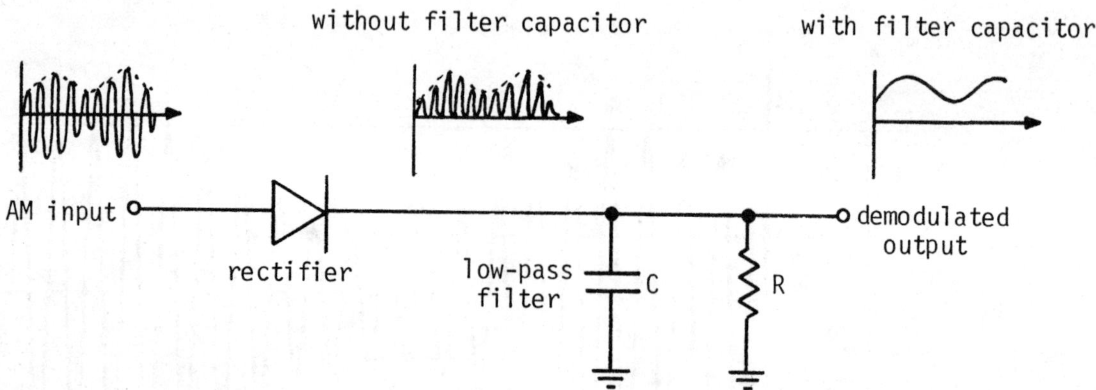

Figure 10.5 Circuit diagram of an envelope detector that performs demodulation.

shown in the figure. The RC time constant is chosen so that the capacitor charges and discharges in unison with the low frequency variation of the envelope of the rectified wave. At the same time, the carrier frequency component is attenuated because of the small impedance ($X_C = 1/\omega C$) of the capacitor at that frequency. Note that there is a dc level in the demodulated output (in addition to a small carrier frequency component still present after filtering, not shown). The dc level is proportional to the overall amplitude of the AM input. Therefore, increasing or decreasing the carrier level will change the dc level of the demodulated output.

The most common application of amplitude modulation is in the transmission of audio frequency signals by broadcast stations. The audio signals are used to modulate a radio frequency (rf) carrier and the composite wave is transmitted by electromagnetic radiation to radio receivers, where it is demodulated and the audio signals are recovered. One disadvantage of this process is the fact that broadcast power must be expended to transmit the large carrier component in the frequency spectrum, even though this component contains none of the "information" (audio content) being transmitted. For this reason, another type of amplitude modulator was developed to produce only the sidebands of the spectrum. None of the information is lost since it is contained entirely in the sidebands. A <u>balanced modulator</u> produces this kind of AM waveform, called a <u>suppressed carrier</u> wave. The frequency spectrum of the suppressed carrier wave consists exclusively of the lower and upper sidebands. It is often called a <u>double-sideband</u> (DSB) signal to distinguish it from still another type of AM signal, the <u>single-sideband</u> (SSB) wave in which only one sideband is transmitted. Conventional amplitude modulation (which is used by commercial broadcast stations) may be referred to as double-sideband, full carrier modulation to distinguish it from suppressed carrier or single-sideband modulation.

The appearance of a suppressed carrier waveform, as viewed for example on an oscilloscope, is noticeably different from that of a conventional AM waveform. This difference is of course to be expected, since the frequency spectrums of the two are different. In the suppressed carrier wave, the positive portion of the modulating signal appears in the positive peaks of the wave and the negative portion in the negative peaks. The mirror images of the positive and negative peaks are also present. This appearance is best appreciated by studying an actual suppressed carrier waveform. Figure 10.6 shows an oscilloscope photograph of a suppressed carrier wave modulated by a single frequency audio wave. Study Figure 10.6 carefully and compare it with Figure 10.2. The disadvantage of suppressed carrier modulation is that it is difficult to demodulate.

In this experiment, we will use the versatile 1596 balanced modulator/demodulator chip, which is capable of performing either suppressed carrier modulation and demodulation, or

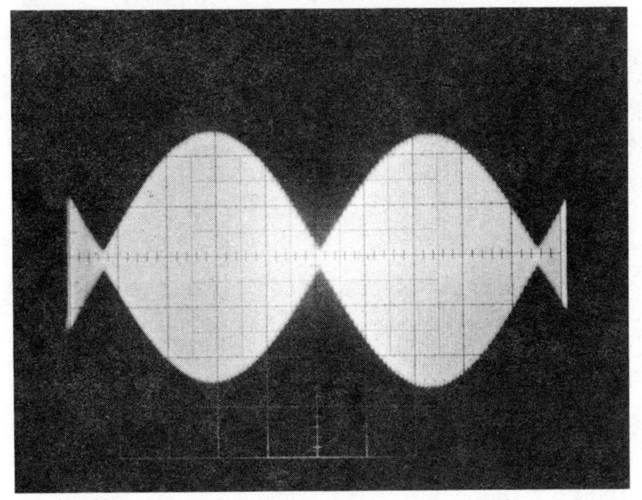

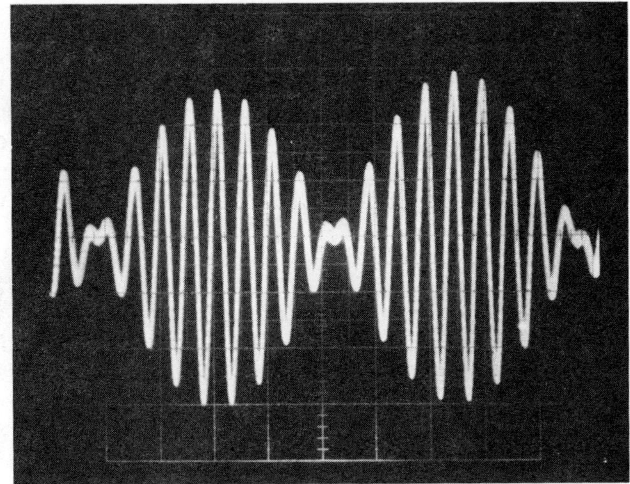

(a) $f_c = 20f_m$ $f_c = 100f_m$

Figure 10.6 Suppressed carrier AM waveforms. f_c = carrier frequency; f_m = modulating frequency.

conventional, double sideband, full carrier modulation and demodulation. The proportion of the carrier frequency in the output spectrum can be controlled by an external adjustment, so it is possible to eliminate it entirely for suppressed carrier operation or fully inject it for full-carrier modulation.

Mathematically, suppressed carrier modulation is equivalent to direct multiplication of the modulating and carrier signals, as shown by equation (5):

$$e = e_c e_m = A(\sin\omega_c t)\sin\omega_m t \qquad (5)$$

where A is a constant (the peak value of the composite signal). Recall the trigonometric identify:

$$\sin a \sin b = \tfrac{1}{2}\left[\cos(a - b) - \cos(a + b)\right] \qquad (6)$$

Letting $a = \omega_c t$ and $b = \omega_m t$, we find:

$$A\sin\omega_c t\sin\omega_m t = \tfrac{1}{2} A[\cos(\omega_c - \omega_m)t - \cos(\omega_c + \omega_m)t] \qquad (7)$$

We see from equation (7) that only the sideband frequencies $(\omega_c - \omega_m)$ and $(\omega_c + \omega_m)$ are present.

The 1596 modulator circuitry uses the carrier input to switch pairs of transistors on and off. By this process, the modulating signal is effectively switched on and off at a rate determined by the carrier frequency. The result is the same as multiplying the modulating signal by a square wave whose frequency is the same as the carrier. Since a square wave can be regarded as being composed of a dominant (fundamental) sine wave plus an infinite number of smaller amplitude odd harmonics, the result is multiplication of the fundamental by the modulating signal, as specified by equation (5). The higher frequency sidebands ($3\omega_c \pm \omega_m$, $5\omega_c \pm \omega_m$, etc.) are also created, but at a much lower level and they are easily filtered out. The advantage of this technique is that unlike conventional, discrete-component balanced modulators, no transformers are required. A full description of the theory of operation of the 1596 modulator/demodulator may be found in Motorola Application Note AN531.

EXERCISES

1. Suppose a known dc level E_1 is connected to the modulating signal input of an AM modulator, and that the amplitude of the modulator output A_1 is then measured. The dc level is then changed to E_2 and a new output amplitude A_2 is measured. This process is repeated several times, each time using a new dc level E and resulting in a new output amplitude A. If the values of A are plotted (on the vertical axis) against the corresponding value of E (on the horizontal axis), what would you expect the graph to look like? Explain.

2. Sketch the waveform representing a 100 kHz sine wave that is amplitude modulated by a 1 kHz waveform when the modulation index is 0.3. Assume the carrier has peak value 1 volt. Show the values of E_1 and E_2 (see Figure 10.3) in your sketch. Sketch the frequency spectrum and label the sideband frequencies.

3. What peak value of the modulating signal $e = A_m \sin\omega_m t$ would result in 100% modulation in Exercise 2?

4. Sketch the modulated waveform you would expect to see if the modulating signal input in Exercise 1 were changed to a square wave. Your sketch should cover a period of time equal to 1 or 2 complete cycles of the square wave.

5. Suppose the 100 kHz carrier of Exercise 1 is modulated by the 1 kHz signal in a suppressed carrier, double-sideband modulator. Sketch the waveform and frequency spectrum for this case.

PROCEDURE

1. Connect the 1496/1596 balanced modulator/demodulator in the circuit shown in Figure 10.7.

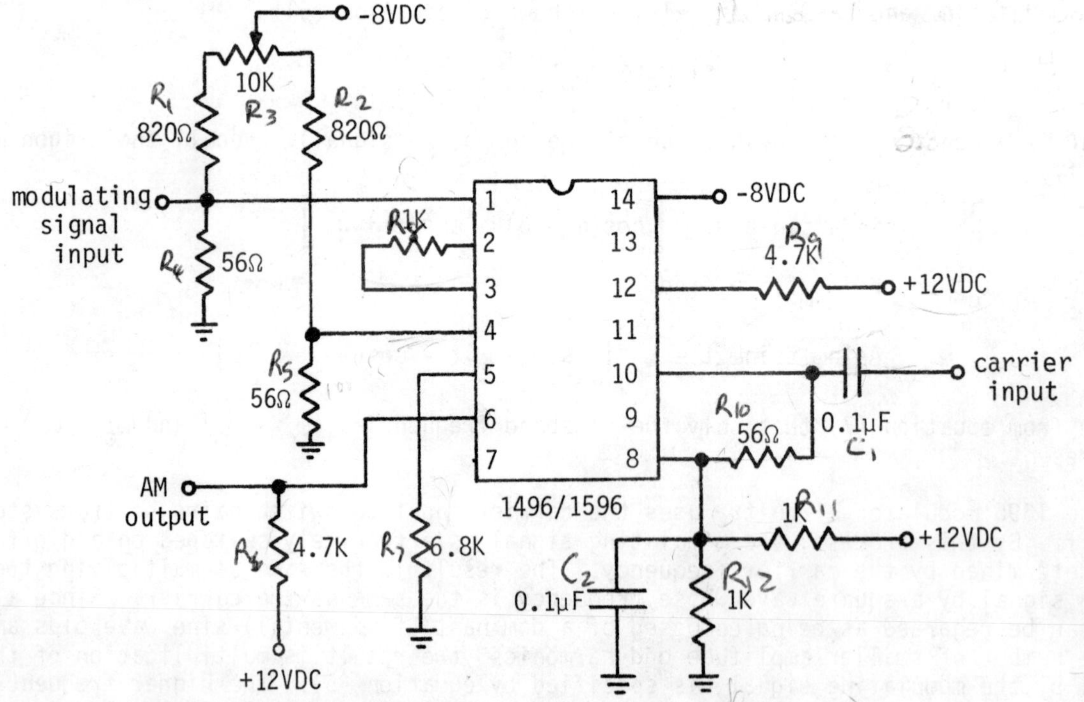

Figure 10.7

2. Using a signal generator with output amplitude set to zero, connect a 100 kHz sine wave to the carrier input. Then increase the 100 kHz input level to 10 mV peak. By monitoring the AM output on an oscilloscope, adjust the 10K potentiometer until a maximum carrier level is observed at the AM output. Record this level.

3. Measure and record the peak-to-peak level of the AM output when each of the following dc levels is connected to the modulating signal input: -0.5 V, -0.4 V, -0.3 V, -0.2 V, -0.1 V, 0.0 V, + 0.1 V, + 0.2 V, + 0.3 V, + 0.4 V, + 0.5 V. These voltages may be derived from a potentiometer connected as a voltage divider across the fixed supply voltages if desired. Be certain to measure the dc level with the input connected.

4. Connect a 1 kHz sine wave signal to the modulation input. With this signal connected, adjust its peak amplitude to 0.3 V. Sketch the waveform observed at the AM output, and make the appropriate measurements necessary to determine the modulation index (see Figure 10.3).

5. Increase the level of the 1 kHz modulating signal until 100% modulated output is observed. Sketch the waveform.

6. To learn the effect of overmodulation on the AM output, increase the level of the 1kHz modulating signal beyond that required for 100% modulation. Sketch the resulting waveform. Then restore 100% modulation.

7. Vary the frequency of the modulating signal input and observe the effect this has on the output AM waveform. Also change the modulating signal input to triangular and square waves and observe the effects on the output.

8. To develop an appreciation for the difference between modulation and summation of waveforms, connect the passive summing network shown in Figure 10.8.

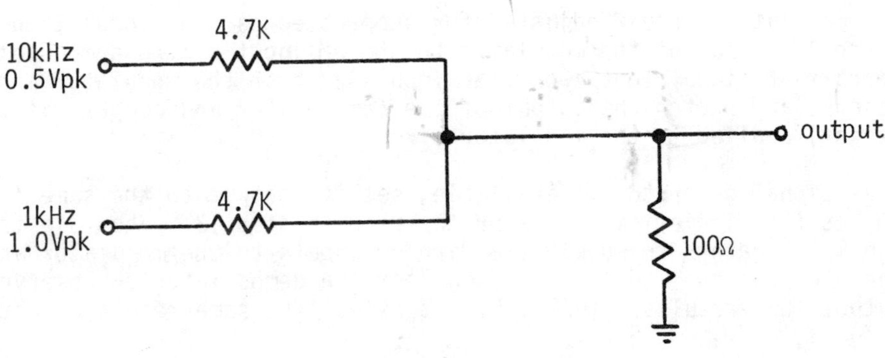

Figure 10.8

Observe the output of the summing network on an oscilloscope and sketch its waveform.

9. Connect the passive demodulator circuit (envelope detector) shown in Figure 10.9 to the AM output of the 1496/1596 modulator. With the carrier input to the modulator set as before, and with a 1 kHz sine wave for the modulating input (adjusted for 100% modulation), observe the output of the detector and sketch its waveform. Now remove the .01 μF capacitor and sketch the output waveform. For sketching purposes, reduce the carrier frequency to about 5 kHz, so that it will be clearly visible in the waveform. Then reset the carrier frequency to 100 kHz and reconnect the capacitor. Determine

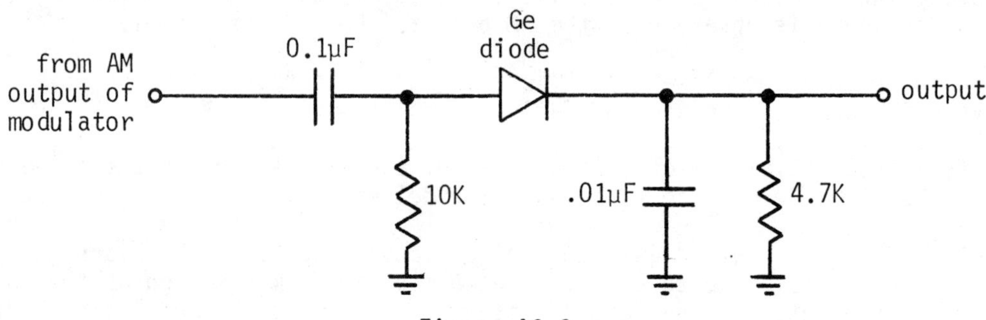

Figure 10.9

the effect of small changes in carrier amplitude on the output of the detector.

10. Disconnect the demodulator circuit and the modulating signal input. Do not change the level setting of the modulating signal generator. To obtain suppressed carrier modulation, observe the AM output on an oscilloscope and adjust the 10K potentiometer until the carrier output level is a minimum.

11. Reconnect the 1 kHz modulating signal to the modulator input and observe the AM output on the oscilloscope. Sketch the waveform, noting in particular how it differs from the waveform observed in step 5.

12. Connect another 1496/1596 chip in the demodulator configuration shown in Figure 10.10. (Alternatively, one such demodulator may be preconnected for sharing by all experimenters.) NOTE: Do not disconnect your modulator circuit, as it will be used in this part of the experiment.

13. With your modulator circuit adjusted for suppressed carrier modulation as in step 11, connect the AM output of the modulator to the AM input of the demodulator. Connect the same carrier signal to the carrier inputs on both the modulator and the demodulator. Observe and sketch the output of the demodulator and compare it with the modulating signal applied to the modulator circuit.

14. If another signal generator is available, set its output to the same frequency and amplitude as the carrier signal generator used in step 13. Then use these two separate signal generators to supply the carrier inputs to the modulator and demodulator (one generator for the modulator and one for the demodulator). Observe the demodulator output that results. (NOTE: Do not expect the same result as obtained in step 13.)

15. Once again using a single carrier frequency generator, repeat step 13 with the modulator adjusted to produce a double-sideband, full carrier, 100% modulated AM wave, as in step 5.

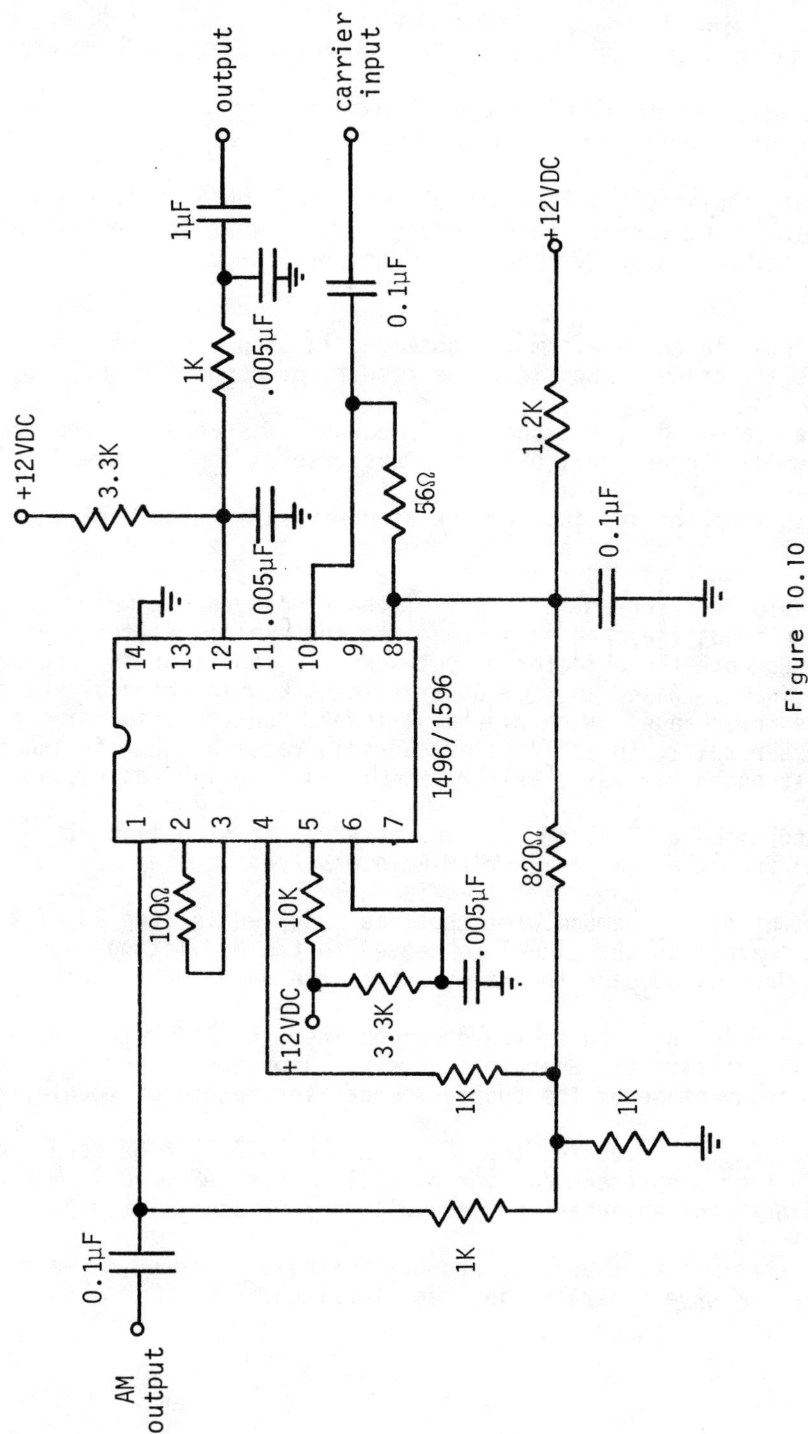
output
carrier
input
1μF
+12VDC
0.1μF
.005μF
1K
+12VDC
3.3K
1.2K
56Ω
.005μF
0.1μF
14
13
12
11
10
9
8
1
2
3
4
5
6
7
1496/1596
820Ω
100Ω
10K
.005μF
+12VDC
3.3K
1K
1K
1K
0.1μF
AM
output

Figure 10.10

QUESTIONS

1. Construct a graph showing peak-to-peak output level versus dc input level to the 1496/1596 modulator, using the data obtained in step 3 of the Procedure. Explain the relationship between these quantities. What should be the ideal relationship?

2. Sketch the waveform observed in step 4 of the Procedure. What is the modulation index for this waveform? Show your calculations, based on values you measured in step 4.

3. Sketch the 100% modulated waveform you observed in step 5 of the Procedure. Based on the carrier level you recorded in Procedure step 2, what is the theoretical level of the sidebands in the frequency spectrum of this 100% modulated wave? Sketch the frequency spectrum.

4. Sketch the overmodulated waveform you observed in step 6 of the Procedure. Why is this an undesirable situation? (Consider the results of demodulation.)

5. Describe and explain the effects on the AM output of changes in the frequency and waveshape of the modulating signal, based on your observations in step 7 of the Procedure.

6. Sketch the output of the summing network as observed in step 8 of the Procedure. In what ways does it differ from an AM waveform? Sketch its frequency spectrum.

7. Sketch the waveforms observed in step 9 of the Procedure. Compare the detector outputs with and without the capacitor connected and explain differences in terms of circuit action. Compare the detector output when the capacitor is connected to the modulating signal input. Based on your observations, how is the dc level of the detector output affected by changes in carrier amplitude? Can you think of a practical application for this result? (Hint: consider a radio receiver that is tuned to a frequency whose signal strength may vary, as for example in a moving vehicle.)

8. Sketch the suppressed carrier AM wave you observed in step 11 of the Procedure. How does it differ from the full-carrier AM wave observed in step 5?

9. Sketch the output of the demodulator that you observed in step 13 of the Procedure. How does it compare with the signal connected to the modulating input of the modulator circuit? How does it compare to the output of the detector circuit of Figure 10.9?

10. Describe and explain the results you observed when the modulator and demodulator were driven by different carrier generators in step 14 of the Procedure. What do you conclude is one disadvantage of the suppressed carrier method of modulation?

11. Repeat question 9 for Procedure step 15. Can the 1496/1596 be used successfully for demodulating either suppressed carrier or full carrier AM waves? Why could the use of this demodulator be disadvantageous for full carrier demodulation?

12. Suggest a way that might be used to produce a single-sideband AM wave using the suppressed carrier AM wave generated in this experiment.

11 Frequency Modulation, Voltage-Controlled Oscillators, and the Phase-Locked Loop

OBJECTIVES

1. To understand the process of frequency modulation and how it differs from amplitude modulation.
2. To construct and verify the operation of a voltage controlled oscillator and learn how it is used as a frequency modulator.
3. To learn how a phase-locked loop can be used to perform demodulation of an FM waveform.
4. To construct and verify the operation of an integrated-circuit FM modulator/demodulator system.
5. To learn the nature and use of a frequency shift keyed (FSK) system in digital data transmission and to verify its operation experimentally.

EQUIPMENT AND MATERIALS REQUIRED

1. Dual trace oscilloscope.
2. ± 15 V DC power supplies.
3. Voltmeter, 0-10 V DC.
4. Low frequency function generator (sine, square, triangular), 0-100 Hz, 0-1 V pk.
5. 566 voltage controlled oscillator.
6. 565 Phase-locked loop.
7. 311 Voltage comparator (111 or 211 may be used).
8. Resistors: 820 Ω, 1K, 4.7K (3), 10K (2), 15K, 22K (2).
9. Potentiometers: 1K, 10K.
10. Capacitors: .001 μF (2), .01 μF (2), .047 μF, 0.1 μF, 1.0 μF, 2 μF, 10 μF.

DISCUSSION

A voltage-controlled oscillator (VCO) is an oscillator whose frequency is controlled by the level of an externally applied voltage. Recall that frequency modulation (FM) is the process of changing the frequency of a waveform in accordance with the instantaneous changes in level of another waveform. Thus, a VCO may be used to perform frequency modulation. Figure 11.1 illustrates the process.

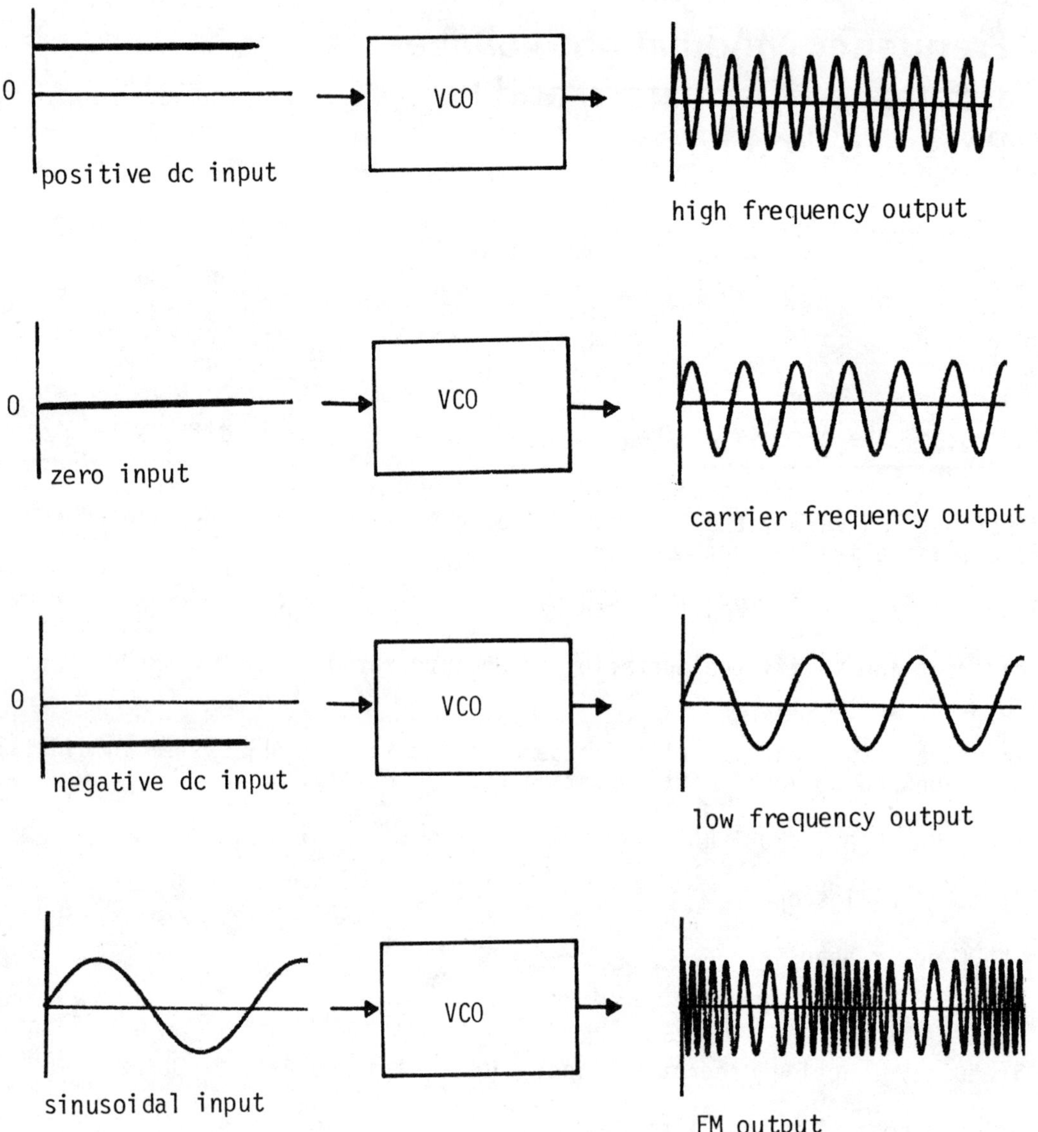

Figure 11.1 Output waveforms generated by a voltage-controlled oscillator (VCO) with dc and ac inputs.

Note in Figure 11.1 that the amplitude of the VCO output remains constant, i.e., output amplitude is independent of input amplitude. Only the output frequency changes. We can regard frequency modulation as a process of increasing or decreasing the frequency of a carrier wave so, as illustrated in the figure, the output signal is the same as the carrier when the input is zero volts (no modulation). The maximum change in the output frequency above or below the carrier frequency is called the (maximum) deviation δ.

The sensitivity of the VCO is the amount of frequency change per volt of input amplitude change. Ideally, the VCO should be linear, in the sense that the output frequency should change in direct proportion to changes in input amplitude. As an example, suppose an FM system has a carrier frequency of 1 MHz and that the modulating signal $e_m = 2\sin\omega_m t$ causes the output frequency to change from 995 kHz to 1.005 MHz. Then the deviation δ is

.005 MHz or 5 kHz and the sensitivity is 5 kHz/2V = 2.5 kHz/V. We should therefore expect a + 0.5 V input to cause a frequency change of (0.5 V)(2.5 kHz/V) = 1.25 kHz and a -0.5V input to cause a frequency change of -1.25 kHz. In other words, input amplitude changes between ± 0.5 V should cause the output frequency to change from 998.75 kHz to 1.00125 MHz.

The modulation index m_f for an FM wave is defined by

$$m_f = \frac{\delta}{f_m} \tag{1}$$

where δ is the maximum deviation and f_m is the frequency of the modulating signal. Note that, unlike AM, the modulation index m_f may be greater than one. The frequency spectrum for an FM wave is more complex than that of an AM wave. The spectrum contains an infinite number of sidebands and its frequency components depend heavily on m_f. In fact, for certain values of m_f the carrier component is completely suppressed.

Besides commercial broadcoasting, a commonly encountered application of frequency modulation is frequency shift keying (FSK). FSK is used for the serial transmission of digital (binary) data. Serial transmission means that the data bits (0's and 1's) are generated one after the other in time. So, for example, transmission of the binary number 0110 would be accomplished by first transmitting a zero, then a one, followed by another one, and finally a zero. This method of binary data transmission is in contrast to the parallel method in which all the bits are produced simultaneously, one bit per signal line. In Experiment 9, where we studied A/D and D/A converters, the bits were transmitted in parallel between the devices. The advantage of serial transmission is that only one signal line is required.

When frequency-shift keying is employed for digital data transmission, a binary one is represented by a high frequency tone and a binary zero by a lower frequency tone. (The frequencies are in the audio range, since telephone lines are commonly used for this type of transmission.) In the terminology of FSK, a zero is referred to as a space and a one is called a mark. In one of several standards, a space is represented by 2025 Hz and a mark by 2225 Hz.

In this experiment we will investigate the 566 voltage-controlled oscillator. The 566 VCO generates square and triangular wave outputs whose center (unmodulated) frequency can be set by an external resistor-capacitor combination (R_1C_1). The output frequency is given by

$$f = \frac{2(V^+ - V_5)}{R_1C_1V^+} \tag{2}$$

where V^+ is the voltage from pin 8 to pin 1 and V_5 is the voltage from pin 5 to pin 1.

From (2) we can show that the theoretical sensitivity S is

$$S = \frac{-2}{R_1C_1V^+} \text{ Hz/V}$$

Note from equation (2) that an increase in the modulating voltage V_5 causes a decrease in the output frequency, contrary to the scheme described earlier and shown in Figure 11.1. Of course, the modulating signal could be inverted before connection to pin 5 if we wanted increasing voltage to cause increasing frequency.

A phase locked loop (PLL) is a device whose principal components are a VCO and a phase comparator. A phase comparator produces an output voltage that is proportional to the phase difference between two signals applied to it. Figure 11.2 shows a block diagram of a phase locked loop.

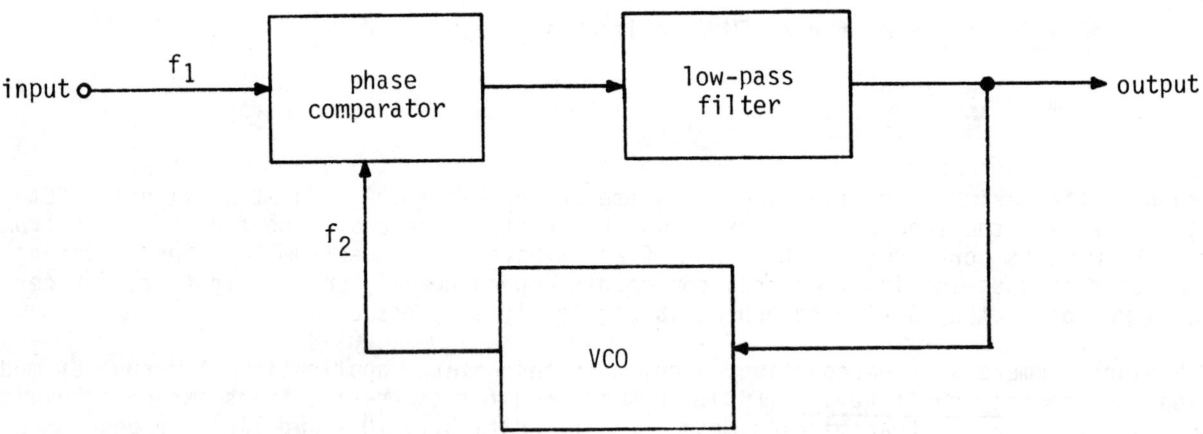

Figure 11.2 Block diagram of a phase locked loop.

When there is a phase difference between the output of the VCO and the input to the PLL, the phase comparator produces a voltage which is filtered and fed back to the input of the VCO. This feedback is arranged so that the voltage applied to the VCO causes it to change the frequency of its output as necessary to bring the phase difference to zero. We should note that a small difference between the two frequencies applied to the phase comparator has the same effect as a phase difference. This is illustrated in Figure 11.3, where two square waves of different frequencies are compared for an "apparent" phase difference over one cycle. (Strictly speaking, phase difference is not defined for two signals having different frequencies.)

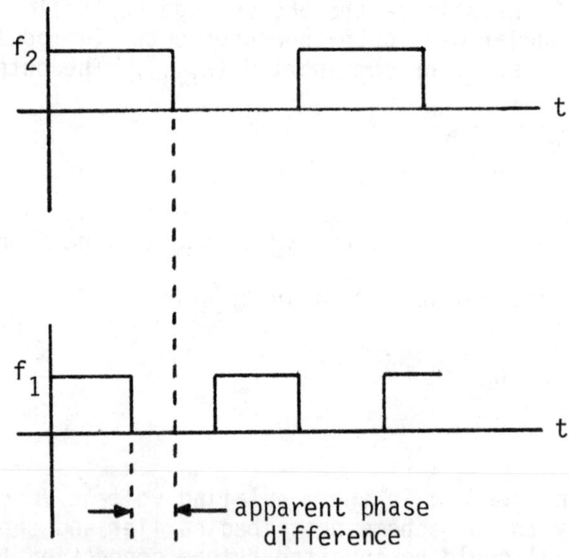

Figure 11.3 The apparent phase difference detected by a phase comparator when the inputs have different frequencies.

In Figure 11.3, f_2 is slightly lower than f_1, giving rise to an apparent phase difference. In the PLL, this phase difference would be detected and cause the frequency of f_2 to be increased until it matched f_1. We say that the PLL has achieved a lock when this is the case. Locking will not occur if the frequency difference is too great.

Among the many uses of a PLL, FM demodulation is one of the most common. If the input shown in Figure 11.2 is an FM signal, continually changing in frequency, then the phase comparator output voltage will also change continuously. This voltage will change in exactly the same way as the input frequency f_1 is changing and will therefore be proportional to the original modulating signal that produced the FM signal. FM demodulation is thus accomplished. The purpose of the low pass filter is to reject any high frequency components (f_1 or f_2) that may appear in the phase comparator output.

In this experiment, we will use the 565 phase locked loop to demodulate an FSK signal produced by a 566 VCO. Since an FSK signal is either the mark frequency or the space frequency, then the output of the PLL will be a series of pulses that occur whenever that frequency changes. The pulses will thus occur in the same sequence as the bits that produced the frequency shifts and we will have recovered the original binary data. We will use a voltage comparator connected to the output of the PLL to improve detection of the pulses.

EXERCISES

1. The carrier frequency in an FM system is 95.7 MHz. When the modulating signal $e_m = 2.5\sin(4\pi \times 10^3 t)$V is applied, the modulated signal frequency varies from 95.68 MHz to 95.72 MHz. What is the deviation? What is the sensitivity? What is the modulation index? What will be the minimum and maximum frequencies of the modulated signal if $e_m = 0.8\sin(2\pi \times 100t)$?

2. A 566 VCO has pin 1 connected to -5 V DC and pin 8 connected to + 5 V DC. If the voltage between pin 5 and pin 1 is 6.4 V and C_1 = .047 μF, what value of R_1 is required to produce a frequency of 2 kHz? (Hint: In this case, $V^+ = 5-(-5) = 10$ V).

3. Using the values of R_1, C_1, and V^+ from exercise 2, calculate the theoretical sensitivity of the 566 VCO. What frequency would be generated if the voltage between pin 5 and pin 1 were set to 8 V? What frequency would be generated if the voltage between pin 5 and ground were set to 2.5 V?

4. An FSK modulator is driven by a clock signal (square wave) that represents a sequence of binary one's and zero's. When the clock is high, a binary one is present, and when it is low, a binary zero. If the clock has frequency 10 Hz, describe the output of the modulator.

PROCEDURE

1. To verify the operation and check the linearity of the 566 VCO, connect the circuit shown in Figure 11.4. The external RC combination that controls the unmodulated output frequency is composed of the 4.7K resistor in series with the 10K variable resistor and the .047 μF capacitor. Monitor the square wave output on an oscilloscope and simultaneously connect it to a frequency counter. Connect pin 5 to the grounded .01 μF capacitor shown in Figure 11.4 and rotate the 10K potentiometer back and forth, observing the effect this has on the output frequency. (Does frequency increase or decrease as the total resistance in series with pin 6 is increased?)

2. With pin 5 still connected to the capacitor, adjust the 10K potentiometer until the output frequency is 2.0 kHz. Measure and record the DC voltage between pin 5 and pin 1. Then, with power off, carefully disconnect the 4.7K resistor from pin 6 and the 10 K potentiometer from the + 5 V supply and measure the total resistance of this series combination. Then reconnect the series combination as before, without disturbing the potentiometer setting.

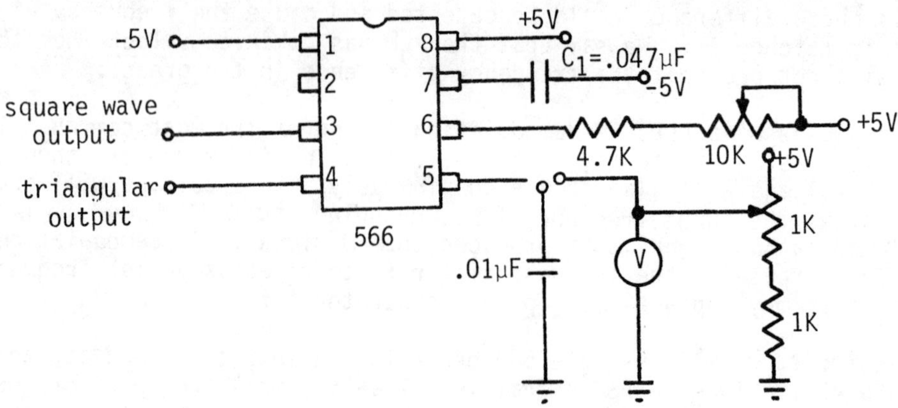

Figure 11.4

3. Connect the 1K potentiometer (in the voltage divider network shown in Figure 11.4) to pin 5. Measure and record the output frequency as the 1K potentiometer is adjusted to produce 8.3 V, 8.4 V, 8.5 V, 8.6 V, ..., 9.1 V between pin 5 and pin 1 (0.1 V increments from 8.3 V to 9.1 V).

4. Remove the voltage divider network and the .01 μF capacitor from pin 5 and connect the network shown in Figure 11.5 in its place.

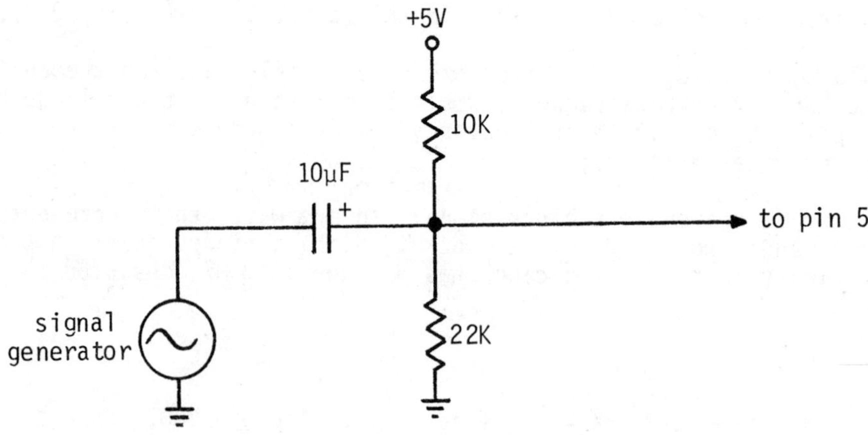

Figure 11.5

5. Adjust the signal generator to produce a sine wave with frequency about 1 Hz and peak value 1 V. Adjust the oscilloscope so that about 5 cycles of the square wave output are displayed and observe the behavior of the waveform. You should see a very graphic demonstration of frequency modulation. Observe the effects of small changes in the frequency and amplitude of the modulating signal produced by the signal generator. Also experiment with changes in the waveform (triangular, square) produced by the signal generator.

6. In the following steps, we will use the 565 PLL to demodulate an FM signal produced by the 566 VCO. Without disturbing your VCO circuit, connect the 565 PLL and 311 (or 111, or 211) voltage comparator in the circuit shown in Figure 11.6. Do not make any connections to the FM input at this time.

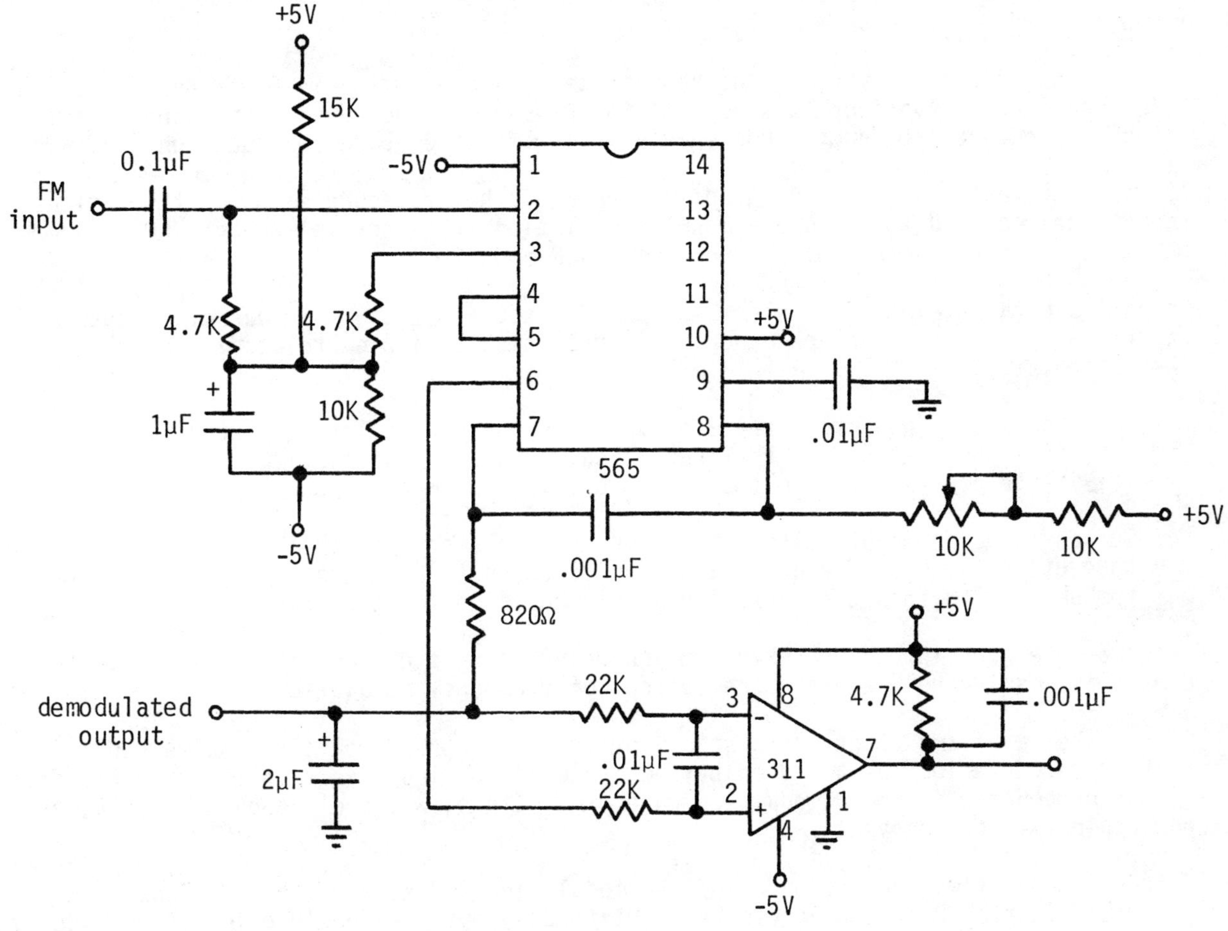

Figure 11.6

7. The variable resistance connected to pin 8 of the PLL is used for external adjustment of the VCO inside the PLL. Pin 4 is the VCO output (which is connected externally via pin 5 to the phase comparator for input). Connect a frequency counter to pin 4 and, with the FM input to pin 2 still disconnected, adjust the 10K potentiometer to obtain 2.1 kHz at pin 4.

8. Set the signal generator that you have connected through the network of Figure 11.5 (and thence to pin 5 of the VCO) so that it produces a 0.5 V peak-to-peak sine wave (with respect to ground) at 100 Hz. This setting is now causing the VCO to generate an FM signal. Connect the square wave output of the VCO (pin 3) to the FM input of Figure 11.6. Use an oscilloscope to monitor the signal generator output and the PLL demodulated output simultaneously. Sketch the waveforms observed.

9. In this step, we will verify the operation of the 566 VCO as an FSK modulator and the 565 PLL as an FSK demodulator. The voltage comparator is used to produce square pulses from the demodulated output of the PLL. Set the signal generator to produce a 0.2 V

peak-to-peak square wave at 10 Hz. This setting is now causing the VCO to generate an alternate sequence of space and mark frequencies. Monitor the square wave output of the signal generator and the voltage comparator output simultaneously on an oscilloscope. Sketch the waveforms observed. Also sketch the waveforms at the demodulated output of the PLL.

QUESTIONS

1. Describe the effect of varying the resistance in series with pin 6 of the 566 VCO, based on your observations in step 1 of the Procedure. How does increasing resistance affect the output frequency? How should it be affected, based on equation (2)?

2. Compare the resistance you measured in step 2 of the Procedure with the theoretical resistance required to produce a 2 kHz oscillation frequency in the 566 VCO. (Remember that V^+ is the total voltage between pin 8 and pin 1.)

3. Using the data obtained from step 3 of the Procedure, plot VCO frequency versus dc input voltage on linear graph paper. How are these variables related? How should they be related?

4. Using the data obtained from step 3 of the Procedure, calculate the sensitivity of the VCO. Compare your value with the theoretical value (see Discussion). How could the sensitivity be determined from the graph you plotted in Question 3?

5. For at least one dc input voltage that you used in step 3 of the Procedure, calculate the theoretical output frequency based on equation (2). Compare your measured output frequency with this theoretically predicted value for that case.

6. Describe the behavior of the waveform you observed in step 5 of the Procedure. How does this display help explain the concept of frequency modulation?

7. Sketch the waveforms observed in step 8 of the Procedure. Do your results confirm that FM demodulation was achieved by the phase locked loop? Are there any other frequency components in the demodulated waveform? If so, where did they come from, and how could they be removed?

8. What was the maximum deviation δ and the modulation index m_f of the FM signal you generated in step 8 of the Procedure? (Hint: use your previously calculated value of VCO sensitivity.)

9. Sketch the waveforms you observed in step 9 of the Procedure. Explain their relationship to each other, especially in regard to phase relation. Using your previously calculated value of VCO sensitivity, determine the space and mark frequencies that you generated when the VCO was used to produce the FSK signal.

Design Project Number 6: Digitally Controlled Modulator

Design an integrated circuit AM modulator that meets the following specifications:

1. Produces a full carrier, double sideband, amplitude modulated waveform having 100% modulation when the modulating signal is a 500 Hz sine wave.

2. Has a carrier frequency that is controlled by an 8-bit digital input. The carrier frequency is adjustable, by changing the digital input, over a minimum range of 5 kHz to 15 kHz.

You may use an external (laboratory) signal generator to produce the 500 Hz modulating signal and you may adjust its amplitude as necessary to obtain 100% modulation. The carrier frequency input to the modulator should originate from a voltage controlled oscillator that generates a square wave. You may use any two fixed-voltage laboratory power supplies. Voltage regulator(s) must be used if additional supply voltages are required.

Construct your circuit and test its performance to determine how well it meets the design specifications. Use a frequency counter to measure the carrier frequency produced for various digital input combinations. You may modify and refine your circuit after testing as necessary to improve its performance. Use potentiometers connected as variable resistors wherever you believe precision-valued components are necessary.

Write a report that contains

1. Your design equations and calculations. Justify and explain all decisions that you make relative to component selection, design tradeoffs, and design changes made after circuit tests. In particular, justify any use of non-standard or precision-valued components.

2. A schematic diagram of your final circuit.

3. Test results that show how well your circuit met the design specifications.

12 Voltage Comparators

OBJECTIVES

1. To learn how voltage comparators can be used to determine the larger of two signals.

2. To measure and understand the significance of certain voltage comparator characteristics, including input offset voltage, response time, rise time, and hysteresis.

3. To learn certain voltage comparator applications, including a window detector, a hysteresis circuit, and a square wave oscillator.

4. To investigate and confirm experimentally the operation of the voltage comparator in the applications listed.

EQUIPMENT AND MATERIALS REQUIRED

1. Dual-trace oscilloscope.

2. ± 5 V DC power supplies.

3. Voltmeter, 0-50 mV DC, 0-5 V DC.

4. Function generator (square and triangular), 1 kHz - 100 kHz, 0-2 V pk.

5. 311 (or 111, or 211) voltage comparator.

6. Resistors: 1K (3), 1.5K, 3.3K, 4.7K, 100K (2), 220K (2), 470K.

7. Potentiometers: 1K (2).

8. Capacitors: 0.1 μF (5).

DISCUSSION

A voltage comparator is a device that is used to determine which is the larger of two input voltages. The ouput of the comparator is one of two different levels, or states, depending on which input is larger. We have seen examples of how voltage comparators are used in several previous experiments.

In its simplest form, a voltage comparator is an operational amplifier with no feedback. The differential gain is therefore the very large open-loop gain that is typical of an operational amplifier. Figure 12.1 shows an operational amplifier used as a voltage comparator.

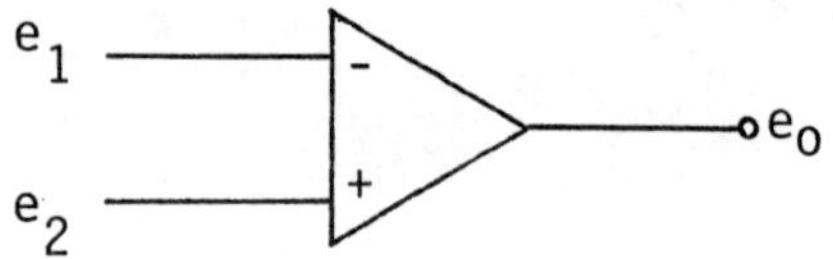

Figure 12.1 An operational amplifier used as a voltage comparator.

If the voltage applied to the inverting input, e_1 in Figure 12.1, is more positive than the voltage applied to the non-inverting input, e_2, then e_1-e_2 is positive and the output of the comparator is negative. Since the open loop gain is very large, e_1 needs only be slightly more positive than e_2 to drive the output all the way to its negative limit. Similarly, if e_2 is more positive than e_1, then the comparator output is driven all the way to its positive limit.

Although a conventional operational amplifier may be used as a voltage comparator in the way we have described, there are integrated circuits manufactured specifically as voltage comparators to meet the special needs of many applications. For example, some voltage comparators are designed so that the output voltage levels can be set independently of the power supply voltages. In many applications, the comparator output drives digital logic circuitry, so it is desirable to have the comparator voltage levels compatible with the digital logic levels (the voltages corresponding to logic 0 and logic 1). In fact, comparators are often used as <u>level shifters</u> to make the logic levels of one type of digital circuitry compatible with another type.

The <u>input offset</u> of a voltage comparator is the minimum differential input voltage required to cause the comparator to switch its output from one level to the other. Clearly, the greater the open-loop gain of the comparator, the smaller the input offset voltage.

Two other important characteristics of a voltage comparator are related to the length of time it takes the output to change state. These are called the <u>response time</u> and the <u>rise time</u>. The response time is the total time required for the output to change from one level to a specified percentage, say 10%, of the other level. In other words, response time is a measure of the delay between the application of an input change and the onset of the output change. Rise time is the total time required for the output to change from 10% of its new value to 90% of its new value. Therefore, the total time required for the output to change is the sum of the response and rise times. Figure 12.2 illustrates these ideas.

<u>Overdrive</u> is the amount of voltage in excess of the differential input voltage that is applied to the comparator to cause it to switch states. Overdrive is illustrated in Figure 12.2. The response time is heavily dependent on the amount of overdrive: the greater the overdrive the faster the response time. The rise time, on the other hand, depends heavily on the bandwidth of the comparator. Rise-time and bandwidth are inversely related, as shown by the following equation:

$$t_r = \frac{0.35}{BW} \tag{1}$$

where t_r is the rise-time in seconds, and BW is the bandwidth of the comparator, in Hz. For dc-coupled comparators (the usual case) the bandwidth is the same as the 3 dB cutoff frequency of the comparator. The significance of equation (1) is that rise time can be reduced (improved) by increasing the bandwidth, but only at the expense of input offset voltage. Recall that the gain-bandwidth product is constant. Therefore, an increase in

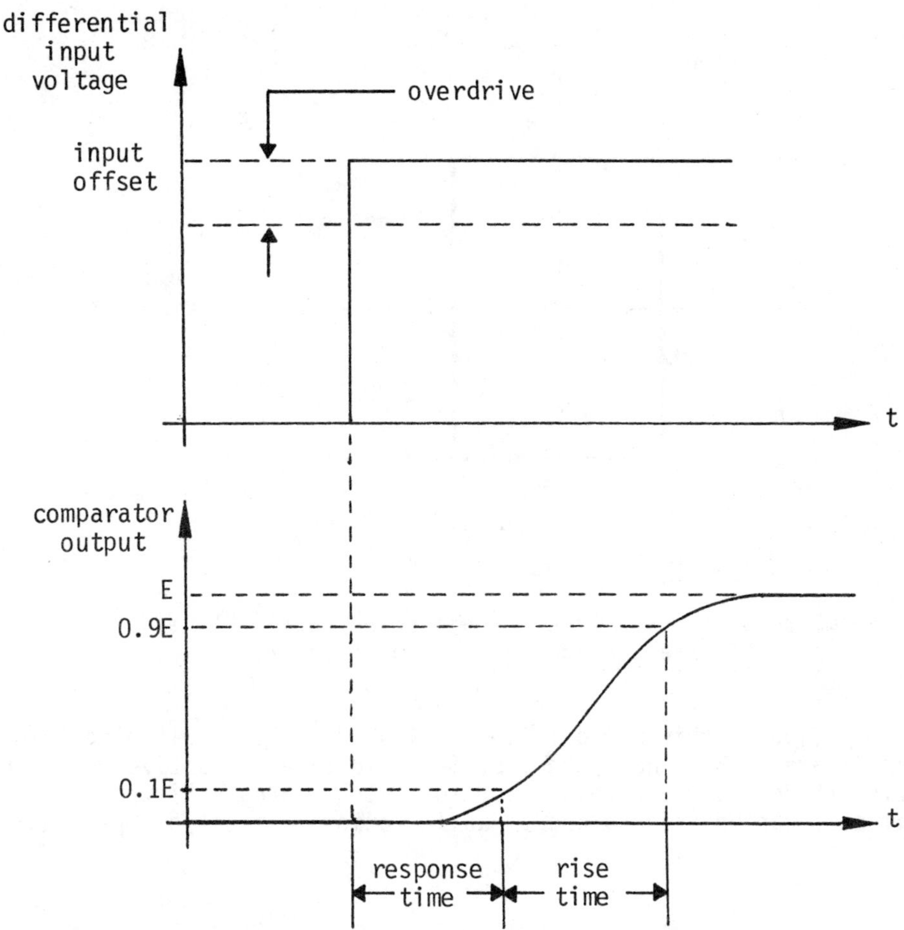

Figure 12.2 Response and rise times of the output of a voltage comparator. The response time depends on the overdrive present in the input.

bandwidth is necessarily accompanied by a decrease in gain, which means an increase in the input offset voltage.

Another specification associated with comparators is hysteresis. Hysteresis is a general term that refers to the fact that many devices respond differently when their input is increasing from the way they do when their input is decreasing. In terms of comparator operation, hysteresis means that the output switches state from one output level to the other when the differential input increases to a certain value, but switches from the second output level to the first for a different input value, when the input is decreasing. Figure 12.3 illustrates hysteresis. In Figure 12.3, e_{01} and e_{02} are the two different output voltage levels. We may assume that the non-inverting input is at some fixed reference voltage (such as ground) and that e_{in} is the voltage applied to the inverting input. When that voltage reaches the upper trigger level (UTL), the comparator output switches to e_{02} volts, as shown in the figure. However, e_{in} must then decrease to the lower trigger level (LTL) before the comparator output will switch to its other value, e_{01} volts. Quantitatively, the total hysteresis is (UTL-LTL) volts.

In some voltage comparator applications, hysteresis is a desirable characteristic. If the input signals have random noise "spikes" or "glitches," we can use hysteresis to make

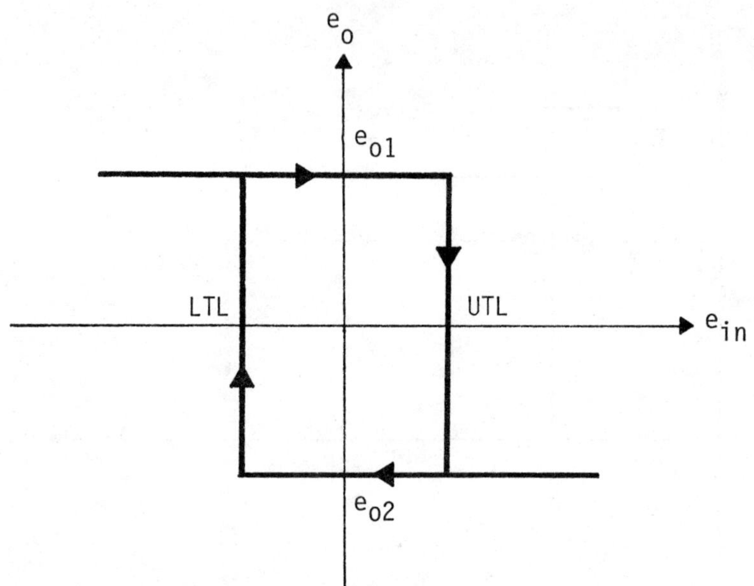

Figure 12.3 Hysteresis in a voltage comparator. LTL and UTL are the lower and upper trigger levels.

the comparator insensitive to this noise in the region around its switching point. For example, if e_{in} in Figure 12.3 is near the LTL, it would take a positive noise pulse equal in magnitude to UTL-LTL to cause the comparator to switch states. Figure 12.4 shows a comparator circuit that can be used to create an intentional amount of hysteresis.

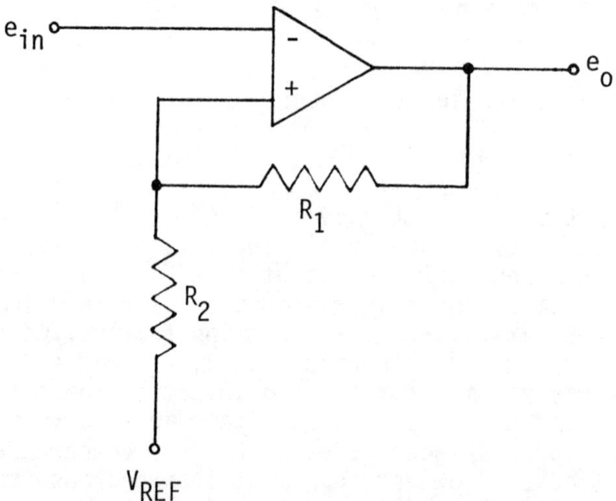

Figure 12.4 A voltage comparator circuit designed to create hysteresis.

In Figure 12.4, a portion of the output voltage (e_{01} or e_{02}, depending on the output state) is fed back to the non-inverting input along with a portion of a fixed reference voltage V_{REF} (which could be 0 V, i.e. ground). Using the superposition principle, we can show that the lower and upper trigger levels (assuming negligible input offset) are given by

$$\text{LTL} = \frac{R_1}{R_1 + R_2} V_{REF} + \frac{R_2}{R_1 + R_2} e_{02} \tag{2}$$

$$\text{UTL} = \frac{R_1}{R_1 + R_2} V_{REF} + \frac{R_2}{R_1 + R_2} e_{01} \tag{3}$$

where e_{01} and e_{02} are as defined in Figure 12.3. If V_{REF} is ground then of course each of the terms multiplying V_{REF} in (2) and (3) are zero, and if e_{01} or e_{02} is 0 V, then the corresponding term multiplying e_{01} or e_{02} is zero.

Two voltage comparators can be used to construct a circuit that switches levels whenever an input voltage falls within a prescribed range. When the input voltage is greater than some lower limit and less than some upper limit, the circuit will produce a high output. This circuit is called a window detector: the high output is produced whenever the input falls within the "window" represented by the two limits. Figure 12.5 shows the window detector circuit.

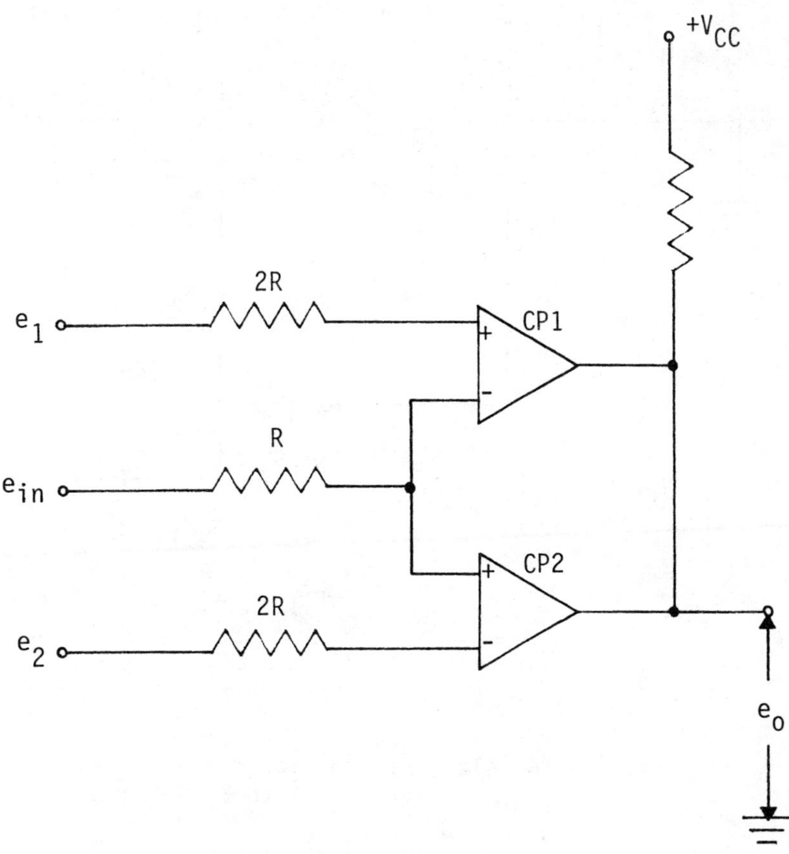

Figure 12.5 Two voltage comparators used to construct a window detector. e_0 is high if $e_2 < e_{in} < e_1$

In Figure 12.5, the voltage e_1 and e_2 are the limits of the window, where $e_1 > e_2$. The resistor values R and 2R have that ratio to minimize input offsets. Note that if $e_2 < e_{in} < e_1$, then the non-inverting input of each comparator is more positive than its inverting input, so the output of each is high. On the other hand, if $e_{in} > e_1$ or $e_{in} < e_2$, then one of the comparator outputs will go low, making the output low.

A final example of an application for voltage comparators is the square wave generator shown in Figure 12.6. Assume the output levels of the comparator are ± V volts. When the output is at + V volts, then + βV is applied to the non-inverting input, where $\beta = R_1/(R_1 + R_2)$. The capacitor charges through R towards + V volts, but when its voltage reaches + βV, the comparator switches its output to -V volts. The voltage applied to the non-inverting input then becomes $-\beta V$ and the capacitor discharges towards -V volts. However, when the capacitor voltage reaches $-\beta V$ the comparator switches again, back to + V volts, and the cycle repeats. Figure 12.7 illustrates the process.

We see that the total time required for the comparator to switch from one state to the other is the time required for the capacitor to charge from $-\beta V$ volts to $+\beta V$ volts (or to discharge from $+\beta V$ to $-\beta V$ volts). The period T of the square wave is thus equal to twice this length of time.

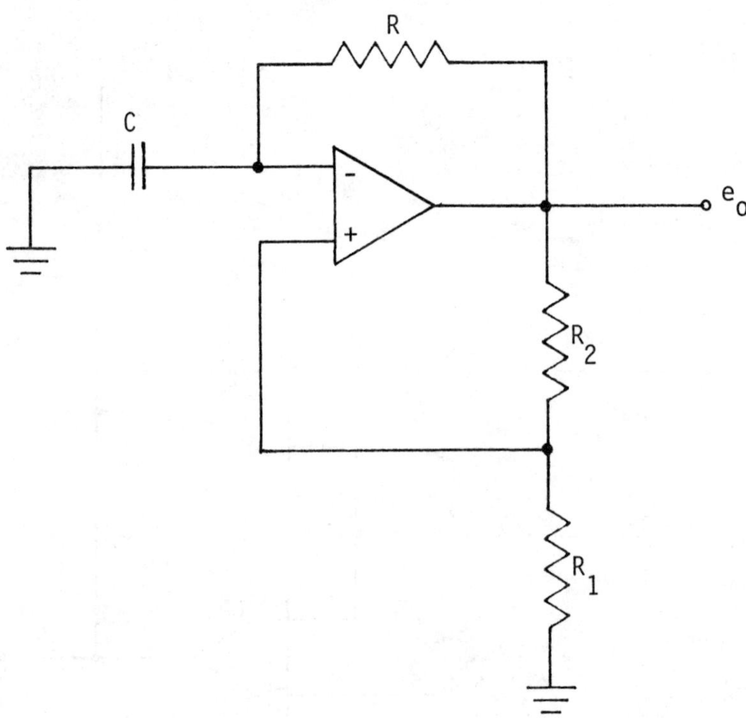

Figure 12.6 A voltage comparator connected as a square wave oscillator.

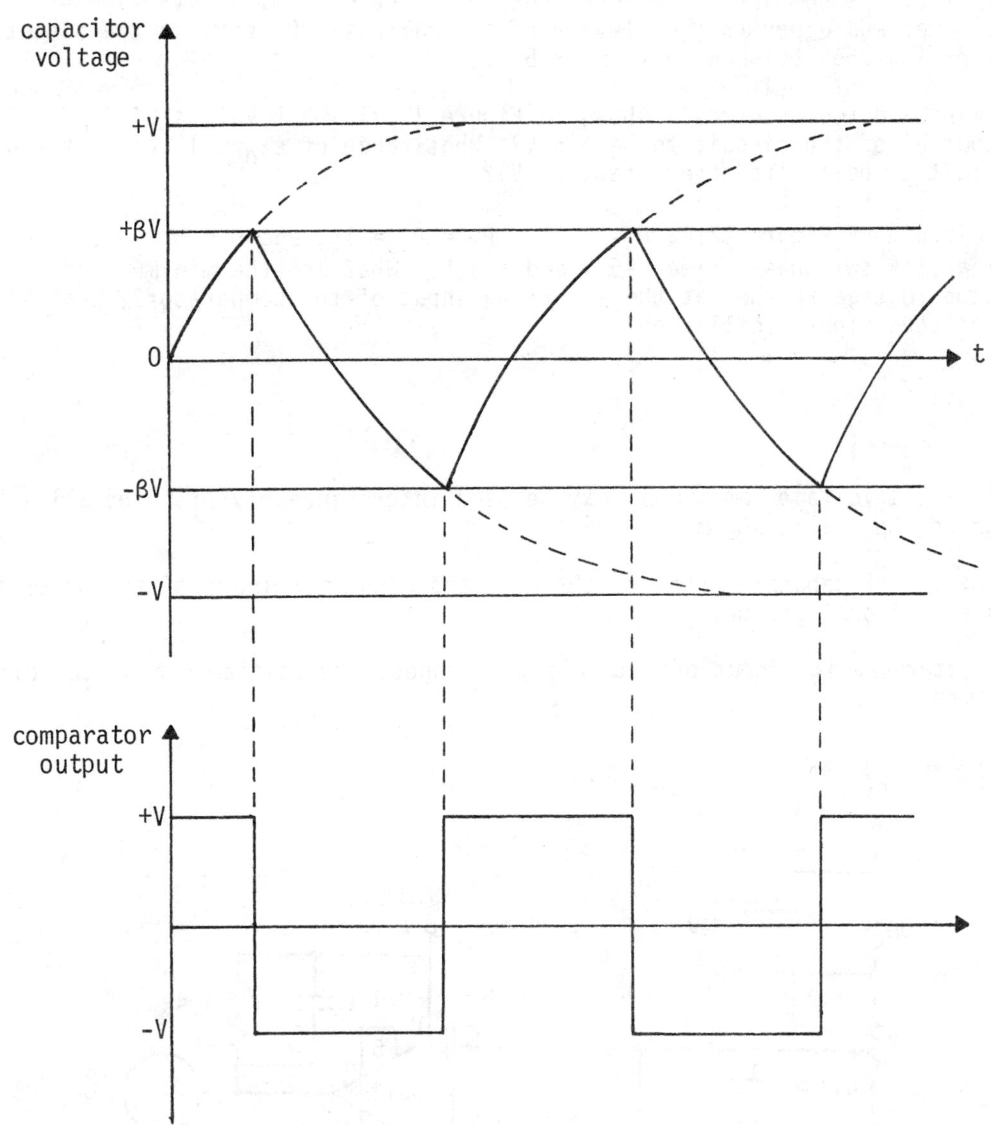

Figure 12.7 The waveforms on the capacitor and at the output of the square wave oscillator shown in Figure 12.6.

It can be shown that

$$T = 2RC\ln\left(\frac{1+\beta}{1-\beta}\right) \tag{4}$$

EXERCISES

1. The gain-bandwidth product of an operational amplifier is 50 x 10^6. When used as voltage comparator, as shown in Figure 12.1, its open-loop gain is 25 x 10^4. What is the rise-time of the comparator's output when it switches from one level to the other? What would be the rise-time if the open-loop gain were twice as large? What are the desirable and undesirable consequences of having the open-loop gain equal to this larger value, from the standpoint of its application as a voltage comparator?

2. A voltage comparator is connected as shown in Figure 12.4 with R_1 = 470K and R_2 = 100K. The output of the comparator switches between 0 and + 5 volts. If V_{REF} = 2 V DC, what are the lower and upper trigger levels of the circuit? Repeat, if the output of the comparator switches between -5 V and + 5 V.

3. In the window detector circuit shown in Figure 12.11, what values of e_{in} will cause the output e_o of the circuit to be + 5 V? What range of e_{in} will cause the output of the circuit to be in its other state (0 V)?

4. In the circuit of Figure 12.6, R = 4.7K $R_1 = R_2$ = 1K, and C = 0.1 μF. The output of the comparator switches between -5 V and + 5 V. What are the minimum and maximum values of the voltage reached at the inverting input of the comparator? What is the frequency of the output oscillation?

PROCEDURE

NOTES:

a. The 111 or 211 voltage comparator may be used interchangeably with the 311 in all circuits where the 311 is shown.

b. Connect a 0.1 μF capacitor between the plus and minus terminals of the power supplies used in all circuits below.

1. To determine the input offset voltage, connect the voltage comparator circuit in Figure 12.8.

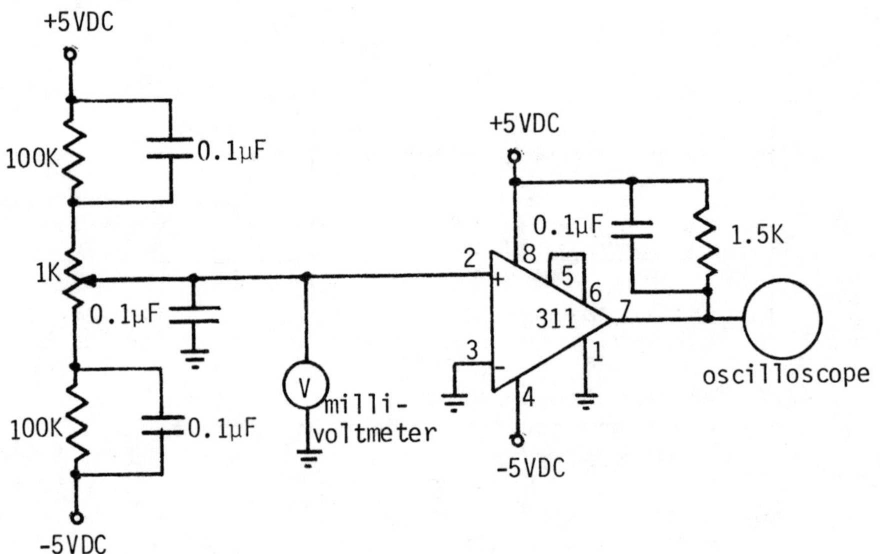

Figure 12.8

The voltage divider network containing the two 100K resistors and the 1K potentiometer is used to obtain a voltage near zero for the input to the comparator. The capacitors are used to suppress oscillations. Adjust the potentiometer as required to cause the output, as observed on the DC-connected oscilloscope, to switch between 5 and 0 volts. Record the millivoltmeter reading when the output switches to + 5 V and again when it switches to 0 V. The output may tend to oscillate when the input is near one of the switching levels; if this is the case, simply continue adjusting the input level until the output has completely switched to one of its two states. It may also be necessary to make small adjustments in the 5 V DC sources supplying the voltage divider in order to obtain an input voltage of the correct polarity.

2. Remove the voltage divider and all the 0.1 μF capacitors in Figure 12.8 (except the power supply bypass capacitor) and connect a 2 V peak triangular wave to the + input of the comparator. Adjust the level of the triangular wave before connecting it to the comparator. The wave should go negative for at least part of each cycle. Connect it through a 0.1 μF capacitor to the + input if necessary to obtain a negative-going portion. Observe the triangular wave and the output of the comparator simultaneously on a dual-trace oscilloscope and sketch the display.

3. To determine the response and rise times of the comparator, connect the circuit shown in Figure 12.9.

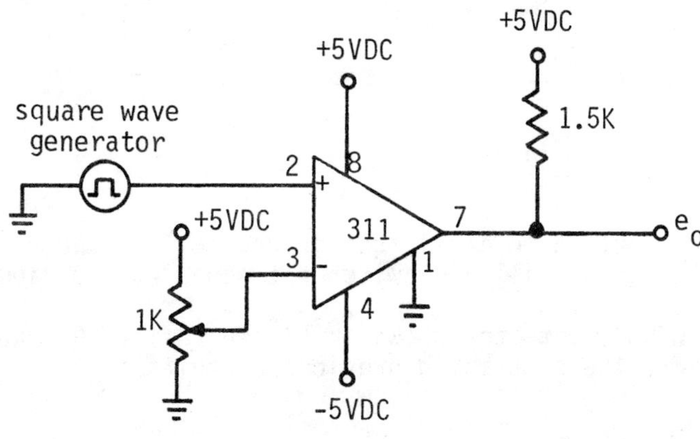

Figure 12.9

Adjust the 1K potentiometer until a 1 V DC level is applied to the minus input of the comparator. Set the square wave generator output to 1.5 V peak at 100 kHz. Observe the square wave and the output of the comparator simultaneously on a dual-trace oscilloscope. Set the horizontal sweep on the oscilloscope as necessary to view one positive-going edge of the output and expand the display so that the response and rise times can be measured, as described in the Discussion. (Sketch the display accurately. Then observe the effect of changes in the amplitude of the square wave on the display.)

4. Change the input to the plus terminal from a square wave to a 2 V peak, 1 kHz triangular wave. Observe one or two cycles of the triangular wave and the comparator output on a dual trace oscilloscope. Sketch the display, noting particularly the levels of the triangular wave at the points where the comparator output switches states.

5. Connect the comparator circuit designed to increase hysteresis shown in Figure 12.10. Pin numbers and power supply connections are the same as in previous circuits. Adjust the 1K potentiometer controlling e_s until e_s = 2.0 V. Monitor e_o on an oscilloscope to determine when the comparator output switches level. Adjust the 1K potentiometer controlling e_{in} until e_{in} is near zero volts. Then gradually increase the level of e_{in} until e_o switches to zero volts. Record the value of e_{in} when this occurs. Continue to increase e_{in} until it is near + 5 V. Then gradually decrease e_{in} until e_o switches to its positive state. Record the value of e_{in} where this occurs.

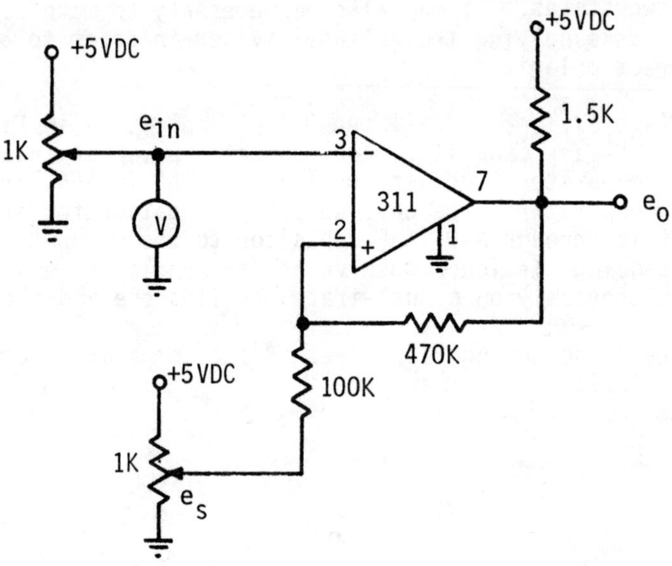

Figure 12.10

6. Change the connection to pin 1 from ground to -5 V and repeat step 6. (Except note that the output will now switch between plus and minus 5 V.)

7. Connect the window detector shown in Figure 12.11. Pin numbers and power supply connections are the same as in previous circuits.

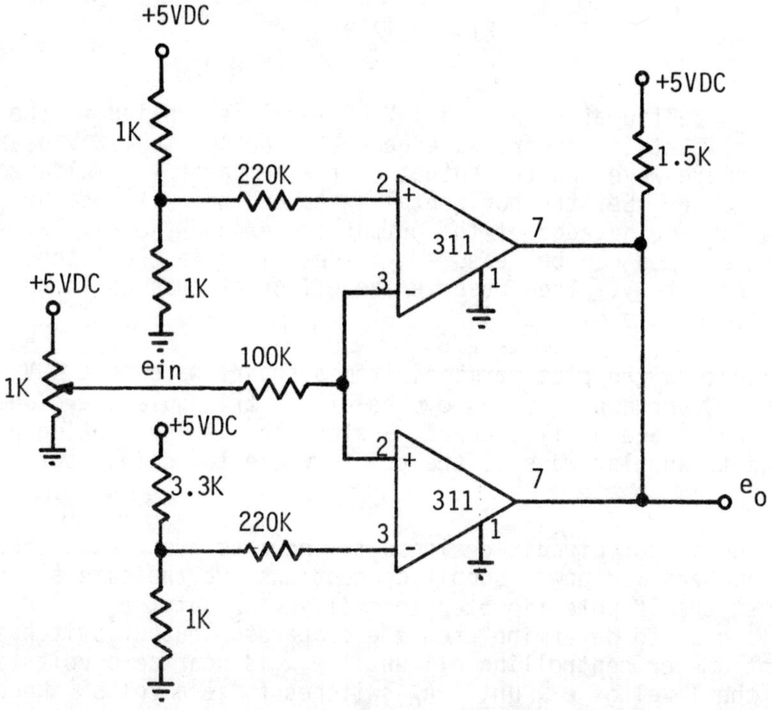

Figure 12.11

Monitor e_o on an oscilloscope. Then increase and decrease the level of e_{in} between 0 and 5 V and observe e_o. Measure and record the range of values of e_{in} that results in e_o being + 5 V and the range over which e_o is 0 V.

8. Change e_{in} to a 5 V peak, 1 kHz triangular wave and observe it simultaneously with e_o on a dual trace oscilloscope. Sketch the display, noting especially the levels of e_{in} where e_o changes state.

9. Connect the astable multivibrator circuit shown in Figure 12.12.

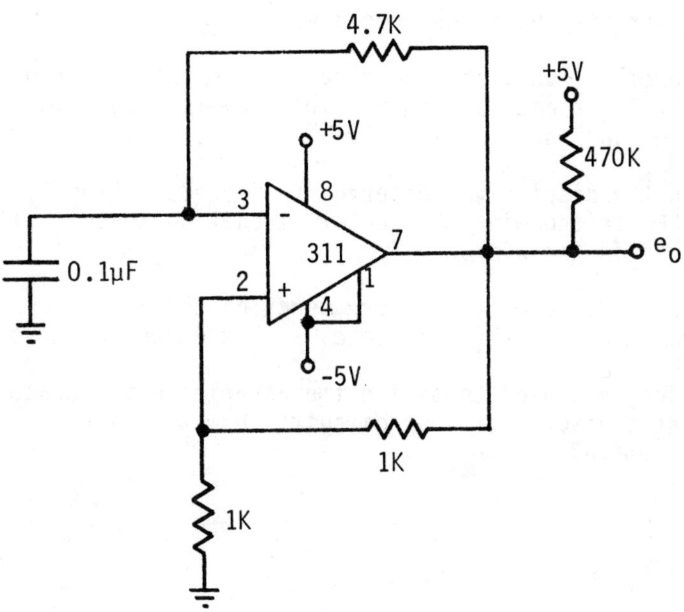

Figure 12.12

Measure and record the frequency of the output. Sketch the waveform generated at the output.

10. Design and construct a 10 kHz astable multivibrator circuit using the 311 voltage comparator. Use the same output resistor (470K) as in Figure 12.12. Measure the frequency of the output of your circuit.

QUESTIONS

1. Calculate the absolute value of the difference between the two input voltages that you found in step 1 were required to switch the output level of the comparator. Compare this calculated value with the manufacturer's specification for input offset voltage. What factors contribute to the input offset that were not compensated for in the experimental procedure?

2. Sketch and explain the display you observed in step 2 of the Procedure. How could the circuit have been altered so that the output switched between + 5 V and -5 V instead of between + 5 V and 0 V?

3. Compare the response time you measured in step 3 of the Procedure with the manufacturer's specification. Note that you should not necessarily expect your value to be the

same. Why? (How much overdrive was present in the input square wave you used to switch the comparator?) Describe the effect of changes in the amplitude of the square wave on the response time.

4. Calculate the bandwidth of the comparator based on the rise time you measured in step 3 of the Procedure. Show your calculations.

5. Sketch and explain the display you observed in step 4 of the Procedure.

6. Calculate the theoretical lower and upper threshold levels of the circuit you built in step 5 of the Procedure. Compare these with the levels you measured in step 5.

7. Repeat question 6 for step 6 of the Procedure.

8. Calculate the theoretical voltage ranges over which the output of the window detector in Procedure step 7 is at each of its possible levels. Compare these ranges with those you determined in Procedure step 7.

9. Sketch and explain the display you observed in Procedure step 8. Describe how you could connect an LED to the window detector circuit so that it illuminates whenever the input voltage is in the window.

10. Calculate the theoretical oscillation frequency of the multivibrator you built in step 9 of the Procedure. Compare this frequency with the frequency you measured in step 9.

11. Show the calculations you used to design the astable multivibrator in Procedure step 10. Draw a schematic diagram of your circuit. How well did your circuit conform to the design specification?

APPENDIX A
Writing Lab Reports

The ability to communicate technical information in clear, concise written English is an exceptionally important skill that should be cultivated by all students of technology. The time and effort spent gathering experimental data is of little value if the results of that effort cannot be interpreted in a way that is meaningful to others, or if conclusions cannot be stated in a way that is both convincing and unambiguous. The technical expertise of an experimenter is often judged solely by the quality of his or her written reports, so time spent learning how to produce a good report can reap significant practical dividends.

The best way to learn how to write good technical reports is to write as many as possible. Through repeated practice, a technical writer learns certain techniques and widely accepted standards for the structure and format of a good report. In the following paragraphs, we summarize these standards and provide guidelines on the correct way to present technical information. The structure we suggest here (i.e., the organization of the topics and the scope of each) is often called that of a "formal" report. You may not be required to submit a complete, formal report that conforms precisely to the suggested format for every experiment you perform. However, you will find useful guidelines for the content of each section in a technical report, no matter how it is organized.

THE TITLE PAGE

The title (cover) page of a report should contain the title of the experiment, the name of the person who wrote the report, the name(s) of any partners who participated in the experiment, the date the experiment was performed, and the date the report is submitted. Although there is no universal standard for the format in which this information is presented, the title of the experiment (and its number, if appropriate) should be more prominent than the other items. Some organizations may have their own standards for the format of a title page.

STATEMENT OF OBJECTIVES

The objectives of the experiment should be clearly stated at the outset. These may be presented in a list or in short narrative form.

Objectives are generally narrow and very limited in scope, so avoid stating them in overly broad terms. Do not generalize your objectives beyond your ability to achieve them in one experiment. For example, it would not be correct to state as an objective "to find the limitations of operational amplifiers." More appropriate would be a statement such as "to verify that the bandwidth of an operational amplifier is decreased when its closed-loop gain is increased."

LIST OF EQUIPMENT AND MATERIALS

The reliability of experimental data depends heavily on the type and quality of the instruments used to obtain it. Experienced investigators can gage the validity of an experiment from a knowledge of the equipment used to perform it. For this reason, it is important to list the make and model numbers of all instruments, power supplies, and other test equipment used. A reader who may wish to duplicate an experiment will require some knowledge of the instrument quality needed to obtain similar experimental results.

Depending on how critical the conclusions drawn from an experimental investigation are to the organization sponsoring it, additional information on the equipment may be required. For example, it may be necessary to report manufacturers' serial numbers on the test equipment used in order to permit the reader to verify calibration records or to perform calibration tests. Electronic components, such as integrated circuit chips, may have to be identified by manufacturer and by any suffixes in the device number that are used to designate a special version of the device. Different versions of a particular chip often have different performance specifications, some of which may be critical to the outcome of an experiment. For example, we find the 741 operational amplifier available from National Semiconductor as a 741, 741A, 741C, or 741E.

The prefix used with a linear integrated circuit number identifies the manufacturer. Some of those commonly in use today are listed below:

Am	Advanced Micro Device
BB	Burr-Brown
μA	Fairchild
HA	Harris
MIC	ITT
ICL, IH	Intersil
MC, MFC	Motorola
LH, LM	National
CA, CD	RCA
SE/NE, N/S	Signetics
SG	Silicon General
UC	Solitron Devices
ECG	Sylvania
SN	Texas Instruments
TOA	Transition

In some less critical experiments, you may not be required to furnish a separate list of equipment and materials. In such cases, the labeling of components on a schematic or block diagram of the experimental setup may be sufficient.

DESCRIPTION OF EXPERIMENTAL PROCEDURE

The Procedure section of a report should contain a brief description, in the writer's own words, of the experimental procedure that he or she followed when the experiment was conducted. This section should contain schematics and/or block diagrams of the test set-up(s), and the written description should refer to these diagrams. All such diagrams should contain labels that identify the components, and each figure should be given a descriptive title. Do not simply write: "Figure 1." Instead, write a title that reveals information about the figure, such as "Schematic diagram of the operational amplifier circuit used to measure gain."

As a guide to the detail that should be used in describing the experimental procedure, assume that the reader has the same knowledge and background as yourself. Then describe the procedure in sufficient depth to enable such a person to perform the experiment using your description. Thus, for example, it would not be necessary to write "The red lead of the voltmeter was connected to the positive terminal of the power supply and the black lead of the voltmeter was connected to the negative terminal." This is too much detail. It is sufficient to say "The supply voltage was measured."

Any particularly complex steps in the procedure should be described in greater detail to eliminate the possibility of misinterpretation. Also, include any special precautions that are necessary. For example, "It was observed that different operational amplifier chips required different values of capacitance to suppress oscillations." Finally, be sure to note any outright errors in experimental technique, including any failures to preadjust or calibrate instruments. For example: "The zero level of the oscilloscope trace was not set before taking the measurements, so it is not known if the results reflect the presence of a dc level."

DATA

The DATA section of a report contains the results of all measurements that were made during the conduct of an experiment. These are called "raw" data. (Incidentally, the word "data" is a plural noun, so "these data" is correct usage. In recent years, common usage has made "data" also acceptable as a singular noun. Therefore, "the data is ..." is also correct.)

Raw data should be presented in data tables. Each table should be given a descriptive title, such as "Gain versus feedback resistance in an operational amplifier circuit." Do not simply write "Table 1." The units of all entries in a data table should be made clearly apparent to the reader. A common method used to show units is to write the units as part of the heading for each column or row of data. For example, a data column might be headed: "Output voltage (mV)." Alternatively, each data entry can be accompanied by its units. This method is often used if the data values have such a large range that some are recorded in, say, microvolts, while others are recorded in volts. For example, the data column shown on the left side is equivalent to that shown on the right:

Output	Output (V)
25 μV	.000025
40 mV	.04
10.6 V	10.6

However, as a general rule, data should be recorded with the same units in which it is measured. Avoid the use of numerical multipliers or divisors in data headings. For example, "Amperes x 10^{-3}" is ambiguous and confusing.

Reduced data are the results of computations that are performed on the raw data. For example, if the voltage V applied to an amplifier and the current I into it are both measured (the raw data), then the input resistance of the amplifier, $R = V/I$, is the corresponding reduced data. Other examples of data reduction include the averaging of a set of raw data values and the computation of percent errors or percent differences in the data. For convenience and ease of interpretation, reduced data is often shown in the same data tables as the raw data from which it is derived. In these cases, it must be made clear in the heading of the column containing reduced data that it in fact contains calculated rather than measured values. For example, a data column containing calculated values of input resistance should be headed: "R_{in} = V/I ohms."

Tables of reduced data should not contain numerical values that have more significant digits than the measured data on which they are based. For example, if V = 2.3 volts and I = 1.7 mA, do not list R_{in}= V/I as 1.3529411 kΩ, as this degree of accuracy is not justified by the raw data.

<u>Graphs</u> can be a very effective way of presenting both raw and reduced data. They are considered part of the DATA section of a report but should not be used to replace tables containing actual measured values.

The horizontal and vertical axes of a graph should always be labelled with the quantities they represent and their units. Place numbers along each axis to make it clear what range of variation is represented by the intervals between major divisions. Provide a descriptive title for each graph, such as "Gain versus frequency for a low-pass filter." Data points should be plotted so they are clearly visible. Do not make tiny "dots" and then obscure or obliterate them with a line. Instead, use a symbol such as ⊙ or ⊗ to represent each experimental data value plotted.

A graph showing experimental data is often drawn incorrectly because the individual who constructed it did not really understand its purpose. The purpose of any graph is to demonstrate a <u>relationship</u> between variables. Often the graph is used to show how well experimental data conform to a theoretical relationship. In these cases, both the theoretical relation and the experimental data should be plotted on the same set of axes. For example, if the experimenter wishes to demonstrate that a set of experimental values of oscillation frequency f and capacitance C conform to the relationship $f = 1/(2\pi\sqrt{6C})$, then the equation $f = 1/2\pi\sqrt{6C})$ should be plotted on the same frequency versus capacitance axes that are used to plot the data. Do not use symbols such as ⊙ to plot theoretical relations.

Sometimes the graph is used simply to show that there is a certain <u>kind</u> of relationship between variables, for example, a linear relationship, rather than to show that it conforms to a specific equation. In these cases, the "best" theoretical relationship should be plotted between the data points. For example, suppose the data consists of voltage across a fixed resistor and current through it, and we want to show that current is linearly related to voltage, that is, that current is directly proportional to voltage. Then a straight line should be fitted between the data points plotted on the graph. The "best" straight line can be precisely defined using a statistical method called regression analysis, but this method is beyond the scope of our discussion. For our purposes, the line should be drawn in a way that best seems to fall between the data points.

One of the most common and most flagrant errors made by students plotting graphs is to draw straight line segments between data points. The graph that results consists of a jagged, zig-zag plot that completely obscures the true relationship between the variables. The relationships between physical quantities in nature are always <u>smooth</u>, that is, their graphs are smooth curves, not zig-zags. It is better to draw no line at all between data points than it is to force an artificial line to fit every data point and thereby portray some unrealistic relationship. Even if a line drawn between data points is smooth, it should not show ambiguous "humps" or "dips" caused by forcing the line to go through every data point.

Figure A.1 shows examples of correct and incorrect graphs drawn through a set of experimental points for which there is a theoretical linear relationship.

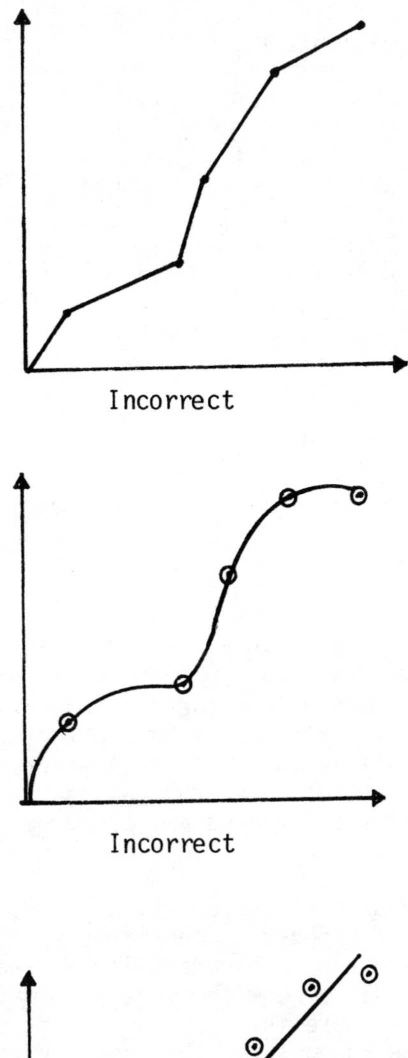

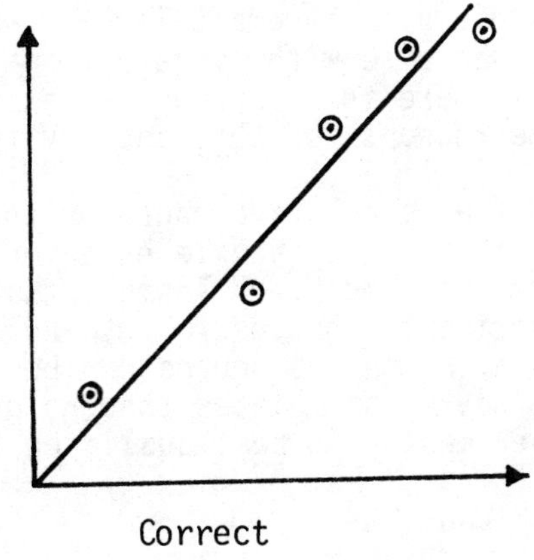

Figure A.1

SAMPLE CALCULATIONS

This section of the report should contain one example of each type of calculation used in data reduction or in theoretical computations. Present each sample calculation with a heading that makes the purpose of the calculation clear. Be sure to include units.

Following are examples:

Calculation of dc input resistance R_{in}

$$E = 12.5 \text{ V}$$

$$I = 0.25 \text{ mA}$$

$$R = E/I = \frac{12.5 \text{ V}}{0.25 \text{ mA}} = 50 \text{ k}\Omega$$

Calculation of theoretical voltage gain A

$$A = \frac{R_f}{R_1} = \frac{100K}{47K} = 2.13$$

Also provide sample calculations showing how you determined values of percent error or percent difference in your data.

ERROR ANALYSIS

The Error-analysis section of the report should list, describe, and justify any sources that the writer believes may have contributed to errors in the experiment. Among these sources are instrument accuracy, component tolerance, and human error. Be specific. State, for example, the manufacturer's specified accuracy of each instrument used, and attempt wherever possible to relate it to the percent errors or differences calculated from the data. Avoid broad, unsubstantiated generalizations such as "The error was caused by resistor tolerance." If resistor tolerances are a source of error, explain, how, why, and how much.

Summarize the results of your error calculations, and, if possible, distinguish between the more important and less important contributors to the error. Look for any trends in the error. For example, is every measurement larger than it should be? Does the percent error tend to increase or decrease with larger or smaller measured values? Attempt to explain any such trends. If there is a serious deviation between theoretical and experimental results, explore the reasons for this and try to present plausible explanations.

Error analysis is admittedly a speculative endeavor that often requires a considerable amount of guesswork. However, it is a worthwhile endeavor since it gives the reader some insight into the conduct of the experiment, an insight that is best supplied by the person who actually performed the experimental procedure. In addition, the very process of contemplating the error and speculating on its source can be very rewarding. Looking for error trends and asking oneself "why?" stimulates thought processes and produces shrewd guesses-- guesses that the experimenter is best qualified to make.

DATA INTERPRETATION AND CONCLUSIONS

This section of the report contains the investigator's written interpretation of the experimental results and the conclusions that he or she has formed based on the results. This is one of the most important parts of a report, but student writers often neglect it. Unless the writer shows that there is concrete evidence to support the relevant theory, or demonstrates that certain important insights have been gained as a result of performing the work, then the reader can only suppose that nothing has been learned.

Avoid the tendency to make sweeping generalizations for which there is no real basis in the data. Do not use the word "prove," as in "These results prove Ohm's Law."

No experimental investigation can absolutely prove any theoretical relation. Say instead "the results tend to confirm the validity of Ohm's Law."

Conclusions should refer to the specific objectives that are given at the beginning of the report. Do not form conclusions that are not relevant to these objectives or for which you have no supporting evidence. For example, it would not be appropriate to conclude that integrated-circuit amplifiers are better than vacuum tube amplifiers based on the results of an experiment designed to measure the gain of an integrated circuit op-amp. Unless there is experimental data in your report that you can use to back up a conclusion, don't state that conclusion.

Sometimes the experimental data reveal unexpected results or suggest that certain new facts may have been discovered. Experimental confirmation of such facts may not have been a specific objective of the experiment, but conclusions involving these results should be stated, provided again that there is experimental evidence to support them. For example: "By comparing the output voltages of the regulator in the last six lines of the data table to those of the first six lines, it can be seen that, contrary to expectation, better regulation occurred in the range of higher load currents." Be careful not to generalize any such discoveries beyond what is justified by the experiment. For example, one could not conclude that "... voltage regulators perform better at high load currents than they do at low load currents."

As in the preceding example, always cite data as evidence for conclusions. Be specific and be quantitative; that is, refer to numerical results that support your conclusions. For example: "The frequency response data show that the gain is down 3 dB at 1.57 kHz, so it may be concluded that the lower cutoff frequency is 1.57 kHz." Another example: "The data table showing voltage versus current in a 4.7 K resistor demonstrates that there is good experimental agreement with Ohm's Law, since no calculation of $R = E/I$ differs from 4.7 K by more than 2.3%"

Finally, this section of the report should contain any recommendations that the writer feels are relevant to the conduct of the experiment. For example, a suggestion on how to change the experimental procedure to reduce measurement error would be an appropriate recommendation. Also appropriate would be a recommendation to investigate further any unexpected result brought to light by the data.

Following is a sample report that illustrates most of the points made in the previous discussion. Study this report and note in particular these points:

1. The report is complete; it contains all of the required sections.

2. The statement of objectives is clear, concise, and appropriately limited in scope. The writer did not, for example, state that the objective was to "investigate operational amplifiers."

3. The Procedure section includes a neatly drawn schematic diagram and does not contain unnecessary detail.

4. The Data section has a data table and graph that are correctly labeled and titled. All units are shown.

5. The Error Analysis refers to specific results in the Data section. It contains a quantitative analysis of the effect of resistor variations and explores other possible error sources in a quantitative way.

6. The Data Interpretation and Conclusions show that the writer examined the experimental results very carefully and arrived at intelligent conclusions based on a quantitative analysis of the data. Specific data is cited to support the conclusions, and reference is made to the original objectives. An appropriate recommendation is made for a way to improve the experimental procedure.

CLOSED LOOP GAIN OF AN
OPERATIONAL AMPLIFIER

By: Manifred Williams

Lab partners:

Alan Dorchester

J. Streich

Date performed: June 4, 1983

Date submitted: June 11, 1983

OBJECTIVES

The objective of this experiment is to confirm experimentally the validity of using the ratio R_f/R_1 to determine the closed-loop voltage gain of an operational amplifier. R_f is the feedback resistance and R_1 is the (input) resistance connected to the inverting input of the amplifier.

EQUIPMENT AND MATERIALS USED

1. Regulated power supplies (2), Sorensen model QRD 40-2, serial numbers 2755, 2761.

2. Dual trace oscilloscope, Tektronix model T935A, serial number B023334.

3. Function generator, Wavetek model 142, serial number 306049.

4. Resistors: 2.2K, 4.7K, 10K (2), 15K, 22K, 33K, 47K. All resistors ½W, 5%.

PROCEDURE

The circuit shown below was connected:

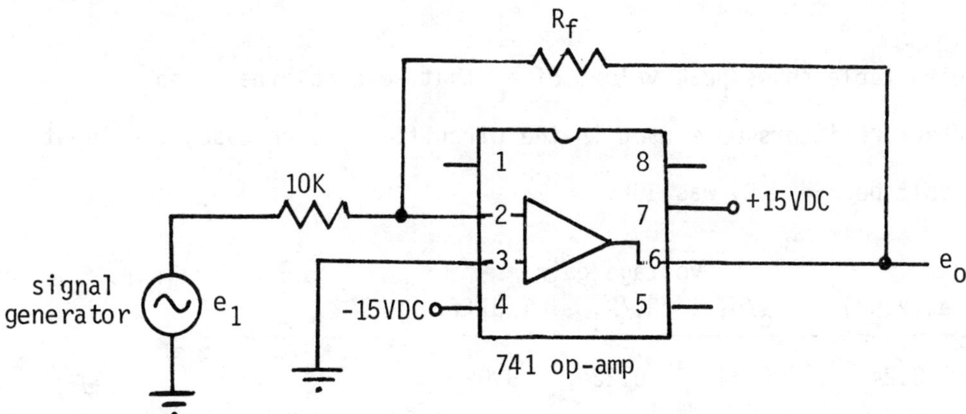

Figure 1 Schematic diagram of the experimental circuit used to determine voltage gain.

The signal generator was adjusted to produce a 1 kHz sine wave with a peak value of 1 volt, as measured on an oscilloscope. The output e_o was observed simultaneously with e_1 on the dual trace oscilloscope.

The peak value of e_o was measured for feedback resistors R_f having nominal (color-coded) values of 2.2K, 4.7K, 10K, 15K, 22K, 33K, and 100K. The peak value of e_1 was maintained at 1V in each case.

DATA

The following table shows peak values of e_o that were obtained when different feedback resistors were used in the circuit. In each case, the input voltage was 1 volt peak and R_1 was 10K.

		voltage gain		
$R_f(\Omega)$	e_o(V,pk)	e_o/e_1	R_f/R_1	% diff.
2.2K	0.24	0.24	0.22	9.09
4.7K	0.50	0.50	0.47	6.38
10K	0.98	0.98	1.00	-2.00
15K	1.65	1.65	1.50	10.00
22K	2.30	2.30	2.20	4.55
33K	3.52	3.52	3.30	6.67
47K	4.81	4.81	4.70	2.34

Table 1. Output voltage and gain versus R_f in the circuit of Figure 1.

It was noted that the output voltage was inverted (180 degrees out of phase) with respect to the input voltage in each case.

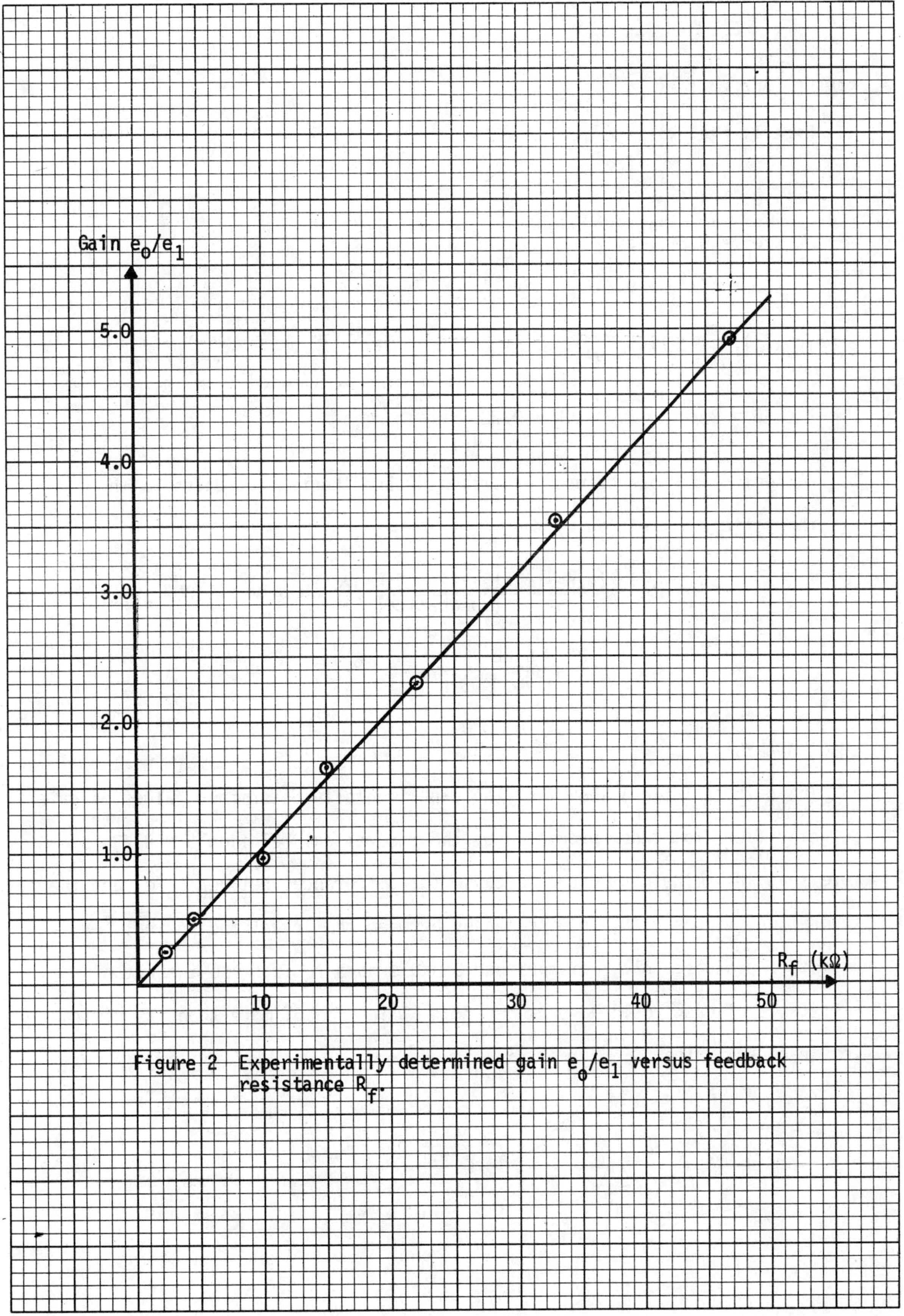

Figure 2 Experimentally determined gain e_o/e_1 versus feedback resistance R_f.

SAMPLE CALCULATIONS

1. Experimental voltage gain G:

e_1 = 1V, pk

e_o = 0.24V, pk

$G = e_o/e_1 = 0.24/1 = .24$

2. Theoretical voltage gain G:

$G = R_f/R_1 = 2.2K/10K = 0.22$

3. Percent difference between experimental and theoretical gains (with theoretical as reference):

$$\% \text{ difference} = \frac{(\text{experimental G}) - (\text{theoretical G})}{(\text{theoretical G})} \times 100\%$$

$$= \frac{0.24 - 0.22}{0.22} \times 100\% = 9.09\%$$

ERROR ANALYSIS

The data in Table 1 show that the experimentally determined values of gain do not differ from those predicted by the theoretical relation $G = R_f/R_1$ by more than 10%. The most probable cause for differences between e_o/e_1 and the theoretical values R_f/R_1 is differences between the actual and nominal resistance values. All resistors had 5% tolerance ratings, so the actual resistance of each could have ranged from 5% less to 5% more than its nominal (color-coded) value.

The following analysis determines the minimum and maximum possible values of R_f/R_1, based on the use of 5% resistors:

Since R_1=10K, its minimum value is R_{1min}=9.5K and its maximum value is R_{1max}=10.5K. Therefore,

$$\frac{R_f}{R_1}_{max} = \frac{R_{fmax}}{R_{1min}} = \frac{1.05R_f}{9.5K} = .1105R_f \times 10^{-3}$$

$$\frac{R_f}{R_1}_{min} = \frac{R_{fmin}}{R_{1max}} = \frac{.95R_f}{10.5K} = .0905R_f \times 10^{-3}$$

For each value of R_f in the data table, e_o/e_1 is less than $(R_f/R_1)_{max}$ and greater than $(R_f/R_1)_{min}$. Therefore, the percent differences shown in the table are not greater than would be expected due to resistance variations.

Another possible cause of differences between e_o/e_1 and R_f/R_1 is the fact that R_f/R_1 is only an approximation of e_o/e_1. The actual relation is

$$\frac{e_o}{e_1} = \frac{-R_f}{R_1} \frac{A}{A+1/\beta)}$$

where $$\beta = \frac{R_1}{R_1+R_f}$$

and A is the open-loop gain of the amplifier. However, the minimum gain A of the 741 op-amp is specified to be 5×10^4, and since the maximum value of $1/\beta$ is (10K+47K)/10K = 5.7, the minimum value of $A/(A+1/\beta)$ is

$$\frac{A}{A+1/\beta} = \frac{5 \times 10^4}{5 \times 10^4+5.7} = .9988$$

This value is so close to 1 that use of the approximation R_f/R_1 can be assumed to have contributed negligible error.

The method used to measure e_1 and e_o may have contributed some error. These values were obtained by measuring peak values on an oscilloscope. The limited resolution of this method and the fact that it is difficult to zero the trace accurately may have influenced the results.

DATA INTERPRETATION AND CONCLUSIONS

As noted in the "Error Analysis" section of this report, the difference between e_o/e_1 and R_f/R_1 did not exceed 10% in any of the recorded data. The error analysis showed that these differences could be attributed to differences between the actual and nominal values of the 5% resistors used, so it is concluded that there is good experimental evidence for using R_f/R_1 to determine the closed-loop gain.

The data table shows that all of the experimentally determined values of e_o/e_1 were higher than the ratio R_f/R_1 would predict, with the exception of the case where R_f=10K. In the latter case, the difference was -2.00%, while all other cases show positive percent differences. These results suggest that the 10K resistor used for R_1 was less than its nominal value. Since the same resistor R_1 was used in every case, this fact would account for the ratio R_f/R_1 being larger than otherwise expected. The one case where e_o/e_1 was less than R_f/R_1 may have been due to R_f being abnormally small, or due to the other error sources described earlier.

The graph of e_o/e_1 versus R_f (Figure 2 of the "Data" section) shows conclusively that the closed-loop gain is linearly related to the value of feedback resistance R_f, as would be expected from the equation

$$\frac{e_o}{e_1} = G = \frac{R_f}{R_1}$$

With R_1 fixed in value, G is directly proportional to R_f. By comparison with the general equation for a straight line, y=mx+b, it can be seen that the slope m of $G=R_f/R_1$ is $1/R_1$. The slope of the line in Figure 1 was measured to be $1.04x10^{-3}$. Therefore, the value of R_1 predicted by this value of slope is

$$R_1 = \frac{1}{m} = \frac{1}{1.04x10^{-3}} = 9.62K$$

This value of R_1 is less than the nominal value of 10K that was used in the data reduction and is further evidence that R_1 was abnormally small.

None of the data reduction or computations in this report reflect the fact that the output was 180° out of phase with the input, as noted in the "Data" section. Since this was the case in every measurement, it can be concluded that supplying input to the inverting terminal and grounding the non-inverting terminal of the amplifier causes a 180° phase inversion. Therefore, all gain computations, e_o/e_1 and R_f/R_1, should have a minus sign in front of them to denote phase inversion.

It is recommended that the experimental procedure include provisions to measure the actual values of the input and feedback resistors R_1 and R_f. Using measured values rather than nominal (color-coded) values in the computations should reduce differences between the theoretical and experimental results. Alternatively, precision resistors could be used instead of 5% resistors.

APPENDIX B
Specifications and Data Sheets

The data sheets on the following pages
are reprinted through the courtesy of:

NATIONAL SEMICONDUCTOR-*pages 172-233, 243-245*
RCA SOLID STATE-*pages 234-238*
SIGNETICS CORPORATION-*pages 239-242*

National Semiconductor

LM311 Voltage Comparator

General Description

The LM311 is a voltage comparator that has input currents more than a hundred times lower than devices like the LM306 or LM710C. It is also designed to operate over a wider range of supply voltages: from standard ±15V op. amp supplies down to the single 5V supply used for IC logic. Its output is compatible with RTL, DTL and TTL as well as MOS circuits. Further, it can drive lamps or relays, switching voltages up to 40V at currents as high as 50 mA.

Features

- Operates from single 5V supply
- Maximum input current: 250 nA
- Maximum offset current: 50 nA
- Differential input voltage range: ±30V
- Power consumption: 135 mW at ±15V

Both the input and the output of the LM311 can be isolated from system ground, and the output can drive loads referred to ground, the positive supply or the negative supply. Offset balancing and strobe capability are provided and outputs can be wire OR'ed. Although slower than the LM306 and LM710C (200 ns response time vs 40 ns) the device is also much less prone to spurious oscillations. The LM311 has the same pin configuration as the LM306 and LM710C. See the "application hints" of the LM311 for application help.

Auxiliary Circuits**

**Note: Pin connections shown on schematic diagram and typical applications are for TO-5 package.

Typical Applications**

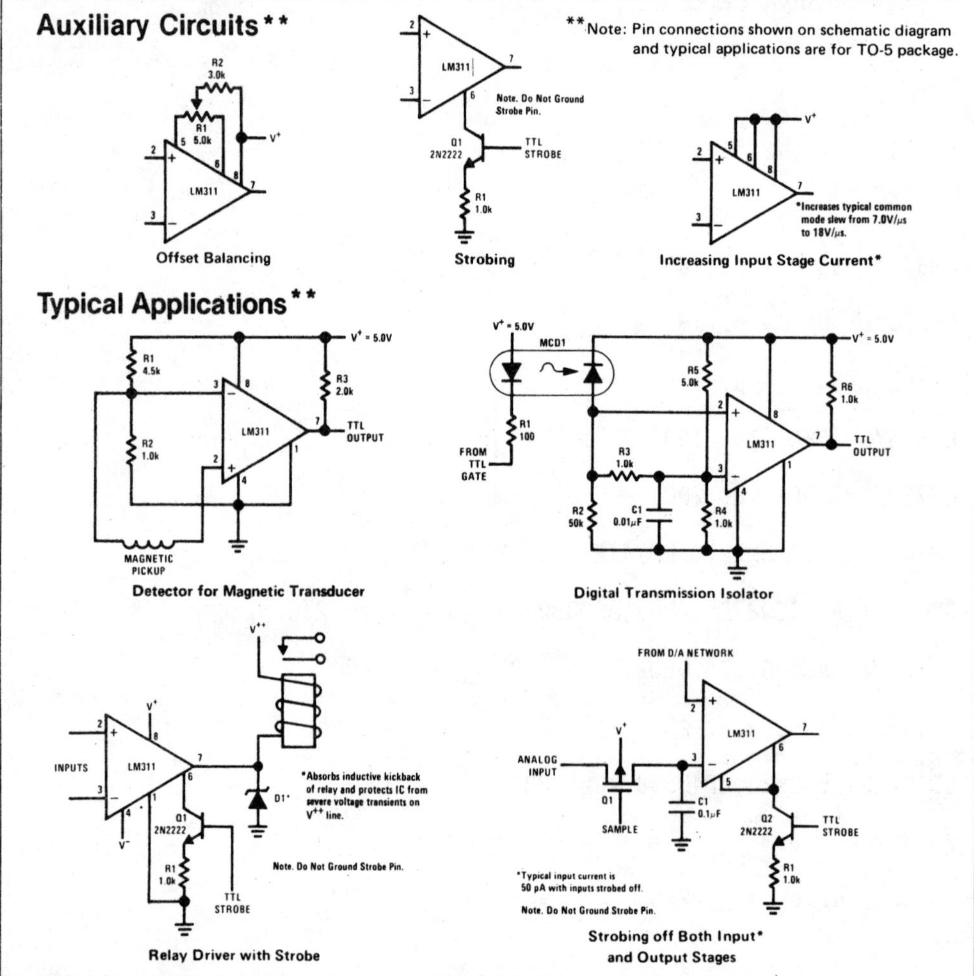

Offset Balancing

Strobing

Increasing Input Stage Current*

Detector for Magnetic Transducer

Digital Transmission Isolator

Relay Driver with Strobe

Strobing off Both Input* and Output Stages

Absolute Maximum Ratings

Total Supply Voltage (V_{84})	36V
Output to Negative Supply Voltage (V_{74})	40V
Ground to Negative Supply Voltage (V_{14})	30V
Differential Input Voltage	±30V
Input Voltage (Note 1)	±15V
Power Dissipation (Note 2)	500 mW
Output Short Circuit Duration	10 sec
Operating Temperature Range	0°C to 70°C
Storage Temperature Range	−65°C to 150°C
Lead Temperature (soldering, 10 sec)	300°C
Voltage at Strobe Pin	$V^+ - 5V$

Electrical Characteristics (Note 3)

PARAMETER	CONDITIONS	MIN	TYP	MAX	UNITS
Input Offset Voltage (Note 4)	$T_A = 25°C$, $R_S \leq 50k$		2.0	7.5	mV
Input Offset Current (Note 4)	$T_A = 25°C$		6.0	50	nA
Input Bias Current	$T_A = 25°C$		100	250	nA
Voltage Gain	$T_A = 25°C$	40	200		V/mV
Response Time (Note 5)	$T_A = 25°C$		200		ns
Saturation Voltage	$V_{IN} \leq -10$ mV, $I_{OUT} = 50$ mA $T_A = 25°C$		0.75	1.5	V
Strobe ON Current	$T_A = 25°C$		3.0		mA
Output Leakage Current	$V_{IN} \geq 10$ mV, $V_{OUT} = 35V$ $T_A = 25°C$, $I_{STROBE} = 3$ mA		0.2	50	nA
Input Offset Voltage (Note 4)	$R_S \leq 50k$			10	mV
Input Offset Current (Note 4)				70	nA
Input Bias Current				300	nA
Input Voltage Range		−14.5	13.8,−14.7	13.0	V
Saturation Voltage	$V^+ \geq 4.5V$, $V^- = 0$ $V_{IN} \leq -10$ mV, $I_{SINK} \leq 8$ mA		0.23	0.4	V
Positive Supply Current	$T_A = 25°C$		5.1	7.5	mA
Negative Supply Current	$T_A = 25°C$		4.1	5.0	mA

Note 1: This rating applies for ±15V supplies. The positive input voltage limit is 30V above the negative supply. The negative input voltage limit is equal to the negative supply voltage or 30V below the positive supply, whichever is less.

Note 2: The maximum junction temperature of the LM311 is 110°C. For operating at elevated temperatures, devices in the TO-5 package must be derated based on a thermal resistance of 150°C/W, junction to ambient, or 45°C/W, junction to case. The thermal resistance of the dual-in-line package is 100°C/W, junction to ambient.

Note 3: These specifications apply for $V_S = \pm 15V$ and the Ground pin at ground, and $0°C < T_A < +70°C$, unless otherwise specified. The offset voltage, offset current and bias current specifications apply for any supply voltage from a single 5V supply up to ±15V supplies.

Note 4: The offset voltages and offset currents given are the maximum values required to drive the output within a volt of either supply with 1 mA load. Thus, these parameters define an error band and take into account the worst-case effects of voltage gain and input impedance.

Note 5: The response time specified (see definitions) is for a 100 mV input step with 5 mV overdrive.

Note 6: Do not short the strobe pin to ground; it should be current driven at 3 to 5 mA.

Typical Performance Characteristics

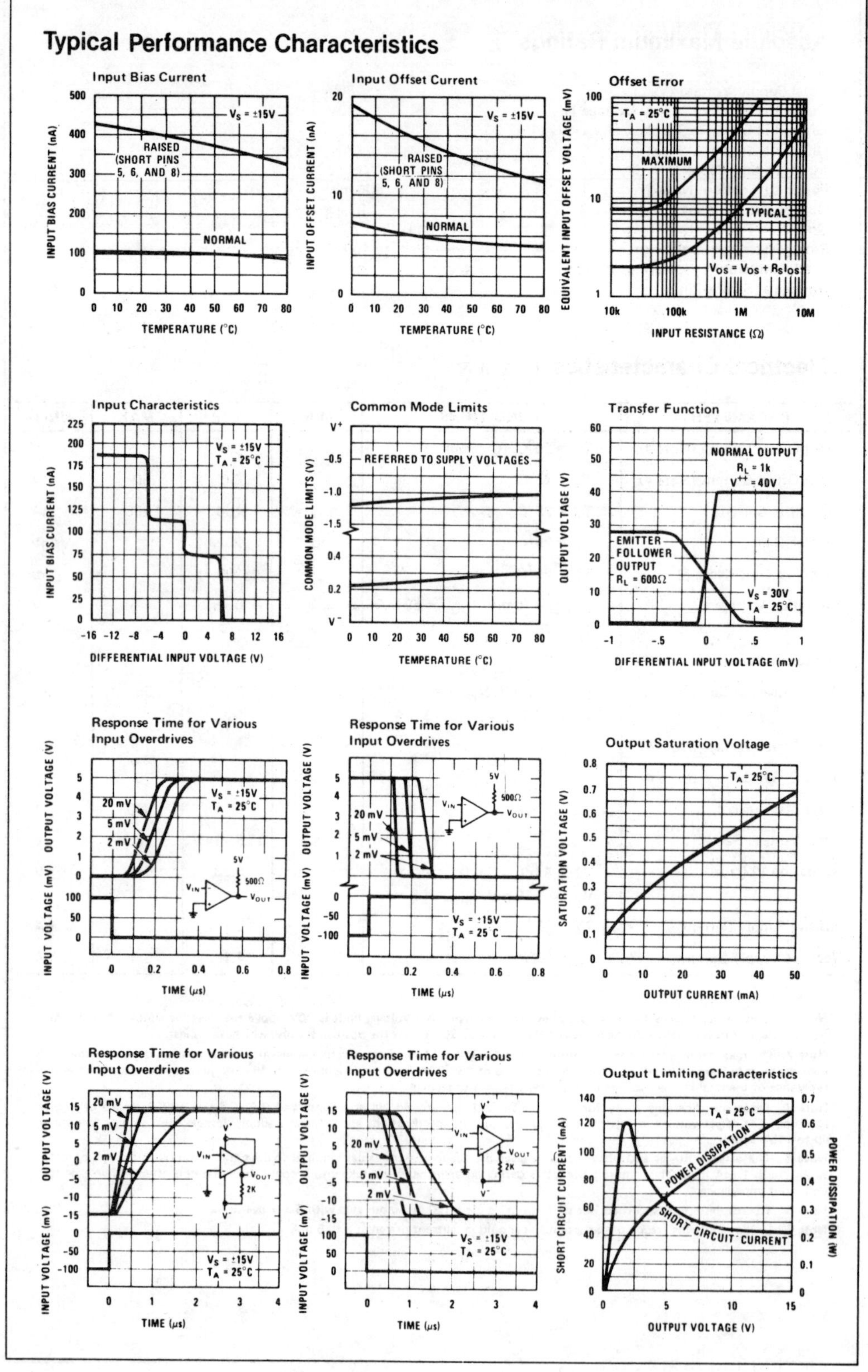

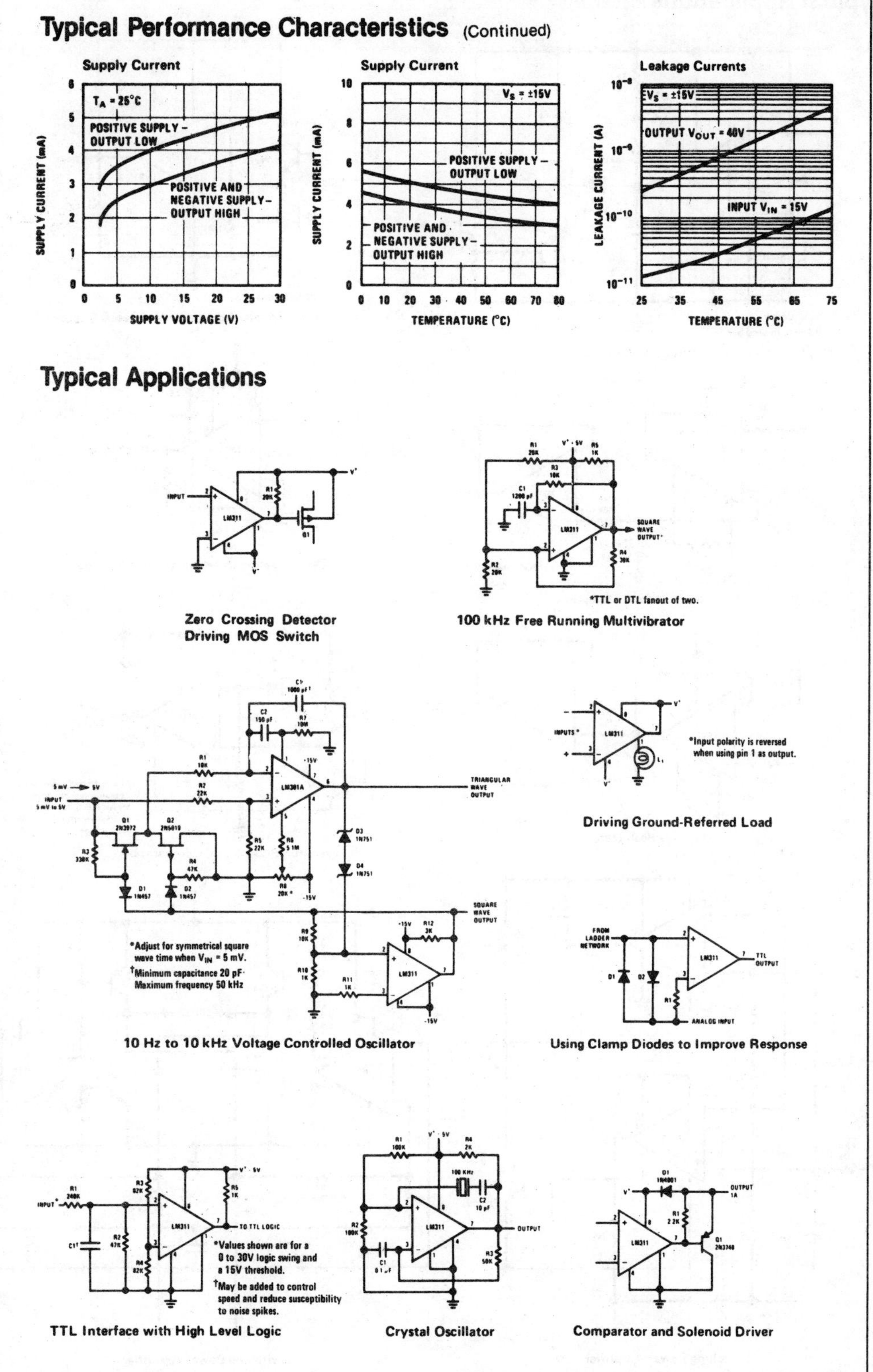

Zero Crossing Detector Driving MOS Switch

100 kHz Free Running Multivibrator

10 Hz to 10 kHz Voltage Controlled Oscillator

Driving Ground-Referred Load

Using Clamp Diodes to Improve Response

TTL Interface with High Level Logic

Crystal Oscillator

Comparator and Solenoid Driver

Typical Applications (Continued)

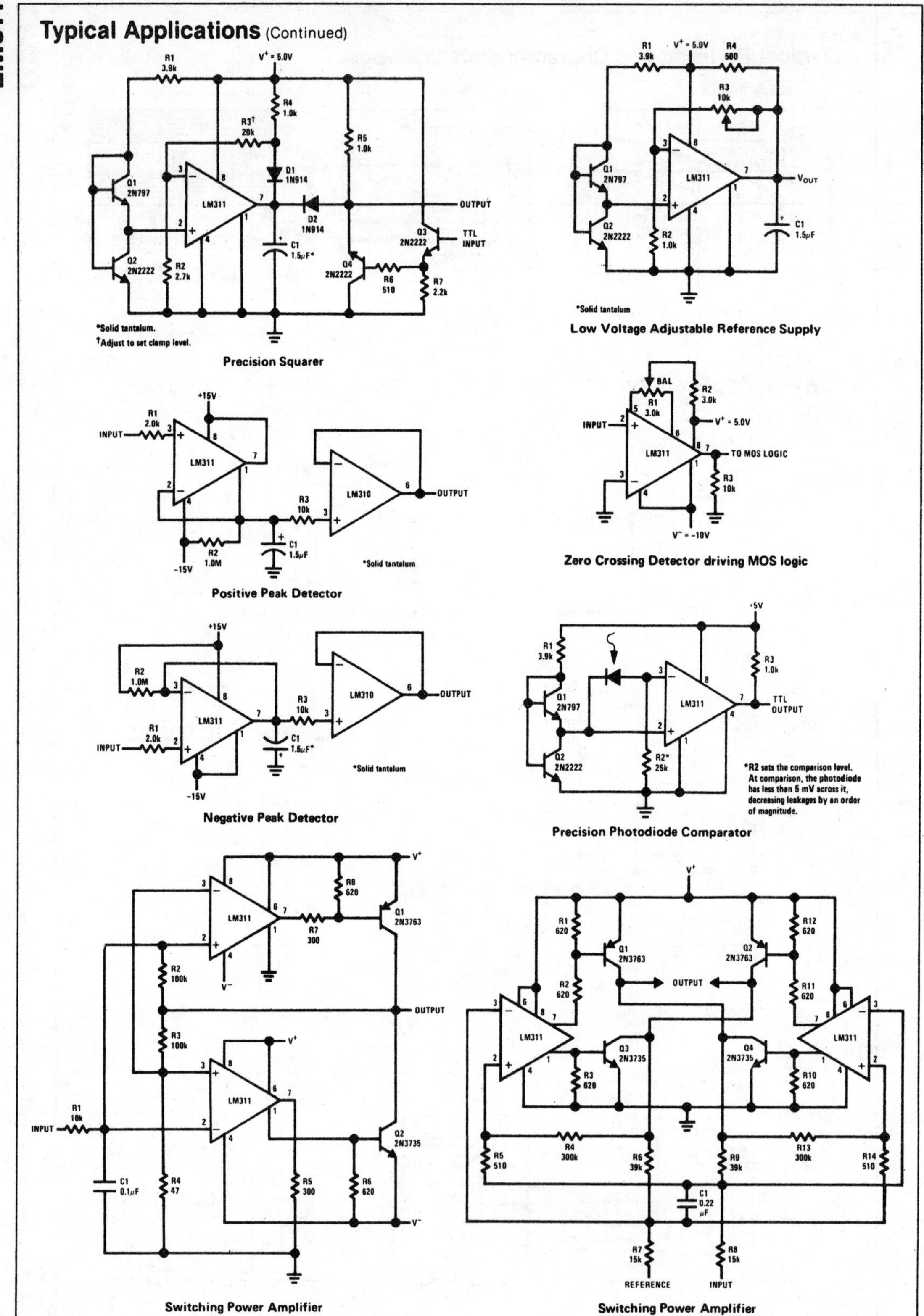

Precision Squarer

Low Voltage Adjustable Reference Supply

Positive Peak Detector

Zero Crossing Detector driving MOS logic

Negative Peak Detector

Precision Photodiode Comparator

Switching Power Amplifier

Switching Power Amplifier

Schematic Diagram

Connection Diagrams *

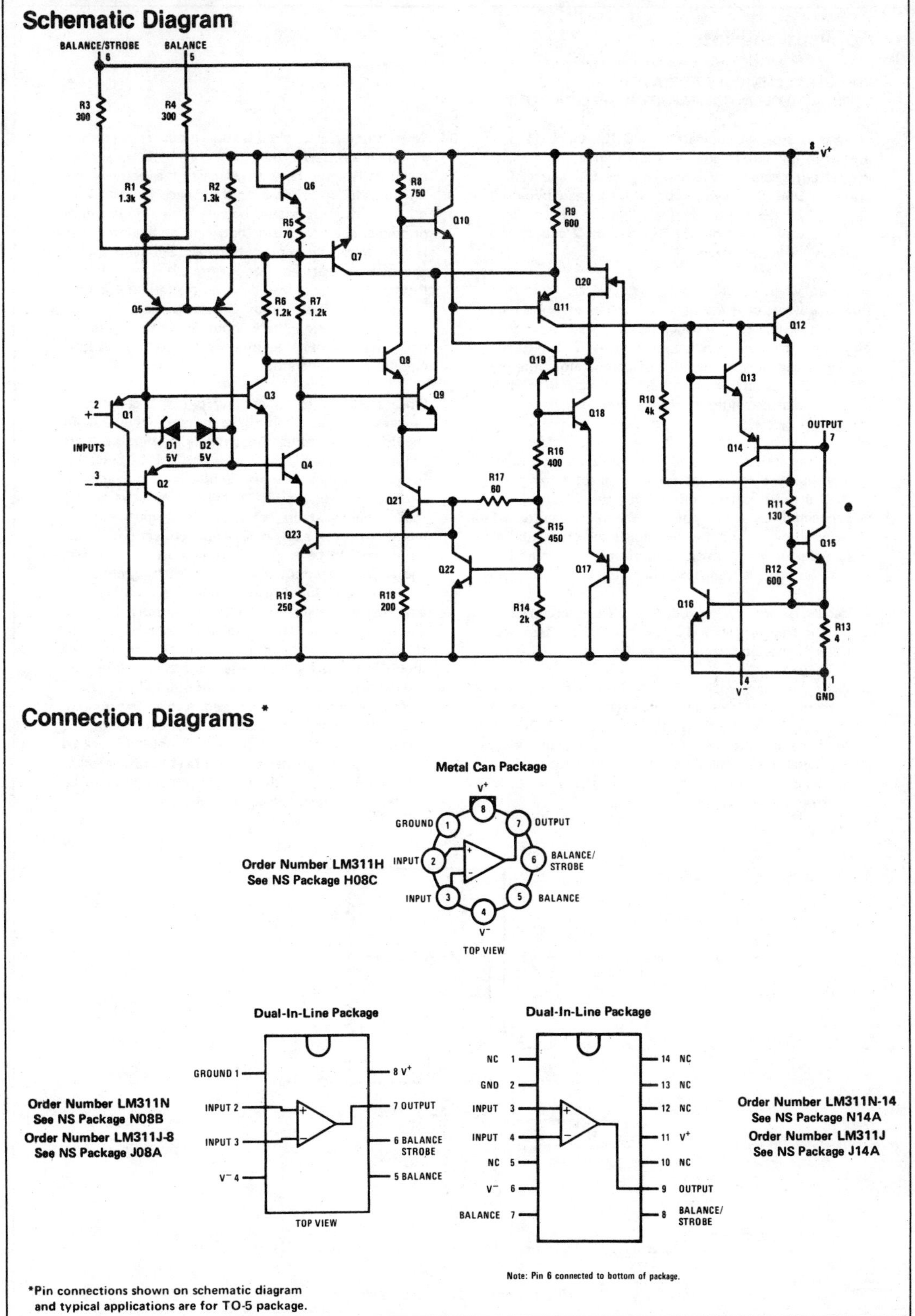

*Pin connections shown on schematic diagram and typical applications are for TO-5 package.

Application Hints

CIRCUIT TECHNIQUES FOR AVOIDING OSCILLATIONS IN COMPARATOR APPLICATIONS

When a high-speed comparator such as the LM111 is used with fast input signals and low source impedances, the output response will normally be fast and stable, assuming that the power supplies have been bypassed (with 0.1 μF disc capacitors), and that the output signal is routed well away from the inputs (pins 2 and 3) and also away from pins 5 and 6.

However, when the input signal is a voltage ramp or a slow sine wave, or if the signal source impedance is high (1 kΩ to 100 kΩ), the comparator may burst into oscillation near the crossing-point. This is due to the high gain and wide bandwidth of comparators like the LM111. To avoid oscillation or instability in such a usage, several precautions are recommended, as shown in *Figure 1* below.

1. The trim pins (pins 5 and 6) act as unwanted auxiliary inputs. If these pins are not connected to a trim-pot, they should be shorted together. If they are connected to a trim-pot, a 0.01 μA capacitor C1 between pins 5 and 6 will minimize the susceptibility to AC coupling. A smaller capacitor is used if pin 5 is used for positive feedback as in *Figure 1*.

2. Certain sources will produce a cleaner comparator output waveform if a 100 pF to 1000 pF capacitor C2 is connected directly across the input pins.

3. When the signal source is applied through a resistive network, R_S, it is usually advantageous to choose an R_S' of substantially the same value, both for DC and for dynamic (AC) considerations. Carbon, tin-oxide, and metal-film resistors have all been used successfully in comparator input circuitry. Inductive wirewound resistors are not suitable.

4. When comparator circuits use input resistors (eg. summing resistors), their value and placement are particularly important. In all cases the body of the resistor should be close to the device or socket. In other words there should be very little lead length or printed-circuit foil run between comparator and resistor to radiate or pick up signals. The same applies to capacitors, pots, etc. For example, if R_S = 10 kΩ, as little as 5 inches of lead between the resistors and the input pins can result in oscillations that are very hard to damp. Twisting these input leads tightly is the only (second best) alternative to placing resistors close to the comparator.

5. Since feedback to almost any pin of a comparator can result in oscillation, the printed-circuit layout should be engineered thoughtfully. Preferably there should be a groundplane under the LM111 circuitry, for example, one side of a double-layer circuit card. Ground foil (or, positive supply or negative supply foil) should extend between the output and the inputs, to act as a guard. The foil connections for the inputs should be as small and compact as possible, and should be essentially surrounded by ground foil on all sides, to guard against capacitive coupling from any high-level signals (such as the output). If pins 5 and 6 are not used, they should be shorted together. If they are connected to a trim-pot, the trim-pot should be located, at most, a few inches away from the LM111, and the 0.01 μF capacitor should be installed. If this capacitor cannot be used, a shielding printed-circuit foil may be advisable between pins 6 and 7. The power supply bypass capacitors should be located within a couple inches of the LM111. (Some other comparators require the power-supply bypass to be located immediately adjacent to the comparator.)

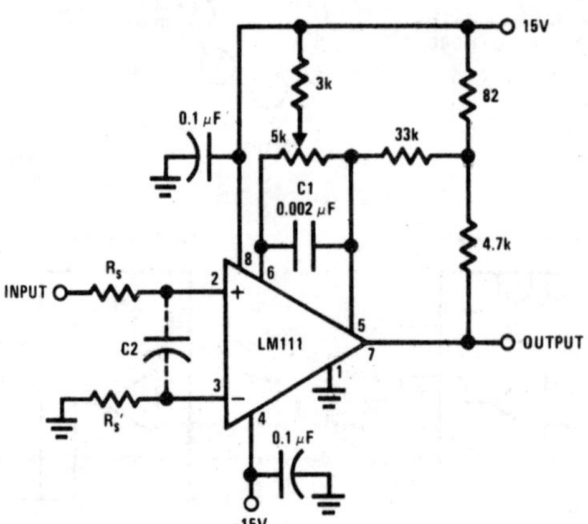

Pin connections shown are for LM111H in 8-lead TO-5 hermetic package

FIGURE 1. Improved Positive Feedback

Application Hints (Continued)

6. It is a standard procedure to use hysteresis (positive feedback) around a comparator, to prevent oscillation, and to avoid excessive noise on the output because the comparator is a good amplifier for its own noise. In the circuit of *Figure 2,* the feedback from the output to the positive input will cause about 3 mV of hysteresis. However, if R_S is larger than 100Ω, such as 50 kΩ, it would not be reasonable to simply increase the value of the positive feedback resistor above 510 kΩ. The circuit of *Figure 3* could be used, but it is rather awkward. See the notes in paragraph 7 below.

7. When both inputs of the LM111 are connected to active signals, or if a high-impedance signal is driving the positive input of the LM111 so that positive feedback would be disruptive, the circuit of *Figure 1* is ideal. The positive feedback is to pin 5 (one of the offset adjustment pins). It is sufficient to cause 1 to 2 mV hysteresis and sharp transitions with input triangle waves from a few Hz to hundreds of kHz. The positive-feedback signal across the 82Ω resistor swings 240 mV below the positive supply. This signal is centered around the nominal voltage at pin 5, so this feedback does not add to the V_{OS} of the comparator. As much as 8 mV of V_{OS} can be trimmed out, using the 5 kΩ pot and 3 kΩ resistor as shown.

8. These application notes apply specifically to the LM111, LM211, LM311, and LF111 families of comparators, and are applicable to all high-speed comparators in general, (with the exception that not all comparators have trim pins).

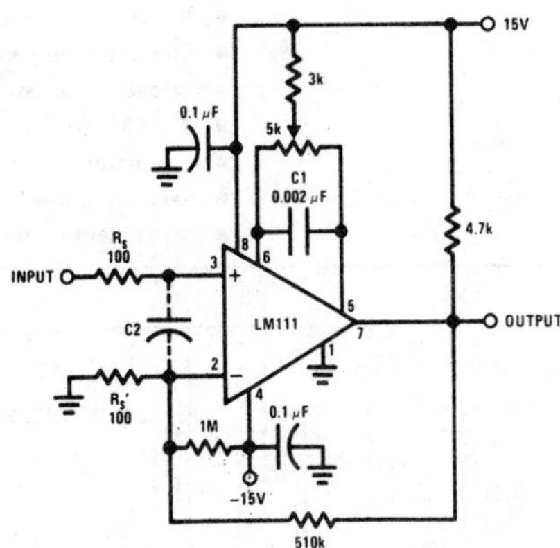

Pin connections shown are for LM111H in 8-lead TO-5 hermetic package

FIGURE 2. Conventional Positive Feedback

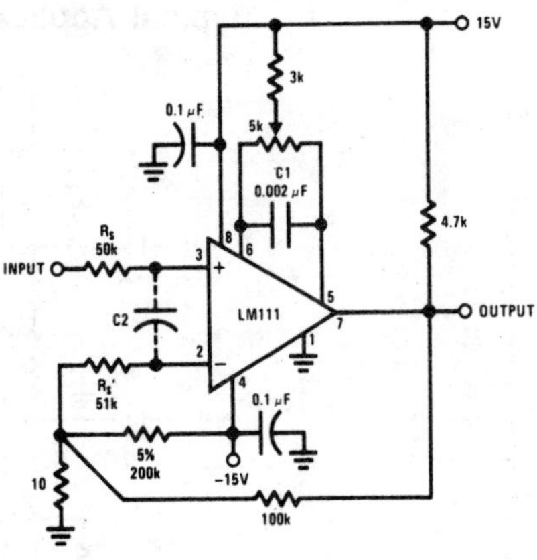

FIGURE 3. Positive Feedback With High Source Resistance

National Semiconductor

Audio/Radio Circuits

LM377 Dual 2 Watt Audio Amplifier

General Description

The LM377 is a monolithic dual power amplifier which offers high quality performance for stereo phonographs, tape players, recorders, and AM-FM stereo receivers, etc.

The LM377 will deliver 2W/channel into 8 or 16Ω loads. The amplifier is designed to operate with a minimum of external components and contains an internal bias regulator to bias each amplifier. Device overload protection consists of both internal current limit and thermal shutdown. For more information, see AN-125. The LM377 is not recommended for new designs; see the LM1877 data sheet for an improved pin-for-pin replacement to the LM377 in audio applications.

Features

- A_{VO} typical 90 dB
- 2W per channel
- 70 dB ripple rejection
- 75 dB channel separation
- Internal stabilization
- Self centered biasing
- 3 MΩ input impedance
- 10–26V operation
- Internal current limiting
- Internal thermal protection

Applications

- Multi-channel audio systems
- Tape recorders and players
- Movie projectors
- Automotive systems
- Stereo phonographs
- Bridge output stages
- AM-FM radio receivers
- Intercoms
- Servo amplifiers
- Instrument systems

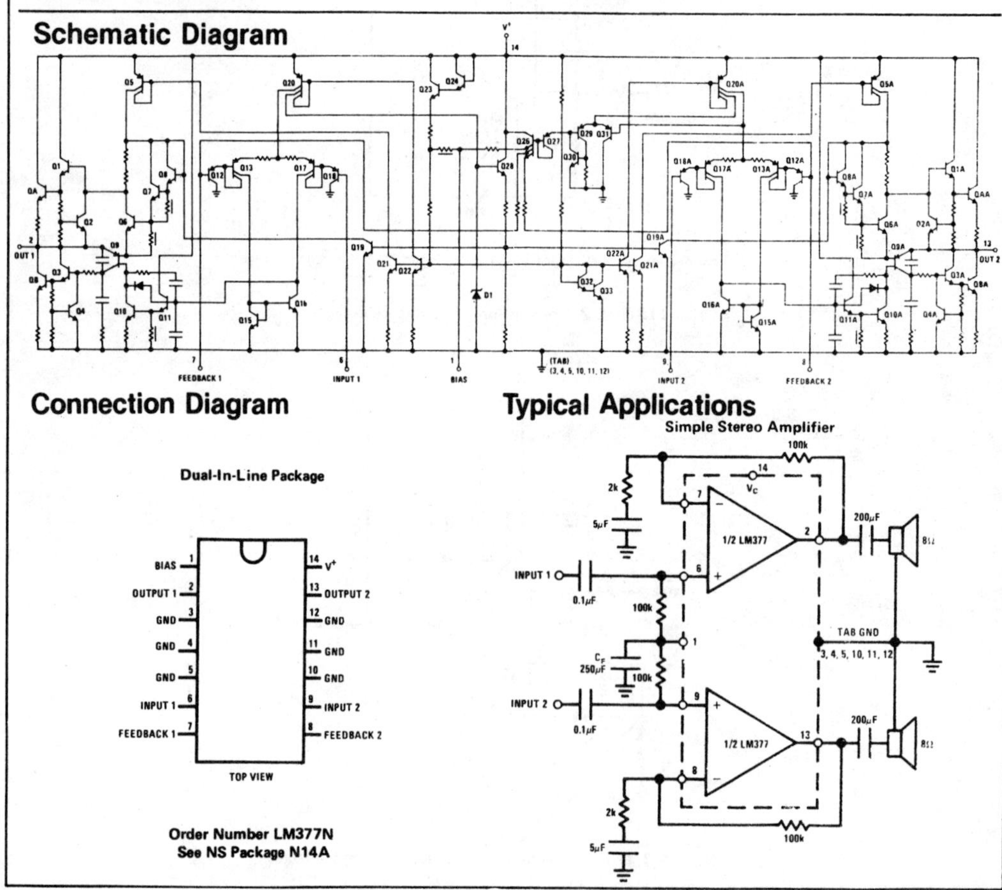

Absolute Maximum Ratings

Supply Voltage	26V
Input Voltage	0V – V_{SUPPLY}
Operating Temperature	0°C to +70°C
Storage Temperature	−65°C to +150°C
Junction Temperature	150°C
Lead Temperature (Soldering, 10 seconds)	300°C

Electrical Characteristics

V_S = 20V, T_{TAB} = 25°C, R_L = 8Ω, A_V = 50 (34 dB), unless otherwise specified.

PARAMETER	CONDITIONS	MIN	TYP	MAX	UNITS
Total Supply Current	P_{OUT} = 0W P_{OUT} = 1.5W/Channel		15 430	50 500	mA mA
DC Output Level			10		V
Supply Voltage		10		26	V
Output Power	T.H.D. = < 5%	2	2.5		W
T.H.D.	P_{OUT} = 0.05W/Channel, f = 1 kHz P_{OUT} = 1W/Channel, f = 1 kHz P_{OUT} = 2W/Channel, f = 1 kHz		0.25 0.07 0.10	 1 	% % %
Offset Voltage			15		mV
Input Bias Current			100		nA
Input Impedance		3			MΩ
Open Loop Gain	R_S = 0Ω	66	90		dB
Output Swing			V_S−6		V_{P-P}
Channel Separation	C_F = 250μF, f = 1 kHz	50	70		dB
Ripple Rejection	f = 120 Hz, C_F = 250μF	60	70		dB
Current Limit			1.5		A
Slew Rate			1.4		V/μs
Equivalent Input Noise Voltage	R_S = 600Ω, 100 Hz – 10 kHz		3		μVrms

Note 1: For operation at ambient temperatures greater than 25°C the LM377 must be derated based on a maximum 150°C junction temperature using a thermal resistance which depends upon device mounting techniques.

Note 2: Dissipation characteristics are shown for four mounting configurations.

a. Infinite sink – 13.4°C/W

b. P.C. board +V_7 sink – 21°C/W. P.C. board is 2 1/2 square inches. Staver V_7 sink is 0.02 inch thick copper and has a radiating surface area of 10 square inches.

c. P.C. board only – 29°C/W. Device soldered to 2 1/2 square inch P.C. board.

d. Free air – 58°C/W.

LM377

Typical Performance Characteristics

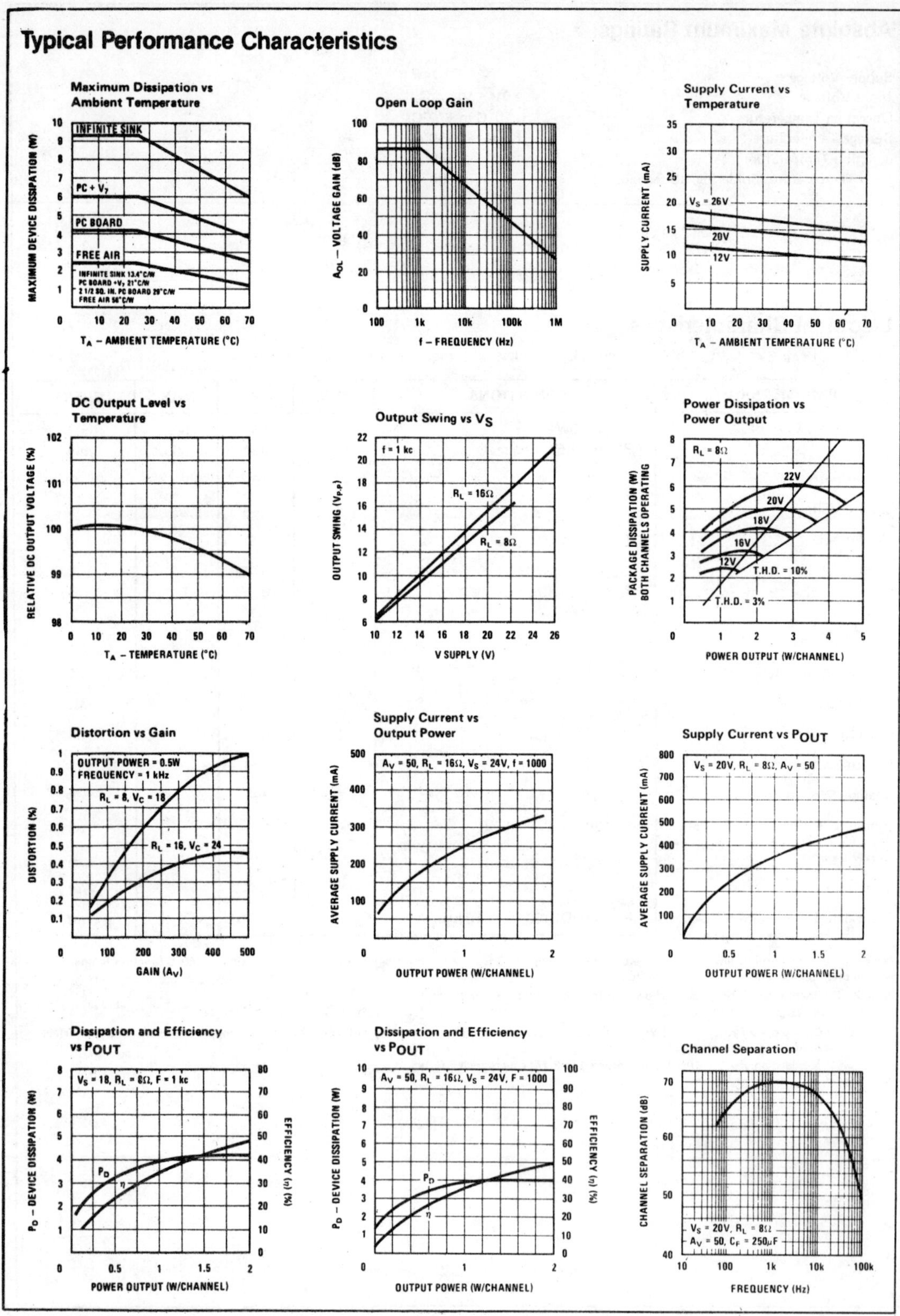

Typical Performance Characteristics (Continued)

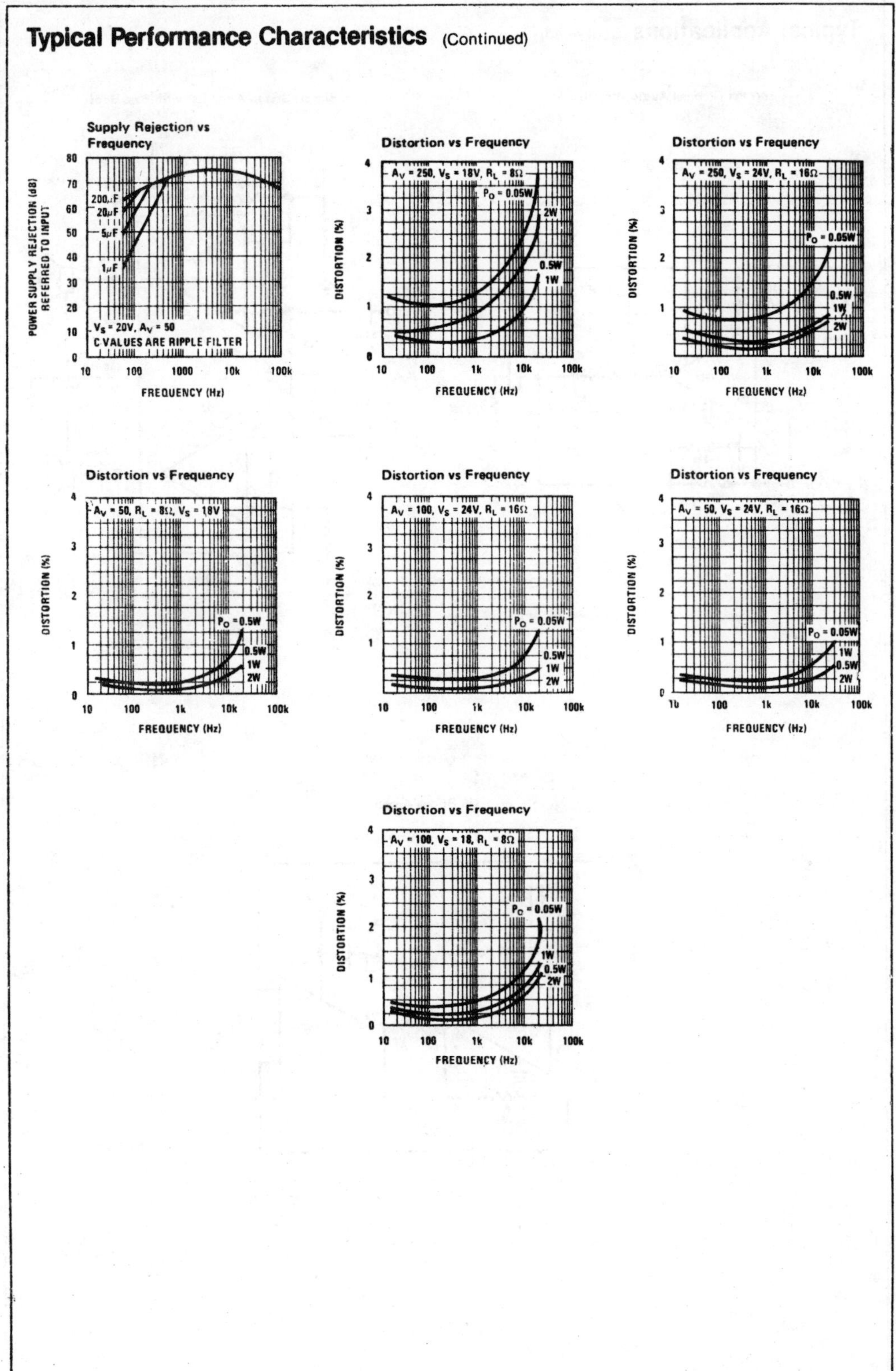

Typical Applications (Continued)

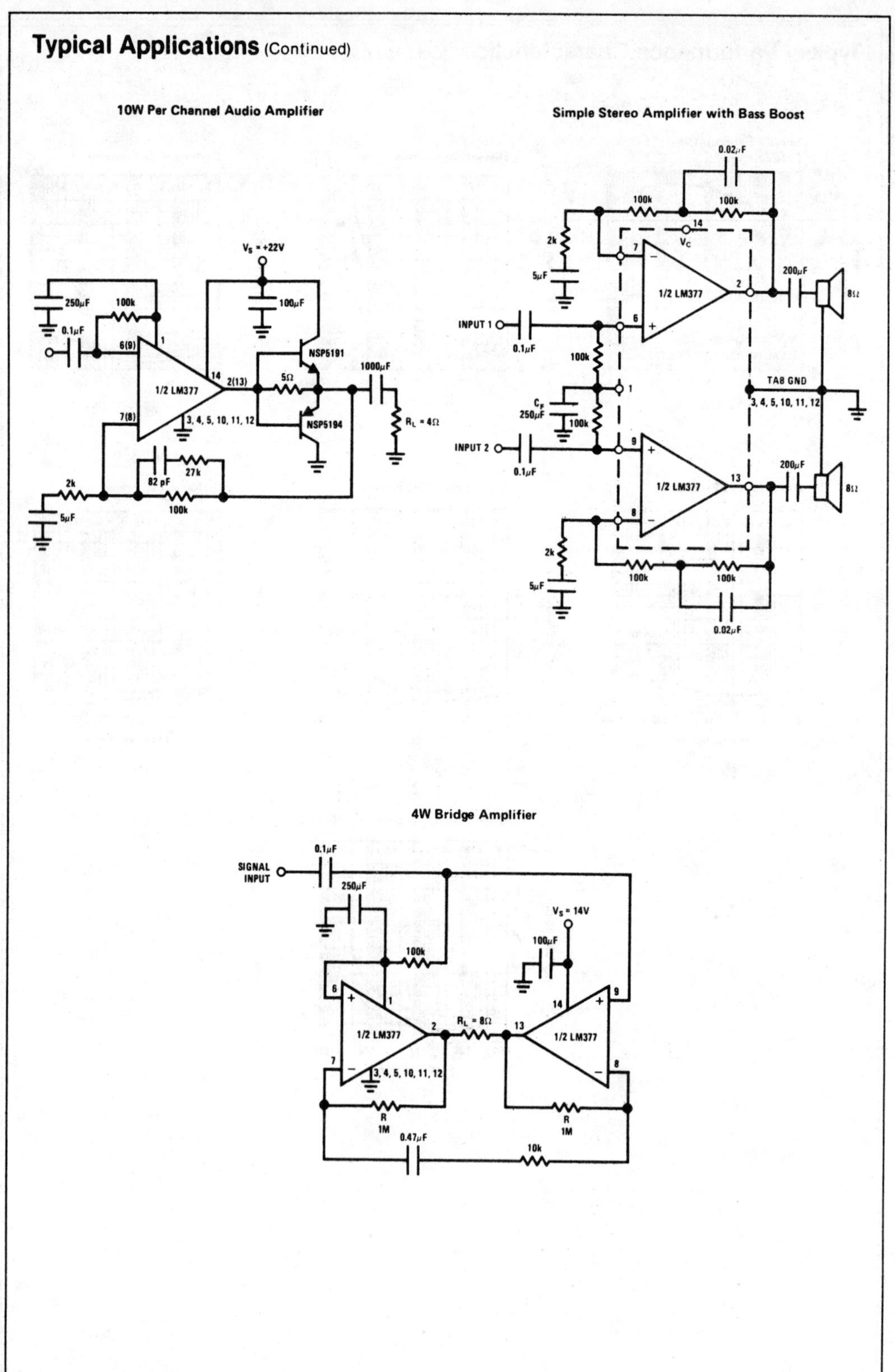

LM555/LM555C Timer

General Description

The LM555 is a highly stable device for generating accurate time delays or oscillation. Additional terminals are provided for triggering or resetting if desired. In the time delay mode of operation, the time is precisely controlled by one external resistor and capacitor. For astable operation as an oscillator, the free running frequency and duty cycle are accurately controlled with two external resistors and one capacitor. The circuit may be triggered and reset on falling waveforms, and the output circuit can source or sink up to 200 mA or drive TTL circuits.

Features

- Direct replacement for SE555/NE555
- Timing from microseconds through hours
- Operates in both astable and monostable modes
- Adjustable duty cycle
- Output can source or sink 200 mA
- Output and supply TTL compatible
- Temperature stability better than 0.005% per °C
- Normally on and normally off output

Applications

- Precision timing
- Pulse generation
- Sequential timing
- Time delay generation
- Pulse width modulation
- Pulse position modulation
- Linear ramp generator

Schematic Diagram

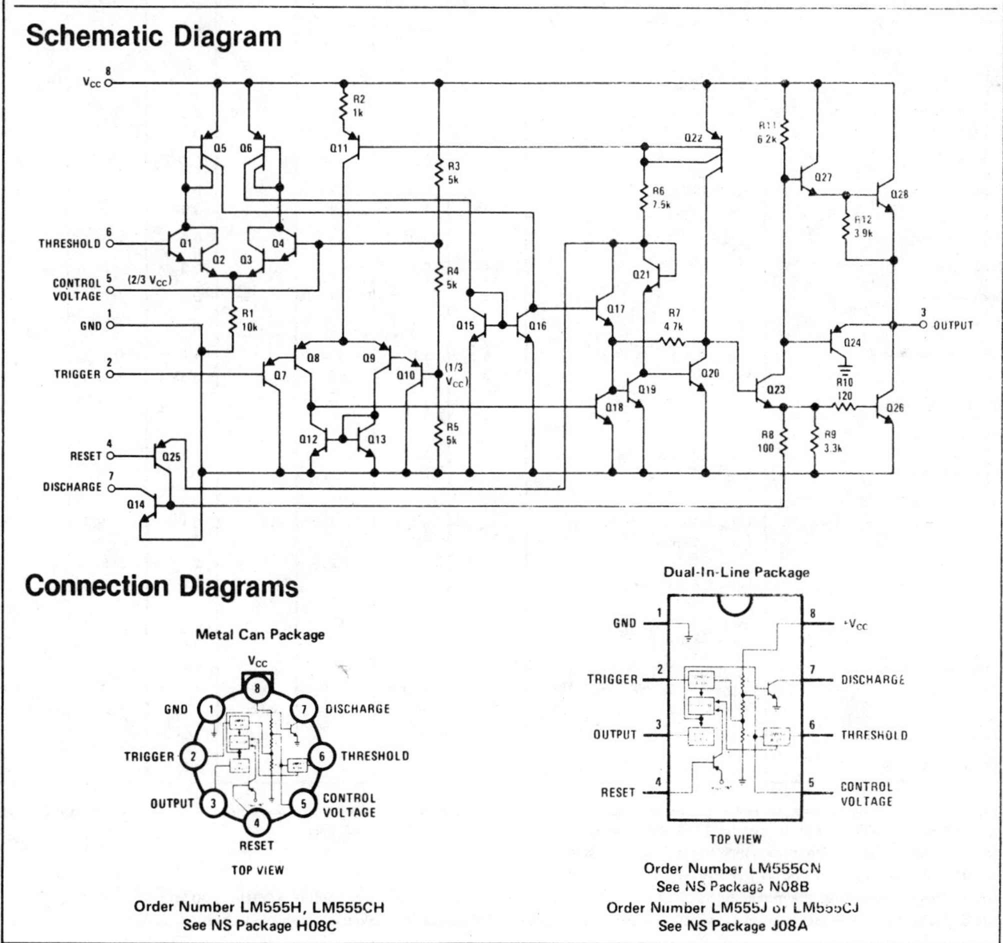

Connection Diagrams

Metal Can Package

GND, TRIGGER, OUTPUT, RESET, CONTROL VOLTAGE, THRESHOLD, DISCHARGE, VCC

TOP VIEW

Order Number LM555H, LM555CH
See NS Package H08C

Dual-In-Line Package

GND, TRIGGER, OUTPUT, RESET, CONTROL VOLTAGE, THRESHOLD, DISCHARGE, +VCC

TOP VIEW

Order Number LM555CN
See NS Package N08B
Order Number LM555J or LM555CJ
See NS Package J08A

Absolute Maximum Ratings

Supply Voltage	+18V
Power Dissipation (Note 1)	600 mW
Operating Temperature Ranges	
LM555C	0°C to +70°C
LM555	−55°C to +125°C
Storage Temperature Range	−65°C to +150°C
Lead Temperature (Soldering, 10 seconds)	300°C

Electrical Characteristics (T_A = 25°C, V_{CC} = +5V to +15V, unless otherwise specified)

PARAMETER	CONDITIONS	LIMITS						UNITS
		LM555			LM555C			
		MIN	TYP	MAX	MIN	TYP	MAX	
Supply Voltage		4.5		18	4.5		16	V
Supply Current	V_{CC} = 5V, R_L = ∞		3	5		3	6	mA
	V_{CC} = 15V, R_L = ∞ (Low State) (Note 2)		10	12		10	15	mA
Timing Error, Monostable								
Initial Accuracy			0.5			1		%
Drift with Temperature	R_A, R_B = 1k to 100 k, C = 0.1μF, (Note 3)		30			50		ppm/°C
Accuracy over Temperature			1.5			1.5		%
Drift with Supply			0.05			0.1		%/V
Timing Error, Astable								
Initial Accuracy			1.5			2.25		%
Drift with Temperature			90			150		ppm/°C
Accuracy over Temperature			2.5			3.0		%
Drift with Supply			0.15			0.30		%/V
Threshold Voltage			0.667			0.667		x V_{CC}
Trigger Voltage	V_{CC} = 15V	4.8	5	5.2		5		V
	V_{CC} = 5V	1.45	1.67	1.9		1.67		V
Trigger Current			0.01	0.5		0.5	0.9	μA
Reset Voltage		0.4	0.5	1	0.4	0.5	1	V
Reset Current			0.1	0.4		0.1	0.4	mA
Threshold Current	(Note 4)		0.1	0.25		0.1	0.25	μA
Control Voltage Level	V_{CC} = 15V	9.6	10	10.4	9	10	11	V
	V_{CC} = 5V	2.9	3.33	3.8	2.6	3.33	4	V
Pin 7 Leakage Output High			1	100		1	100	nA
Pin 7 Sat (Note 5)								
Output Low	V_{CC} = 15V, I_7 = 15 mA		150			180		mV
Output Low	V_{CC} = 4.5V, I_7 = 4.5 mA		70	100		80	200	mV
Output Voltage Drop (Low)	V_{CC} = 15V							
	I_{SINK} = 10 mA		0.1	0.15		0.1	0.25	V
	I_{SINK} = 50 mA		0.4	0.5		0.4	0.75	V
	I_{SINK} = 100 mA		2	2.2		2	2.5	V
	I_{SINK} = 200 mA		2.5			2.5		V
	V_{CC} = 5V							
	I_{SINK} = 8 mA		0.1	0.25				V
	I_{SINK} = 5 mA					0.25	0.35	V
Output Voltage Drop (High)	I_{SOURCE} = 200 mA, V_{CC} = 15V		12.5			12.5		V
	I_{SOURCE} = 100 mA, V_{CC} = 15V	13	13.3		12.75	13.3		V
	V_{CC} = 5V	3	3.3		2.75	3.3		V
Rise Time of Output			100			100		ns
Fall Time of Output			100			100		ns

Note 1: For operating at elevated temperatures the device must be derated based on a +150°C maximum junction temperature and a thermal resistance of +45°C/W junction to case for TO-5 and +150°C/W junction to ambient for both packages.

Note 2: Supply current when output high typically 1 mA less at V_{CC} = 5V.

Note 3: Tested at V_{CC} = 5V and V_{CC} = 15V.

Note 4: This will determine the maximum value of $R_A + R_B$ for 15V operation. The maximum total ($R_A + R_B$) is 20 MΩ.

Note 5: No protection against excessive pin 7 current is necessary providing the package dissipation rating will not be exceeded.

Typical Performance Characteristics

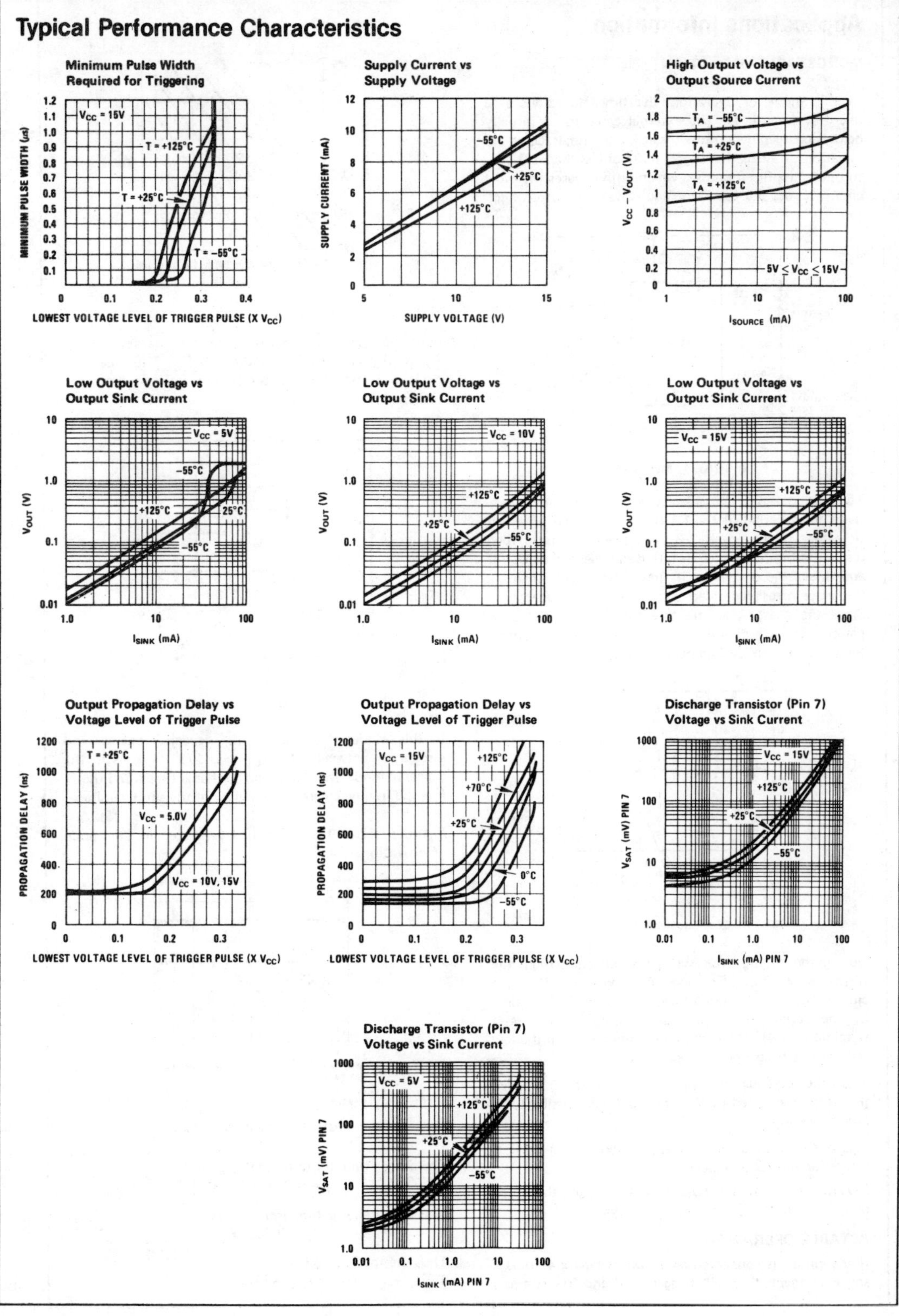

Applications Information

MONOSTABLE OPERATION

In this mode of operation, the timer functions as a one-shot (*Figure 1*). The external capacitor is initially held discharged by a transistor inside the timer. Upon application of a negative trigger pulse of less than 1/3 V_{CC} to pin 2, the flip-flop is set which both releases the short circuit across the capacitor and drives the output high.

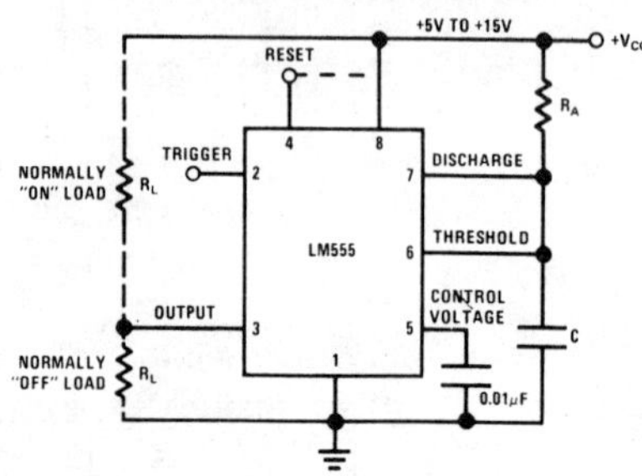

FIGURE 1. Monostable

The voltage across the capacitor then increases exponentially for a period of $t = 1.1\ R_A C$, at the end of which time the voltage equals 2/3 V_{CC}. The comparator then resets the flip-flop which in turn discharges the capacitor and drives the output to its low state. *Figure 2* shows the waveforms generated in this mode of operation. Since the charge and the threshold level of the comparator are both directly proportional to supply voltage, the timing internal is independent of supply.

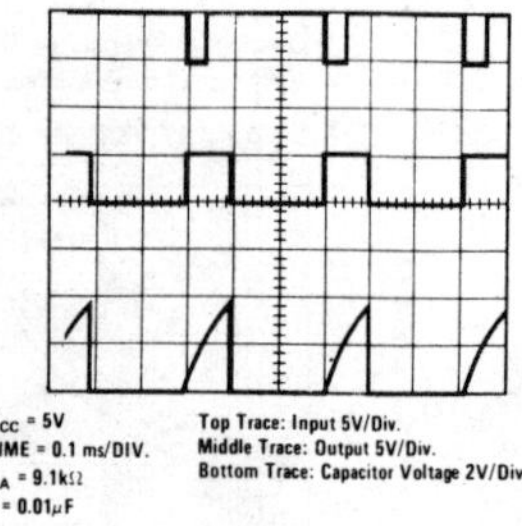

FIGURE 2. Monostable Waveforms

During the timing cycle when the output is high, the further application of a trigger pulse will not effect the circuit. However the circuit can be reset during this time by the application of a negative pulse to the reset terminal (pin 4). The output will then remain in the low state until a trigger pulse is again applied.

When the reset function is not in use, it is recommended that it be connected to V_{CC} to avoid any possibility of false triggering.

Figure 3 is a nomograph for easy determination of R, C values for various time delays.

NOTE: In monostable operation, the trigger should be driven high before the end of timing cycle.

ASTABLE OPERATION

If the circuit is connected as shown in *Figure 4* (pins 2 and 6 connected) it will trigger itself and free run as a

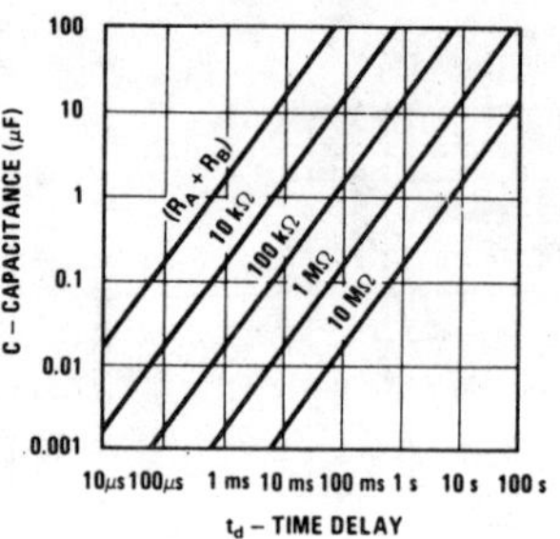

FIGURE 3. Time Delay

multivibrator. The external capacitor charges through $R_A + R_B$ and discharges through R_B. Thus the duty cycle may be precisely set by the ratio of these two resistors.

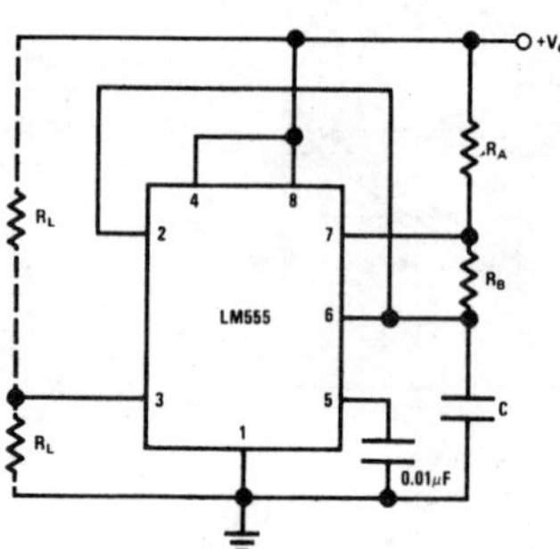

FIGURE 4. Astable

In this mode of operation, the capacitor charges and discharges between 1/3 V_{CC} and 2/3 V_{CC}. As in the triggered mode, the charge and discharge times, and therefore the frequency are independent of the supply voltage.

Figure 5 shows the waveforms generated in this mode of operation.

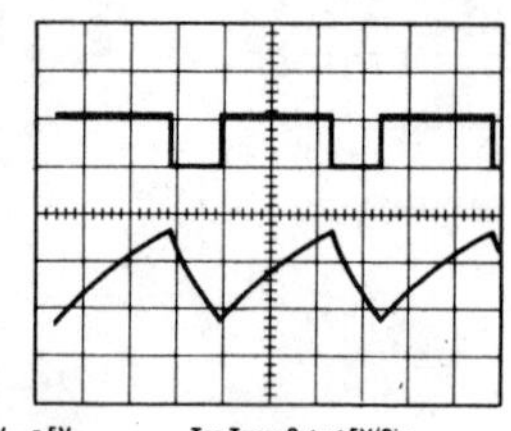

FIGURE 5. Astable Waveforms

The charge time (output high) is given by:

$$t_1 = 0.693\ (R_A + R_B)\ C$$

And the discharge time (output low) by:

$$t_2 = 0.693\ (R_B)\ C$$

Thus the total period is:

$$T = t_1 + t_2 = 0.693\ (R_A + 2R_B)\ C$$

Applications Information (Continued)

The frequency of oscillation is:

$$f = \frac{1}{T} = \frac{1.44}{(R_A + 2 R_B) C}$$

Figure 6 may be used for quick determination of these RC values.

The duty cycle is: $D = \frac{R_B}{R_A + 2R_B}$

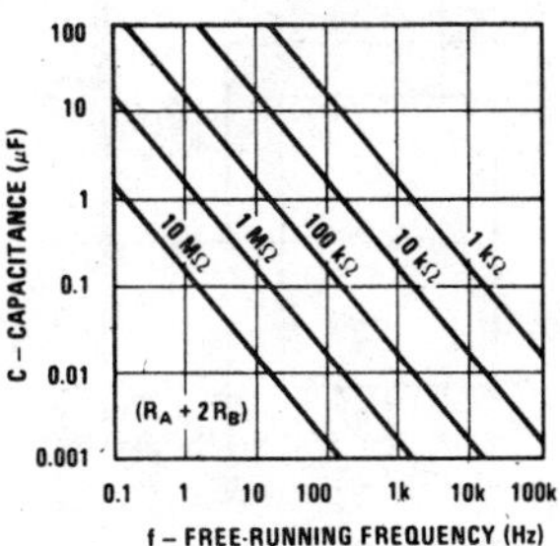

FIGURE 6. Free Running Frequency

FREQUENCY DIVIDER

The monostable circuit of *Figure 1* can be used as a frequency divider by adjusting the length of the timing cycle. *Figure 7* shows the waveforms generated in a divide by three circuit.

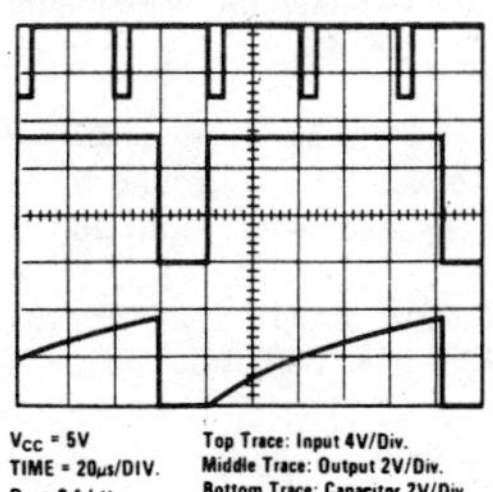

FIGURE 7. Frequency Divider

PULSE WIDTH MODULATOR

When the timer is connected in the monostable mode and triggered with a continuous pulse train, the output pulse width can be modulated by a signal applied to pin 5. *Figure 8* shows the circuit, and in *Figure 9* are some waveform examples.

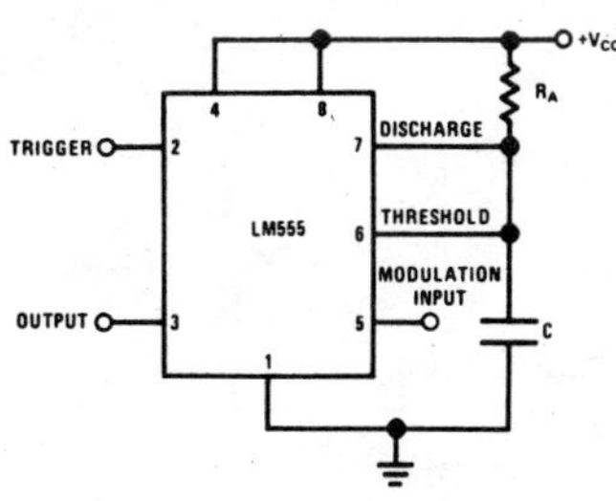

FIGURE 8. Pulse Width Modulator

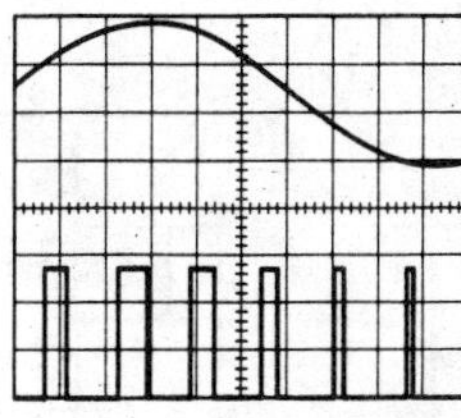

FIGURE 9. Pulse Width Modulator

PULSE POSITION MODULATOR

This application uses the timer connected for astable operation, as in *Figure 10*, with a modulating signal again applied to the control voltage terminal. The pulse position varies with the modulating signal, since the threshold voltage and hence the time delay is varied. *Figure 11* shows the waveforms generated for a triangle wave modulation signal.

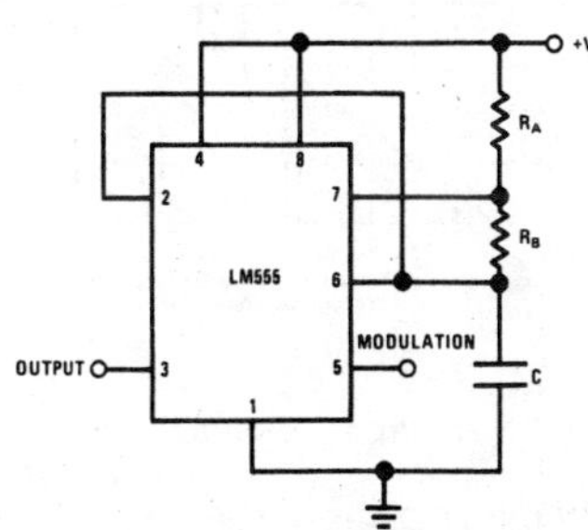

FIGURE 10. Pulse Position Modulator

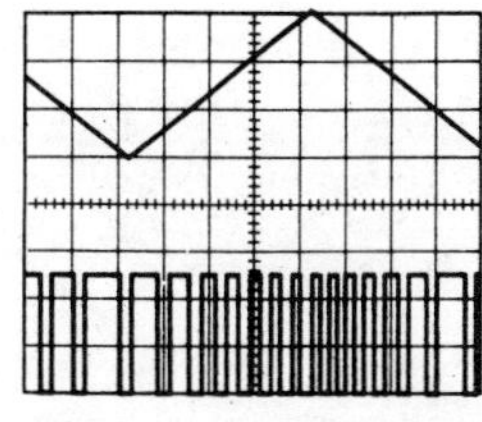

FIGURE 11. Pulse Position Modulator

LINEAR RAMP

When the pullup resistor, R_A, in the monostable circuit is replaced by a constant current source, a linear ramp is

Applications Information (Continued)

generated. *Figure 12* shows a circuit configuration that will perform this function.

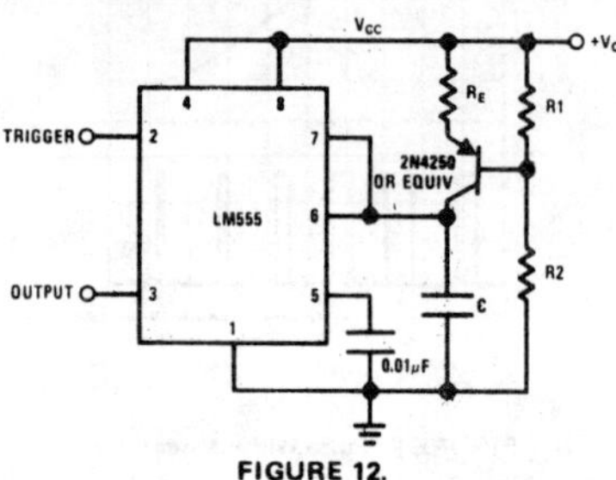

FIGURE 12.

Figure 13 shows waveforms generated by the linear ramp.

The time interval is given by:

$$T = \frac{2/3\ V_{CC}\ R_E\ (R_1 + R_2)\ C}{R_1\ V_{CC} - V_{BE}\ (R_1 + R_2)}$$

$$V_{BE} \simeq 0.6V$$

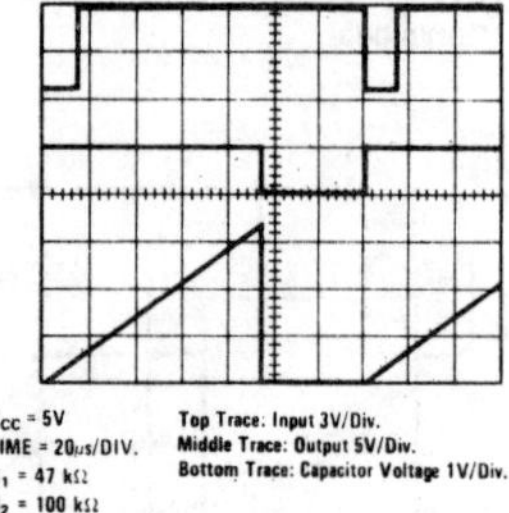

FIGURE 13. Linear Ramp

50% DUTY CYCLE OSCILLATOR

For a 50% duty cycle, the resistors R_A and R_B may be connected as in *Figure 14*. The time period for the output high is the same as previous, $t_1 = 0.693\ R_A\ C$. For the output low it is t_2 =

$$[(R_A\ R_B)/(R_A + R_B)]\ C Ln \left[\frac{R_B - 2R_A}{2R_B - R_A}\right]$$

Thus the frequency of oscillation is $f = \frac{1}{t_1 + t_2}$

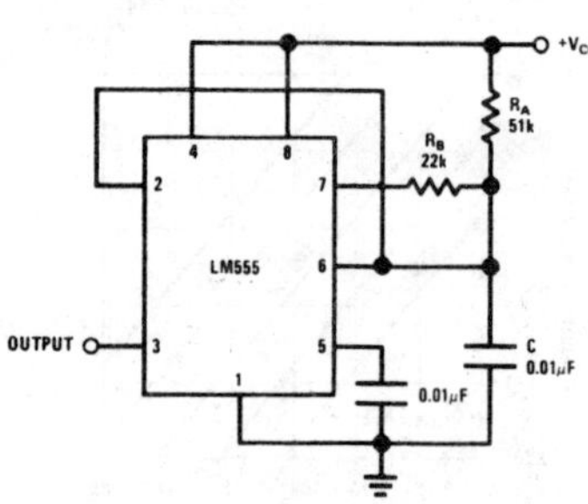

FIGURE 14. 50% Duty Cycle Oscillator

Note that this circuit will not oscillate if R_B is greater than 1/2 R_A because the junction of R_A and R_B cannot bring pin 2 down to 1/3 V_{CC} and trigger the lower comparator.

ADDITIONAL INFORMATION

Adequate power supply bypassing is necessary to protect associated circuitry. Minimum recommended is 0.1μF in parallel with 1μF electrolytic.

Lower comparator storage time can be as long as 10μs when pin 2 is driven fully to ground for triggering. This limits the monostable pulse width to 10μs minimum.

Delay time reset to output is 0.47μs typical. Minimum reset pulse width must be 0.3μs, typical.

Pin 7 current switches within 30 ns of the output (pin 3) voltage.

LM556/LM556C Dual Timer

General Description

The LM556 Dual timing circuit is a highly stable controller capable of producing accurate time delays or oscillation. The 556 is a dual 555. Timing is provided by an external resistor and capacitor for each timing function. The two timers operate independently of each other sharing only V_{CC} and ground. The circuits may be triggered and reset on falling waveforms. The output structures may sink or source 200 mA.

Features

- Direct replacement for SE556/NE556
- Timing from microseconds through hours
- Operates in both astable and monostable modes
- Replaces two 555 timers
- Adjustable duty cycle
- Output can source or sink 200 mA
- Output and supply TTL compatible
- Temperature stability better than 0.005% per °C
- Normally on and normally off output

Applications

- Precision timing
- Pulse generation
- Sequential timing
- Time delay generation
- Pulse width modulation
- Pulse position modulation
- Linear ramp generator

Schematic Diagram

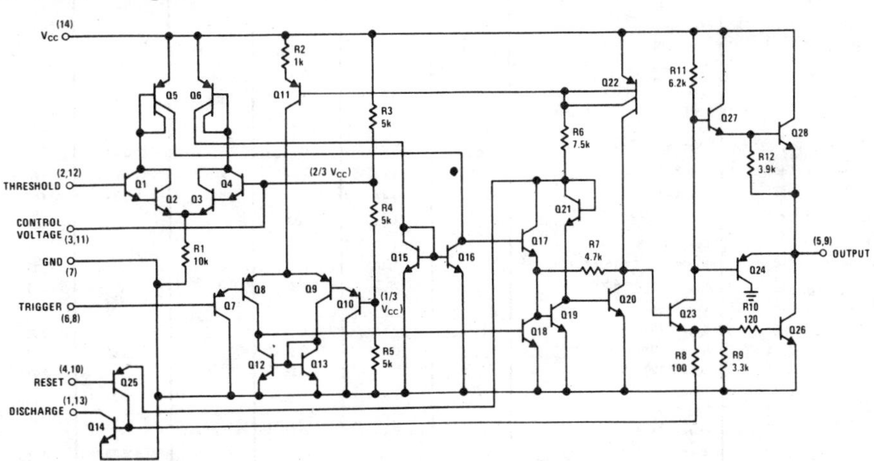

Connection Diagram

Dual-In-Line Package

Vcc 14, DISCHARGE 13, THRESHOLD 12, CONTROL VOLTAGE 11, RESET 10, OUTPUT 9, TRIGGER 8

COMP, FLIP-FLOP, COMP

FLIP-FLOP, COMP, COMP

DISCHARGE 1, THRESHOLD 2, CONTROL VOLTAGE 3, RESET 4, OUTPUT 5, TRIGGER 6, GND 7

TOP VIEW

Order Number LM556CN
See NS Package N14A

Order Number LM556J or LM556CJ
See NS Package J14A

Absolute Maximum Ratings

Supply Voltage	+18V
Power Dissipation (Note 1)	600 mW
Operating Temperature Ranges	
LM556C	0°C to +70°C
LM556	−55°C to +125°C
Storage Temperature Range	−65°C to +150°C
Lead Temperature (Soldering, 10 seconds)	300°C

Electrical Characteristics (T_A = 25°C, V_{CC} = +5V to +15V, unless otherwise specified)

PARAMETER	CONDITIONS	LM556			LM556C			UNITS
		MIN	TYP	MAX	MIN	TYP	MAX	
Supply Voltage		4.5		18	4.5		16	V
Supply Current (Each Timer Section)	V_{CC} = 5V, R_L = ∞		3	5		3	6	mA
	V_{CC} = 15V, R_L = ∞ (Low State) (Note 2)		10	11		10	14	mA
Timing Error, Monostable								
Initial Accuracy			0.5			0.75		%
Drift With Temperature	R_A, R_B = 1k to 100k, C = 0.1µF, (Note 3)		30			50		ppm/°C
Accuracy Over Temperature			1.5			1.5		%
Drift with Supply			0.05			0.1		%/V
Timing Error, Astable								
Initial Accuracy			1.5			2.25		%
Drift With Temperature			90			150		ppm/°C
Accuracy Over Temperature			2.5			3.0		%
Drift With Supply			0.15			0.30		%/V
Trigger Voltage	V_{CC} = 15V	4.8	5	5.2	4.5	5	0.5	V
	V_{CC} = 5V	1.45	1.67	1.9	1.25	1.67	2.0	V
Trigger Current			0.1	0.5		0.2	1.0	µA
Reset Voltage	(Note 4)	0.4	0.5	1	0.4	0.5	1	V
Reset Current			0.1	0.4		0.1	0.6	mA
Threshold Current	(Note 5)	•	0.03	0.1		0.03	0.1	µA
Control Voltage Level And Threshold Voltage	V_{CC} = 15V	9.6	10	10.4	9	10	11	V
	V_{CC} = 5V	2.9	3.33	3.8	2.6	3.33	4	V
Pin 1, 13 Leakage Output High			1	100		1	100	nA
Pin 1, 13 Sat	(Note 6)							
Output Low	V_{CC} = 15V, I = 15 mA		150	240		180	300	mV
Output Low	V_{CC} = 4.5V, I = 4.5 mA		70	100		80	200	mV
Output Voltage Drop (Low)	V_{CC} = 15V							
	I_{SINK} = 10 mA		0.1	0.15		0.1	0.25	V
	I_{SINK} = 50 mA		0.4	0.5		0.4	0.75	V
	I_{SINK} = 100 mA		2	2.25		2	2.75	V
	I_{SINK} = 200 mA		2.5			2.5		V
	V_{CC} = 5V							
	I_{SINK} = 8 mA		0.1	0.25				V
	I_{SINK} = 5 mA					0.25	0.35	V
Output Voltage Drop (High)	I_{SOURCE} = 200 mA, V_{CC} = 15V		12.5			12.5		V
	I_{SOURCE} = 100 mA, V_{CC} = 15V	13	13.3		12.75	13.3		V
	V_{CC} = 5V	3	3.3		2.75	3.3		V
Rise Time of Output			100			100		ns
Fall Time of Output			100			100		ns
Matching Characteristics	(Note 7)							
Initial Timing Accuracy			0.05	0.2		0.1	2.0	%
Timing Drift With Temperature			±10			±10		ppm/°C
Drift With Supply Voltage			0.1	0.2		0.2	0.5	%/V

Note 1: For operating at elevated temperatures the device must be derated based on a +150°C maximum junction temperature and a thermal resistance of +150°C/W junction to ambient for both packages.

Note 2: Supply current when output high typically 1 mA less at V_{CC} = 5V.

Note 3: Tested at V_{CC} = 5V and V_{CC} = 15V.

Note 4: As reset voltage lowers, timing is inhibited and then the output goes low.

Note 5: This will determine the maximum value of $R_A + R_B$ for 15V operation. The maximum total ($R_A + R_B$) is 20 MΩ.

Note 6: No protection against excessive pin 1, 13 current is necessary providing the package dissipation rating will not be exceeded.

Note 7: Matching characteristics refer to the difference between performance characteristics of each timer section.

LM556/LM556C

Typical Performance Characteristics

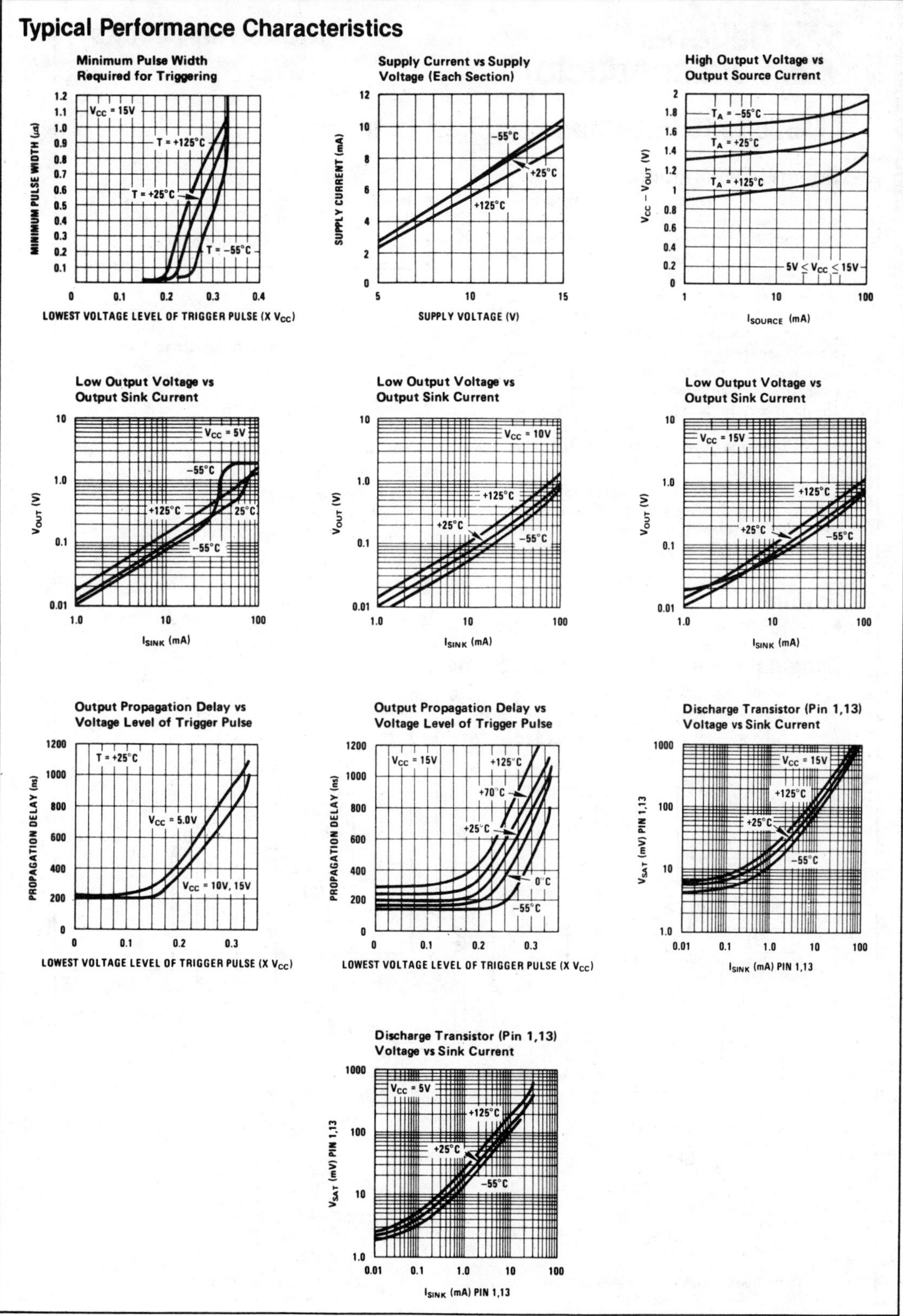

Industrial Blocks

LM565/LM565C Phase Locked Loop

General Description

The LM565 and LM565C are general purpose phase locked loops containing a stable, highly linear voltage controlled oscillator for low distortion FM demodulation, and a double balanced phase detector with good carrier suppression. The VCO frequency is set with an external resistor and capacitor, and a tuning range of 10:1 can be obtained with the same capacitor. The characteristics of the closed loop system—bandwidth, response speed, capture and pull in range—may be adjusted over a wide range with an external resistor and capacitor. The loop may be broken between the VCO and the phase detector for insertion of a digital frequency divider to obtain frequency multiplication.

The LM565H is specified for operation over the −55°C to +125°C military temperature range. The LM565CH and LM565CN are specified for operation over the 0°C to +70°C temperature range.

Features

- 200 ppm/°C frequency stability of the VCO
- Power supply range of ±5 to ±12 volts with 100 ppm/% typical
- 0.2% linearity of demodulated output
- Linear triangle wave with in phase zero crossings available
- TTL and DTL compatible phase detector input and square wave output
- Adjustable hold in range from ±1% to > ±60%.

Applications

- Data and tape synchronization
- Modems
- FSK demodulation
- FM demodulation
- Frequency synthesizer
- Tone decoding
- Frequency multiplication and division
- SCA demodulators
- Telemetry receivers
- Signal regeneration
- Coherent demodulators.

Schematic and Connection Diagrams

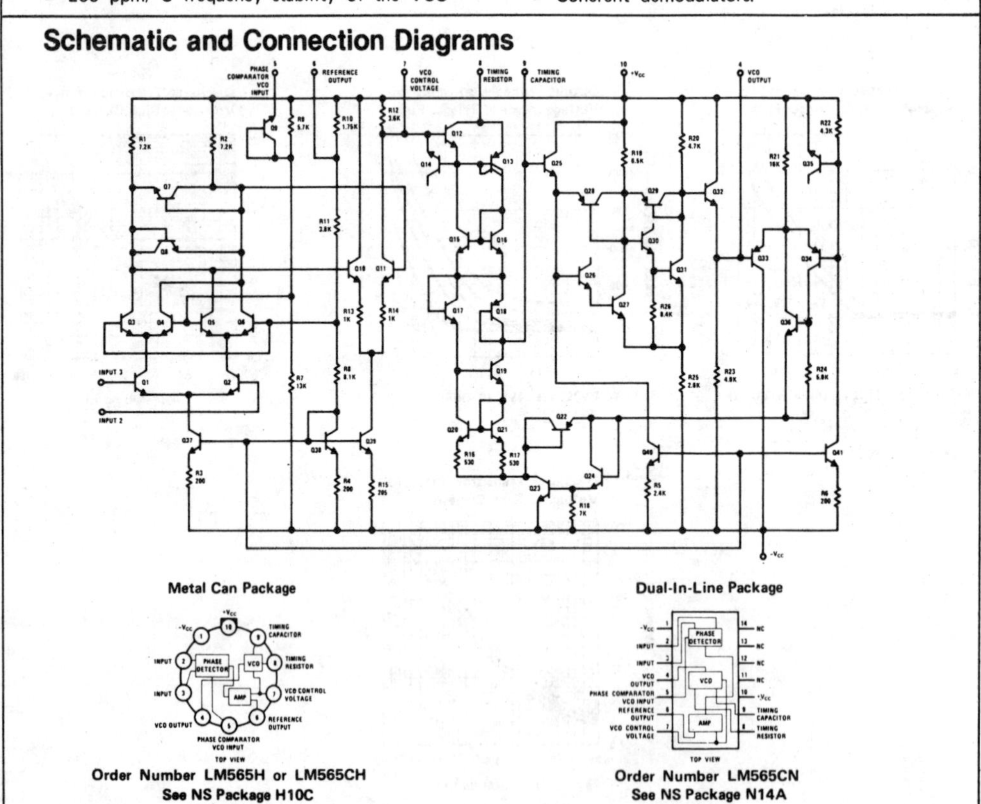

Metal Can Package

Order Number LM565H or LM565CH
See NS Package H10C

Dual-In-Line Package

Order Number LM565CN
See NS Package N14A

Absolute Maximum Ratings

Supply Voltage	±12V
Power Dissipation (Note 1)	300 mW
Differential Input Voltage	±1V
Operating Temperature Range LM565H	−55°C to +125°C
LM565CH, LM565CN	0°C to 70°C
Storage Temperature Range	−65°C to +150°C
Lead Temperature (Soldering, 10 sec)	300°C

Electrical Characteristics (AC Test Circuit, T_A = 25°C, V_C = ±6V)

PARAMETER	CONDITIONS	LM565			LM565C			UNITS
		MIN	TYP	MAX	MIN	TYP	MAX	
Power Supply Current			8.0	12.5		8.0	12.5	mA
Input Impedance (Pins 2, 3)	$-4V < V_2, V_3 < 0V$	7	10			5		kΩ
VCO Maximum Operating Frequency	C_o = 2.7 pF	300	500		250	500		kHz
Operating Frequency Temperature Coefficient			−100	300		−200	500	ppm/°C
Frequency Drift with Supply Voltage			0.01	0.1		0.05	0.2	%/V
Triangle Wave Output Voltage		2	2.4	3	2	2.4	3	V_{p-p}
Triangle Wave Output Linearity			0.2	0.75		0.5	1	%
Square Wave Output Level		4.7	5.4		4.7	5.4		V_{p-p}
Output Impedance (Pin 4)			5			5		kΩ
Square Wave Duty Cycle		45	50	55	40	50	60	%
Square Wave Rise Time			20	100		20		ns
Square Wave Fall Time			50	200		50		ns
Output Current Sink (Pin 4)		0.6	1		0.6	1		mA
VCO Sensitivity	f_o = 10 kHz	6400	6600	6800	6000	6600	7200	Hz/V
Demodulated Output Voltage (Pin 7)	±10% Frequency Deviation	250	300	350	200	300	400	mV_{pp}
Total Harmonic Distortion	±10% Frequency Deviation		0.2	0.75		0.2	1.5	%
Output Impedance (Pin 7)			3.5			3.5		kΩ
DC Level (Pin 7)		4.25	4.5	4.75	4.0	4.5	5.0	V
Output Offset Voltage $\lvert V_7 - V_6 \rvert$			30	100		50	200	mV
Temperature Drift of $\lvert V_7 - V_6 \rvert$			500			500		μV/°C
AM Rejection		30	40			40		dB
Phase Detector Sensitivity K_D		0.6	.68	0.9	0.55	68	0.95	V/radian

Note 1: The maximum junction temperature of the LM565 is 150°C, while that of the LM565C and LM565CN is 100°C. For operation at elevated temperatures, devices in the TO-5 package must be derated based on a thermal resistance of 150°C/W junction to ambient or 45°C/W junction to case. Thermal resistance of the dual-in-line package is 100°C/W.

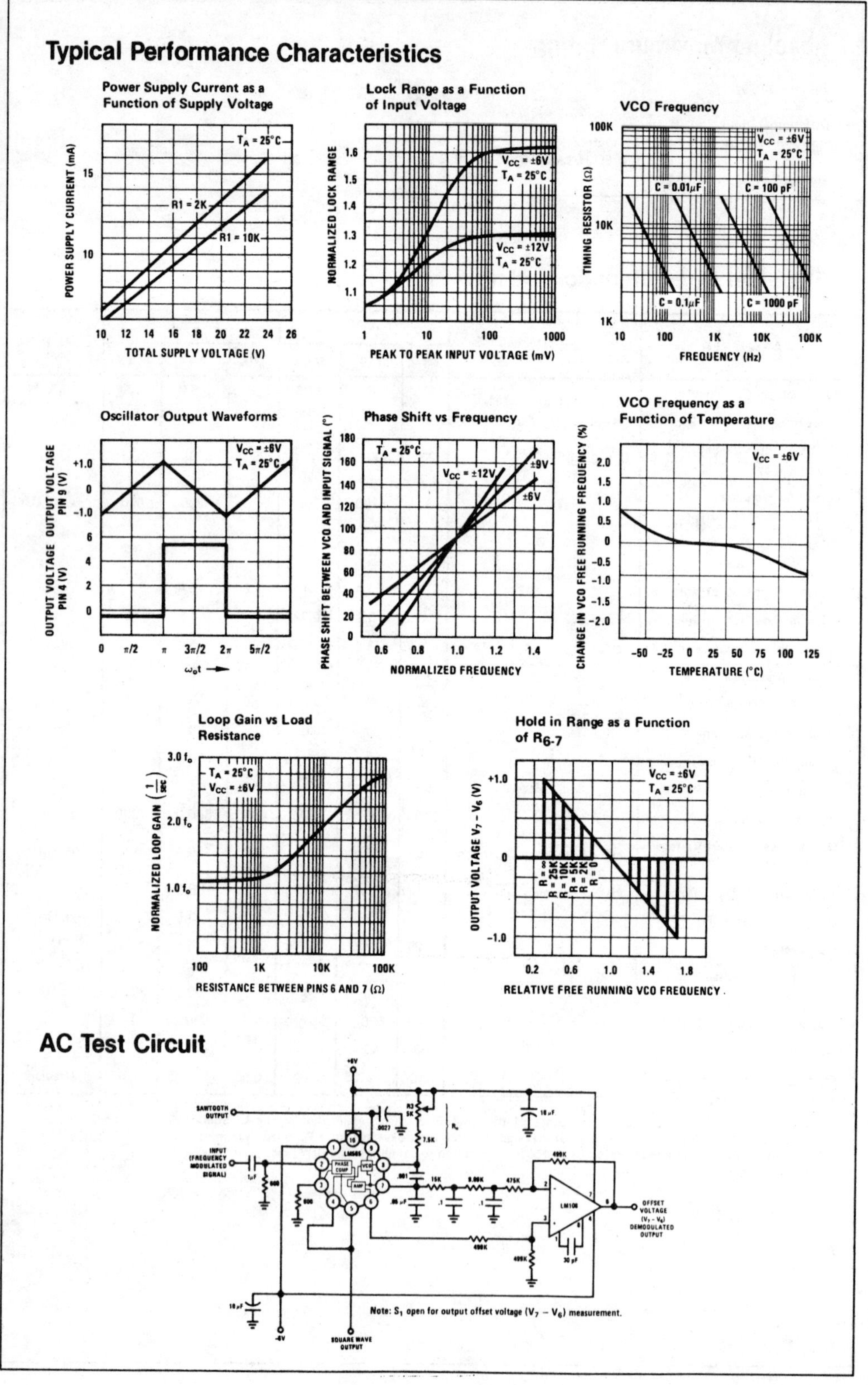
Typical Performance Characteristics
Power Supply Current as a Function of Supply Voltage
T_A = 25°C
R1 = 2K
R1 = 10K
POWER SUPPLY CURRENT (mA)
15
10
10 12 14 16 18 20 22 24 26
TOTAL SUPPLY VOLTAGE (V)
Lock Range as a Function of Input Voltage
V_{CC} = ±6V T_A = 25°C
V_{CC} = ±12V T_A = 25°C
NORMALIZED LOCK RANGE
1.6 1.5 1.4 1.3 1.2 1.1
1 10 100 1000
PEAK TO PEAK INPUT VOLTAGE (mV)
VCO Frequency
V_{CC} = ±6V T_A = 25°C
C = 0.01μF
C = 100 pF
C = 0.1μF
C = 1000 pF
TIMING RESISTOR (Ω)
100K 10K 1K
10 100 1K 10K 100K
FREQUENCY (Hz)
Oscillator Output Waveforms
V_{CC} = ±6V T_A = 25°C
OUTPUT VOLTAGE PIN 9 (V)
+1.0 0 -1.0
OUTPUT VOLTAGE PIN 4 (V)
6 4 2 0
0 π/2 π 3π/2 2π 5π/2
$\omega_o t$
Phase Shift vs Frequency
T_A = 25°C
V_{CC} = ±12V
±9V
±6V
PHASE SHIFT BETWEEN VCO AND INPUT SIGNAL (°)
180 160 140 120 100 80 60 40 20 0
0.6 0.8 1.0 1.2 1.4
NORMALIZED FREQUENCY
VCO Frequency as a Function of Temperature
V_{CC} = ±6V
CHANGE IN VCO FREE RUNNING FREQUENCY (%)
2.0 1.5 1.0 0.5 0 -0.5 -1.0 -1.5 -2.0
-50 -25 0 25 50 75 100 125
TEMPERATURE (°C)
Loop Gain vs Load Resistance
T_A = 25°C
V_{CC} = ±6V
NORMALIZED LOOP GAIN (1/sec)
3.0 f_o
2.0 f_o
1.0 f_o
100 1K 10K 100K
RESISTANCE BETWEEN PINS 6 AND 7 (Ω)
Hold in Range as a Function of R_{6-7}
V_{CC} = ±6V T_A = 25°C
OUTPUT VOLTAGE $V_7 - V_6$ (V)
+1.0 0 -1.0
R = ∞
R = 25K
R = 10K
R = 5K
R = 2K
R = 0
0.2 0.6 1.0 1.4 1.8
RELATIVE FREE RUNNING VCO FREQUENCY
AC Test Circuit
+6V
SAWTOOTH OUTPUT
INPUT (FREQUENCY MODULATED SIGNAL)
LM565
PHASE COMP
VCO
AMP
LM108
OFFSET VOLTAGE ($V_7 - V_6$) DEMODULATED OUTPUT
SQUARE WAVE OUTPUT
-6V
Note: S_1 open for output offset voltage ($V_7 - V_6$) measurement.

Typical Applications

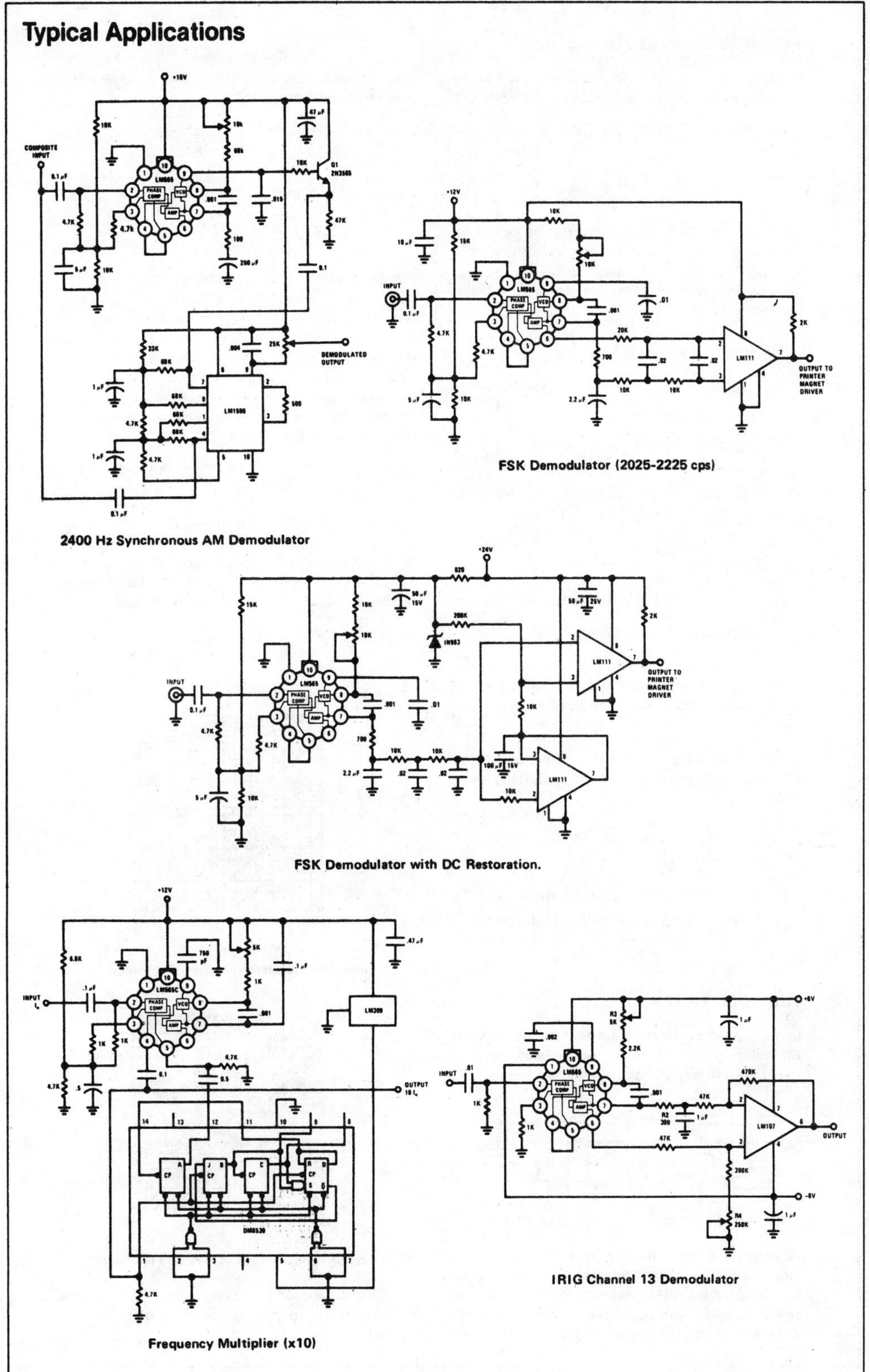

2400 Hz Synchronous AM Demodulator

FSK Demodulator (2025-2225 cps)

FSK Demodulator with DC Restoration.

Frequency Multiplier (x10)

IRIG Channel 13 Demodulator

Applications Information

In designing with phase locked loops such as the LM565, the important parameters of interest are:

FREE RUNNING FREQUENCY

$$f_o \cong \frac{1}{3.7\, R_O C_O}$$

LOOP GAIN: relates the amount of phase change between the input signal and the VCO signal for a shift in input signal frequency (assuming the loop remains in lock). In servo theory, this is called the "velocity error coefficient".

$$\text{Loop gain} = K_o K_D \left(\frac{1}{\text{sec}}\right)$$

$$K_o = \text{oscillator sensitivity} \left(\frac{\text{radians/sec}}{\text{volt}}\right)$$

$$K_D = \text{phase detector sensitivity} \left(\frac{\text{volts}}{\text{radian}}\right)$$

The loop gain of the LM565 is dependent on supply voltage, and may be found from:

$$K_o K_D = \frac{33.6\, f_o}{V_c}$$

f_o = VCO frequency in Hz

V_c = total supply voltage to circuit.

Loop gain may be reduced by connecting a resistor between pins 6 and 7; this reduces the load impedance on the output amplifier and hence the loop gain.

HOLD IN RANGE: the range of frequencies that the loop will remain in lock after initially being locked.

$$f_H = \pm \frac{8\, f_o}{V_c}$$

f_o = free running frequency of VCO

V_c = total supply voltage to the circuit.

THE LOOP FILTER

In almost all applications, it will be desirable to filter the signal at the output of the phase detector (pin 7) this filter may take one of two forms:

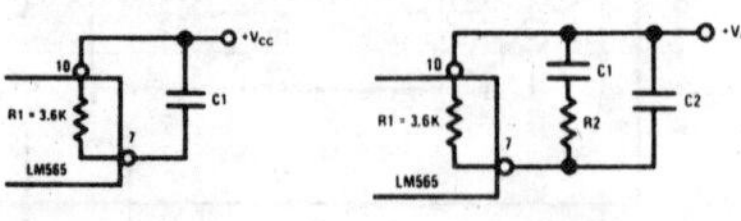

Simple Lag Filter **Lag-Lead Filter**

A simple lag filter may be used for wide closed loop bandwidth applications such as modulation following where the frequency deviation of the carrier is fairly high (greater than 10%), or where wideband modulating signals must be followed.

The natural bandwidth of the closed loop response may be found from:

$$f_n = \frac{1}{2\pi}\sqrt{\frac{K_o K_D}{R_1 C_1}}$$

Associated with this is a damping factor:

$$\delta = \frac{1}{2}\sqrt{\frac{1}{R_1 C_1 K_o K_D}}$$

For narrow band applications where a narrow noise bandwidth is desired, such as applications involving tracking a slowly varying carrier, a lead lag filter should be used. In general, if $1/R_1C_1 < K_oK_d$, the damping factor for the loop becomes quite small resulting in large overshoot and possible instability in the transient response of the loop. In this case, the natural frequency of the loop may be found from

$$f_n = \frac{1}{2\pi}\sqrt{\frac{K_o K_D}{\tau_1 + \tau_2}}$$

$$\tau_1 + \tau_2 = (R_1 + R_2)\, C_1$$

R_2 is selected to produce a desired damping factor δ, usually between 0.5 and 1.0. The damping factor is found from the approximation:

$$\delta \simeq \pi \tau_2 f_n$$

These two equations are plotted for convenience.

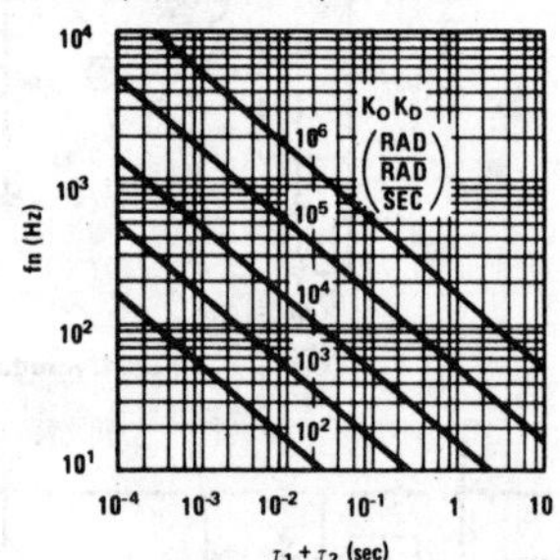

Filter Time Constant vs Natural Frequency

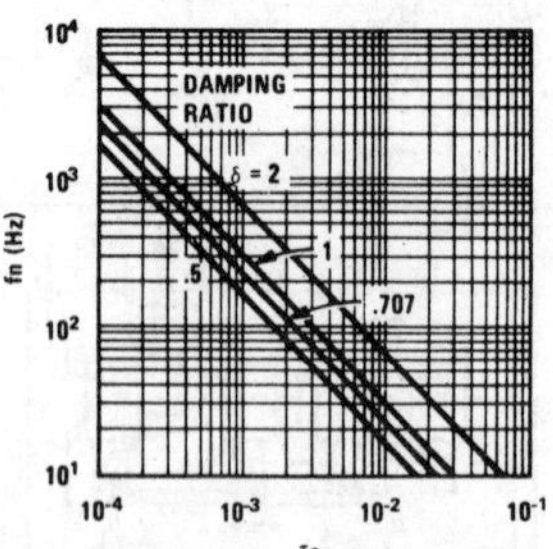

Damping Time Constant vs Natural Frequency

Capacitor C_2 should be much smaller than C_1 since its function is to provide filtering of carrier. In general $C_2 \leq 0.1\, C_1$.

National Semiconductor

Industrial Blocks

LM566/LM566C

LM566/LM566C Voltage Controlled Oscillator

General Description

The LM566/LM566C are general purpose voltage controlled oscillators which may be used to generate square and triangular waves, the frequency of which is a very linear function of a control voltage. The frequency is also a function of an external resistor and capacitor.

The LM566 is specified for operation over the −55°C to +125°C military temperature range. The LM566C is specified for operation over the 0°C to +70°C temperature range.

Features

- Wide supply voltage range: 10 to 24 volts
- Very linear modulation characteristics
- High temperature stability
- Excellent supply voltage rejection
- 10 to 1 frequency range with fixed capacitor
- Frequency programmable by means of current, voltage, resistor or capacitor.

Applications

- FM modulation
- Signal generation
- Function generation
- Frequency shift keying
- Tone generation

Schematic and Connection Diagrams

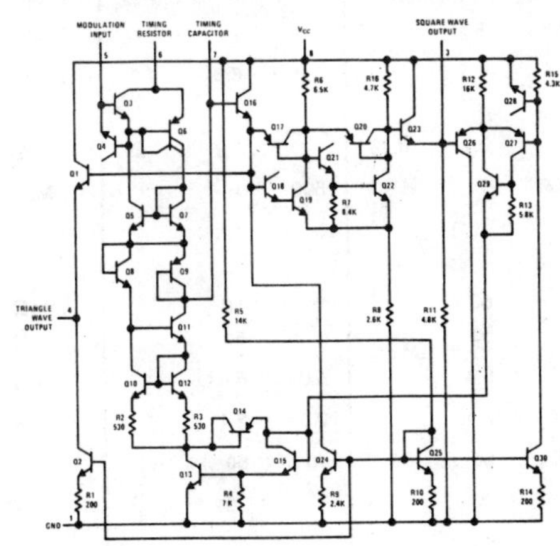

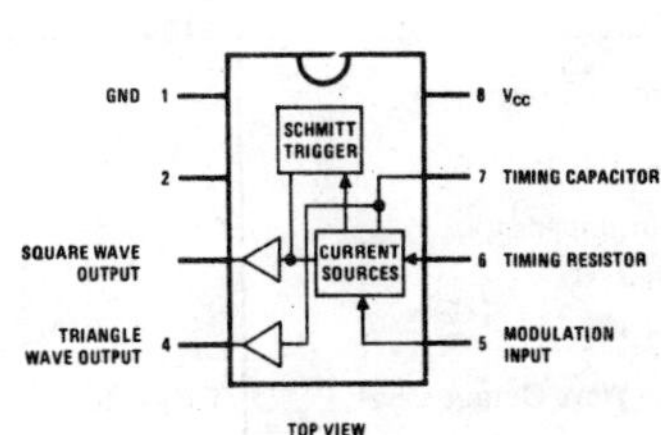

Order Number LM566CN
See NS Package N08B

Typical Application

1 kHz and 10 kHz TTL Compatible Voltage Controlled Oscillator

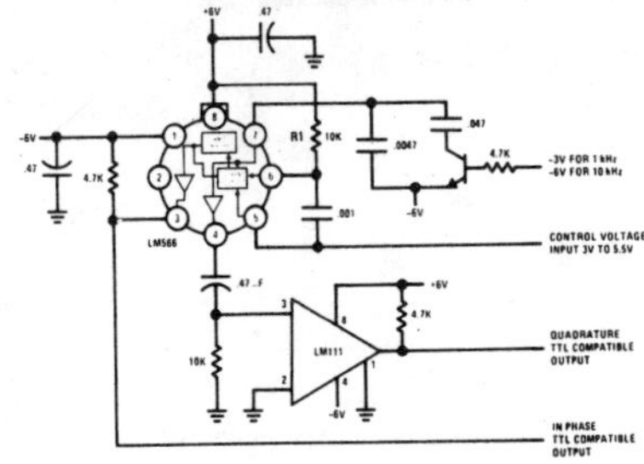

Applications Information

The LM566 may be operated from either a single supply as shown in this test circuit, or from a split (±) power supply. When operating from a split supply, the square wave output (pin 4) is TTL compatible (2 mA current sink) with the addition of a 4.7 kΩ resistor from pin 3 to ground.

A .001 μF capacitor is connected between pins 5 and 6 to prevent parasitic oscillations that may occur during VCO switching.

$$f_O = \frac{2(V^+ - V_5)}{R_1 C_1 V^+}$$

where

$2K < R_1 < 20K$

and V_5 is voltage between pin 5 and pin 1

Absolute Maximum Ratings

Power Supply Voltage		26V
Power Dissipation (Note 1)		300 mW
Operating Temperature Range	LM566	−55°C to +125°C
	LM566C	0°C to 70°C
Lead Temperature (Soldering, 10 sec)		300°C

Electrical Characteristics V_{CC} = 12V, T_A = 25°C, AC Test Circuit

PARAMETER	CONDITIONS	LM566			LM566C			UNITS
		MIN	TYP	MAX	MIN	TYP	MAX	
Maximum Operating Frequency	R0 = 2k C0 = 2.7 pF		1			1		MHz
Input Voltage Range Pin 5		3/4 V_{CC}		V_{CC}	3/4 V_{CC}		V_{CC}	
Average Temperature Coefficient of Operating Frequency			100			200		ppm/°C
Supply Voltage Rejection	10–20V		0.1	1		0.1	2	%/V
Input Impedance Pin 5		0.5	1		0.5	1		MΩ
VCO Sensitivity	For Pin 5, From 8–10V, f_O = 10 kHz	6.4	6.6	6.8	6.0	6.6	7.2	kHz/V
FM Distortion	±10% Deviation		0.2	0.75		0.2	1.5	%
Maximum Sweep Rate		800	1		500	1		MHz
Sweep Range			10:1			10:1		
Output Impedance								
Pin 3			50			50		Ω
Pin 4			50			50		Ω
Square Wave Output Level	R_{L1} = 10k	5.0	5.4		5.0	5.4		Vp-p
Triangle Wave Output Level	R_{L2} = 10k	2.0	2.4		2.0	2.4		Vp-p
Square Wave Duty Cycle		45	50	55	40	50	60	%
Square Wave Rise Time			20			20		ns
Square Wave Fall Time			50			50		ns
Triangle Wave Linearity	+1V Segment at 1/2 V_{CC}		0.2	0.75		0.5	1	%

Note 1: The maximum junction temperature of the LM566 is 150°C, while that of the LM566C is 100°C. For operating at elevated junction temperatures, devices in the TO-5 package must be derated based on a thermal resistance of 150°C/W. The thermal resistance of the dual-in-line package is 100°C/W.

Typical Performance Characteristics

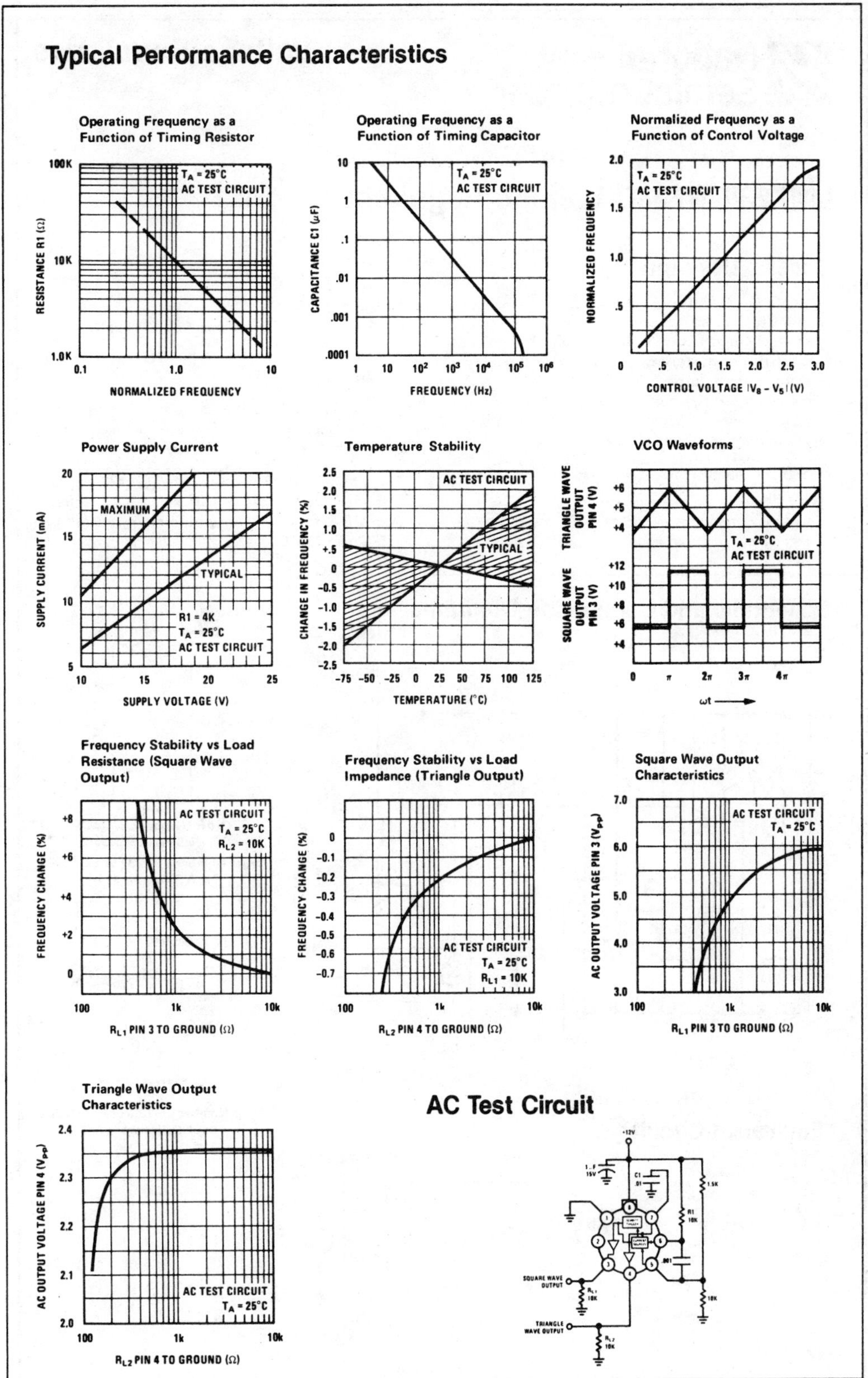

Voltage Regulators

LM723/LM723C Voltage Regulator

General Description

The LM723/LM723C is a voltage regulator designed primarily for series regulator applications. By itself, it will supply output currents up to 150 mA; but external transistors can be added to provide any desired load current. The circuit features extremely low standby current drain, and provision is made for either linear or foldback current limiting. Important characteristics are:

- 150 mA output current without external pass transistor
- Output currents in excess of 10A possible by adding external transistors
- Input voltage 40V max
- Output voltage adjustable from 2V to 37V
- Can be used as either a linear or a switching regulator.

The LM723/LM723C is also useful in a wide range of other applications such as a shunt regulator, a current regulator or a temperature controller.

The LM723C is identical to the LM723 except that the LM723C has its performance guaranteed over a 0°C to 70°C temperature range, instead of −55°C to +125°C.

Schematic and Connection Diagrams *

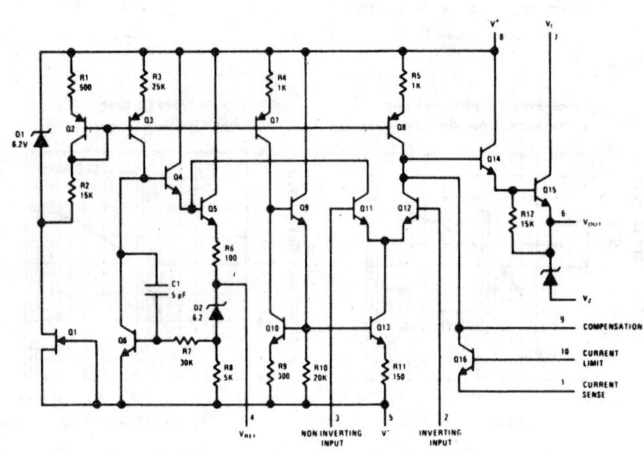

Dual-In-Line Package

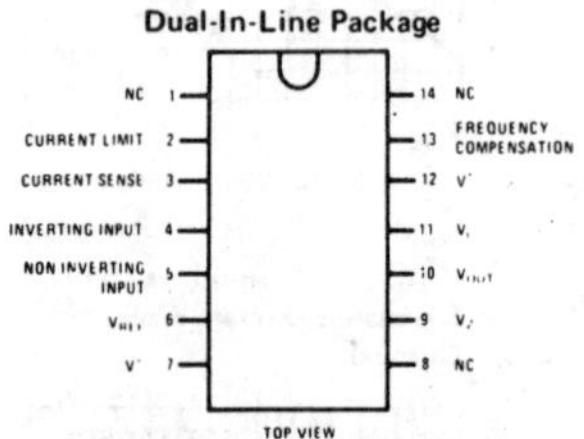

Order Number LM723CN
See NS Package N14A
Order Number LM723J or LM723CJ
See NS Package J14A

Metal Can Package

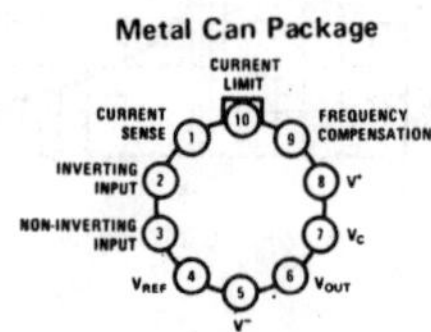

Note: Pin 5 connected to case.

TOP VIEW

Order Number LM723H or LM723CH
See NS Package H10C

Equivalent Circuit *

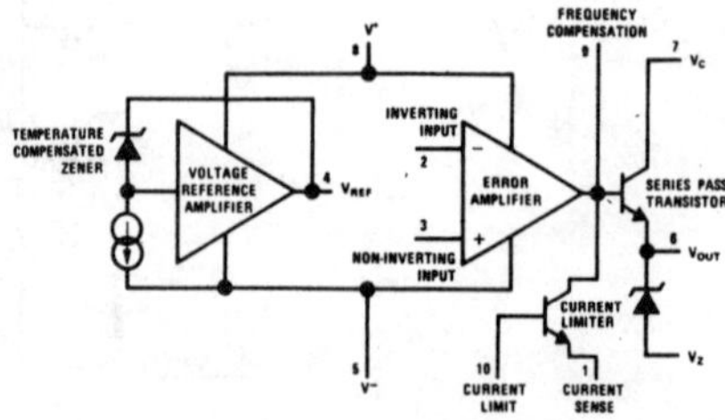

*Pin numbers refer to metal can package.

Absolute Maximum Ratings

Pulse Voltage from V^+ to V^- (50 ms)	50V
Continuous Voltage from V^+ to V^-	40V
Input-Output Voltage Differential	40V
Maximum Amplifier Input Voltage (Either Input)	7.5V
Maximum Amplifier Input Voltage (Differential)	5V
Current from V_Z	25 mA
Current from V_{REF}	15 mA
Internal Power Dissipation Metal Can (Note 1)	800 mW
Cavity DIP (Note 1)	900 mW
Molded DIP (Note 1)	660 mW
Operating Temperature Range LM723	-55°C to +125°C
LM723C	0°C to +70°C
Storage Temperature Range Metal Can	-65°C to +150°C
DIP	-55°C to +125°C
Lead Temperature (Soldering, 10 sec)	300°C

Electrical Characteristics (Note 2)

PARAMETER	CONDITIONS	LM723			LM723C			UNITS
		MIN	TYP	MAX	MIN	TYP	MAX	
Line Regulation	V_{IN} = 12V to V_{IN} = 15V		.01	0.1		.01	0.1	% V_{OUT}
	$-55°C \leq T_A \leq +125°C$			0.3				% V_{OUT}
	$0°C \leq T_A \leq +70°C$						0.3	% V_{OUT}
	V_{IN} = 12V to V_{IN} = 40V		.02	0.2		0.1	0.5	% V_{OUT}
Load Regulation	I_L = 1 mA to I_L = 50 mA		.03	0.15		.03	0.2	% V_{OUT}
	$-55°C \leq T_A \leq +125°C$			0.6				% V_{OUT}
	$0°C \leq T_A \leq = +70°C$						0.6	% V_{OUT}
Ripple Rejection	f = 50 Hz to 10 kHz, C_{REF} = 0		74			74		dB
	f = 50 Hz to 10 kHz, C_{REF} = 5 μF		86			86		dB
Average Temperature Coefficient of Output Voltage	$-55°C \leq T_A \leq +125°C$		.002	.015				%/°C
	$0°C \leq T_A \leq +70°C$					.003	.015	%/°C
Short Circuit Current Limit	R_{SC} = 10Ω, V_{OUT} = 0		65			65		mA
Reference Voltage		6.95	7.15	7.35	6.80	7.15	7.50	V
Output Noise Voltage	BW = 100 Hz to 10 kHz, C_{REF} = 0		20			20		μVrms
	BW = 100 Hz to 10 kHz, C_{REF} = 5 μF		2.5			2.5		μVrms
Long Term Stability			0.1			0.1		%/1000 hrs
Standby Current Drain	I_L = 0, V_{IN} = 30V		1.3	3.5		1.3	4.0	mA
Input Voltage Range		9.5		40	9.5		40	V
Output Voltage Range		2.0		37	2.0		37	V
Input-Output Voltage Differential		3.0		38	3.0		38	V

Note 1: See derating curves for maximum power rating above 25°C.

Note 2: Unless otherwise specified, T_A = 25°C, $V_{IN} = V^+ = V_C$ = 12V, V^- = 0, V_{OUT} = 5V, I_L = 1 mA, R_{SC} = 0, C_1 = 100 pF, C_{REF} = 0 and divider impedance as seen by error amplifier ≤ 10 kΩ connected as shown in Figure 1. Line and load regulation specifications are given for the condition of constant chip temperature. Temperature drifts must be taken into account separately for high dissipation conditions.

Note 3: L_1 is 40 turns of No. 20 enameled copper wire wound on Ferroxcube P36/22-3B7 pot core or equivalent with 0.009 in. air gap.

Note 4: Figures in parentheses may be used if R1/R2 divider is placed on opposite input of error amp.

Note 5: Replace R1/R2 in figures with divider shown in Figure 13.

Note 6: V^+ must be connected to a +3V or greater supply.

Note 7: For metal can applications where V_Z is required, an external 6.2 volt zener diode should be connected in series with V_{OUT}.

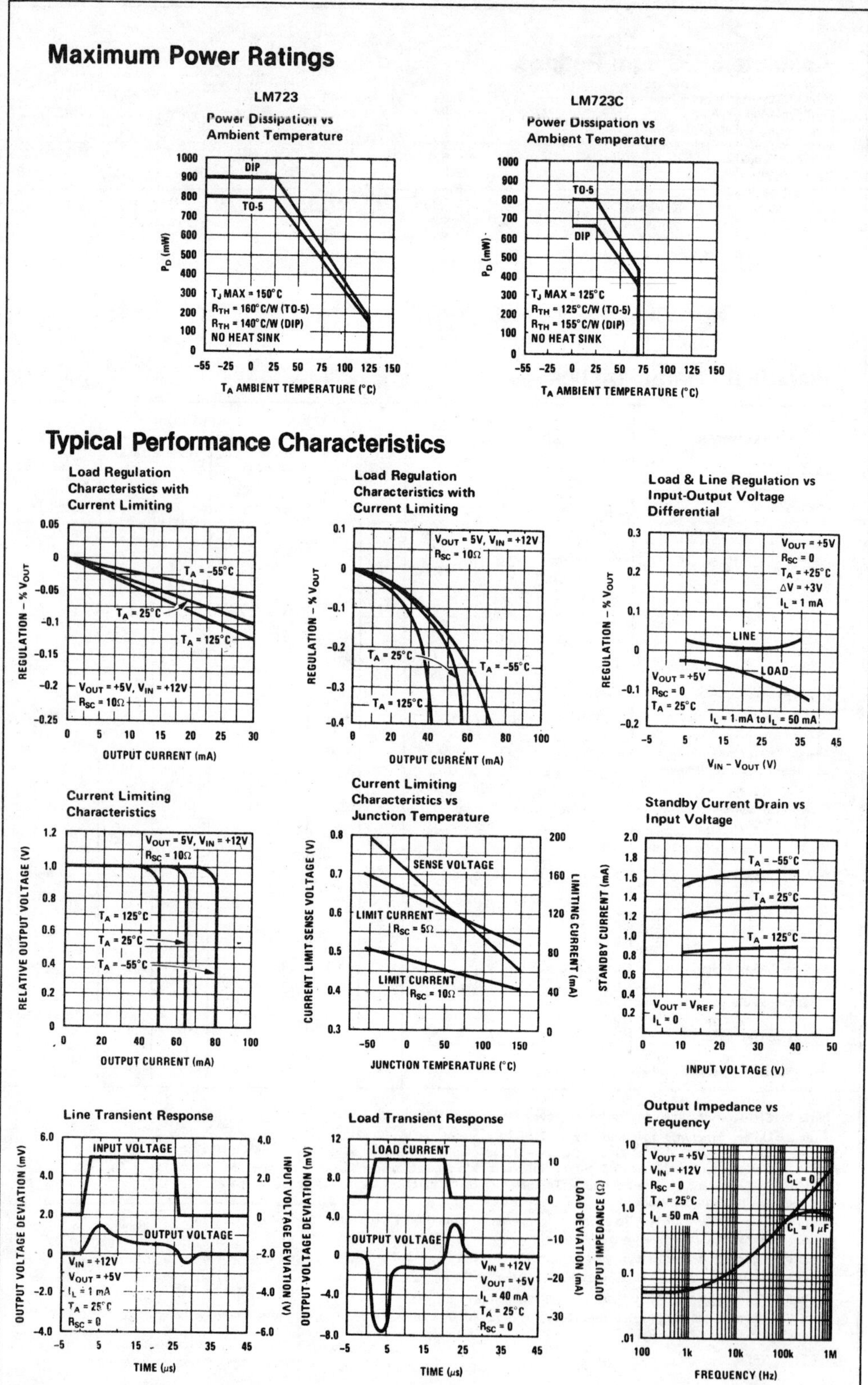
Maximum Power Ratings
LM723
Power Dissipation vs Ambient Temperature
DIP
TO-5
P_D (mW)
T_J MAX = 150°C
R_{TH} = 160°C/W (TO-5)
R_{TH} = 140°C/W (DIP)
NO HEAT SINK
T_A AMBIENT TEMPERATURE (°C)
LM723C
Power Dissipation vs Ambient Temperature
TO-5
DIP
T_J MAX = 125°C
R_{TH} = 125°C/W (TO-5)
R_{TH} = 155°C/W (DIP)
NO HEAT SINK
T_A AMBIENT TEMPERATURE (°C)
Typical Performance Characteristics
Load Regulation Characteristics with Current Limiting
T_A = −55°C
T_A = 25°C
T_A = 125°C
V_{OUT} = +5V, V_{IN} = +12V
R_{SC} = 10Ω
REGULATION – % V_{OUT}
OUTPUT CURRENT (mA)
Load Regulation Characteristics with Current Limiting
V_{OUT} = 5V, V_{IN} = +12V
R_{SC} = 10Ω
T_A = 25°C
T_A = −55°C
T_A = 125°C
REGULATION – % V_{OUT}
OUTPUT CURRENT (mA)
Load & Line Regulation vs Input-Output Voltage Differential
V_{OUT} = +5V
R_{SC} = 0
T_A = +25°C
ΔV = +3V
I_L = 1 mA
LINE
LOAD
V_{OUT} = +5V
R_{SC} = 0
T_A = 25°C
I_L = 1 mA to I_L = 50 mA
REGULATION – % V_{OUT}
V_{IN} – V_{OUT} (V)
Current Limiting Characteristics
V_{OUT} = 5V, V_{IN} = +12V
R_{SC} = 10Ω
T_A = 125°C
T_A = 25°C
T_A = −55°C
RELATIVE OUTPUT VOLTAGE (V)
OUTPUT CURRENT (mA)
Current Limiting Characteristics vs Junction Temperature
SENSE VOLTAGE
LIMIT CURRENT
R_{SC} = 5Ω
LIMIT CURRENT
R_{SC} = 10Ω
CURRENT LIMIT SENSE VOLTAGE (V)
LIMITING CURRENT (mA)
JUNCTION TEMPERATURE (°C)
Standby Current Drain vs Input Voltage
T_A = −55°C
T_A = 25°C
T_A = 125°C
V_{OUT} = V_{REF}
I_L = 0
STANDBY CURRENT (mA)
INPUT VOLTAGE (V)
Line Transient Response
INPUT VOLTAGE
OUTPUT VOLTAGE
V_{IN} = +12V
V_{OUT} = +5V
I_L = 1 mA
T_A = 25°C
R_{SC} = 0
OUTPUT VOLTAGE DEVIATION (mV)
INPUT VOLTAGE DEVIATION (V)
TIME (μs)
Load Transient Response
LOAD CURRENT
OUTPUT VOLTAGE
V_{IN} = +12V
V_{OUT} = +5V
I_L = 40 mA
T_A = 25°C
R_{SC} = 0
OUTPUT VOLTAGE DEVIATION (mV)
LOAD DEVIATION (mA)
TIME (μs)
Output Impedance vs Frequency
V_{OUT} = +5V
V_{IN} = +12V
R_{SC} = 0
T_A = 25°C
I_L = 50 mA
C_L = 0
C_L = 1 μF
OUTPUT IMPEDANCE (Ω)
FREQUENCY (Hz)

TABLE I RESISTOR VALUES (kΩ) FOR STANDARD OUTPUT VOLTAGE

POSITIVE OUTPUT VOLTAGE	APPLICABLE FIGURES	FIXED OUTPUT ±5%		OUTPUT ADJUSTABLE ±10% (Note 5)			NEGATIVE OUTPUT VOLTAGE	APPLICABLE FIGURES	FIXED OUTPUT ±5%		5% OUTPUT ADJUSTABLE ±10%		
	(Note 4)	R1	R2	R1	P1	R2			R1	R2	R1	P1	R2
+3.0	1, 5, 6, 9, 12 (4)	4.12	3.01	1.8	0.5	1.2	+100	7	3.57	102	2.2	10	91
+3.6	1, 5, 6, 9, 12 (4)	3.57	3.65	1.5	0.5	1.5	+250	7	3.57	255	2.2	10	240
+5.0	1, 5, 6, 9, 12 (4)	2.15	4.99	.75	0.5	2.2	-6 (Note 6)	3, (10)	3.57	2.43	1.2	0.5	.75
+6.0	1, 5, 6, 9, 12 (4)	1.15	6.04	0.5	0.5	2.7	-9	3, 10	3.48	5.36	1.2	0.5	2.0
+9.0	2, 4, (5, 6, 12, 9)	1.87	7.15	.75	1.0	2.7	-12	3, 10	3.57	8.45	1.2	0.5	3.3
+12	2, 4, (5, 6, 9, 12)	4.87	7.15	2.0	1.0	3.0	-15	3, 10	3.65	11.5	1.2	0.5	4.3
+15	2, 4, (5, 6, 9, 12)	7.87	7.15	3.3	1.0	3.0	-28	3, 10	3.57	24.3	1.2	0.5	10
+28	2, 4, (5, 6, 9, 12)	21.0	7.15	5.6	1.0	2.0	-45	8	3.57	41.2	2.2	10	33
+45	7	3.57	48.7	2.2	10	39	-100	8	3.57	97.6	2.2	10	91
+75	7	3.57	78.7	2.2	10	68	-250	8	3.57	249	2.2	10	240

TABLE II FORMULAE FOR INTERMEDIATE OUTPUT VOLTAGES

Outputs from +2 to +7 volts [Figures 1, 5, 6, 9, 12, (4)] $V_{OUT} = \left[V_{REF} \times \frac{R2}{R1 + R2}\right]$	**Outputs from +4 to +250 volts** [Figure 7] $V_{OUT} = \left[\frac{V_{REF}}{2} \times \frac{R2 - R1}{R1}\right]$; R3 = R4	**Current Limiting** $I_{LIMIT} = \frac{V_{SENSE}}{R_{SC}}$
Outputs from +7 to +37 volts [Figures 2, 4, (5, 6, 9, 12)] $V_{OUT} = \left[V_{REF} \times \frac{R1 + R2}{R2}\right]$	**Outputs from -6 to -250 volts** [Figures 3, 8, 10] $V_{OUT} = \left[\frac{V_{REF}}{2} \times \frac{R1 + R2}{R1}\right]$; R3 = R4	**Foldback Current Limiting** $I_{KNEE} = \left[\frac{V_{OUT}\,R3}{R_{SC}\,R4} + \frac{V_{SENSE}\,(R3 + R4)}{R_{SC}\,R4}\right]$ $I_{SHORT\,CKT} = \left[\frac{V_{SENSE}}{R_{SC}} \times \frac{R3 + R4}{R4}\right]$

Typical Applications

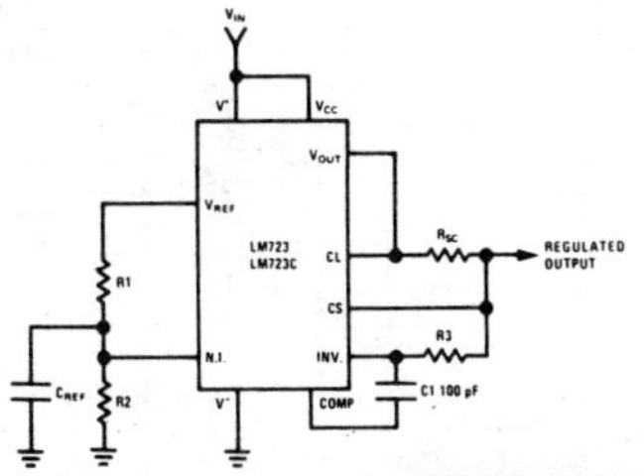

Note: $R3 = \frac{R1\,R2}{R1 + R2}$ for minimum temperature drift.

TYPICAL PERFORMANCE

Regulated Output Voltage	5V
Line Regulation (ΔV_{IN} = 3V)	0.5 mV
Load Regulation (ΔI_L = 50 mA)	1.5 mV

FIGURE 1. Basic Low Voltage Regulator (V_{OUT} = 2 to 7 Volts)

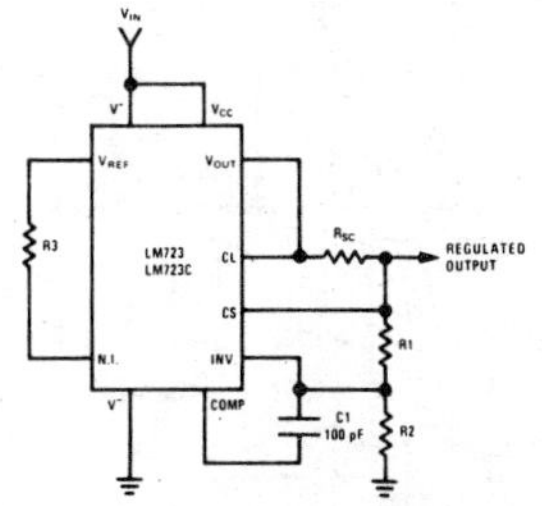

Note: $R3 = \frac{R1\,R2}{R1 + R2}$ for minimum temperature drift.

R3 may be eliminated for minimum component count.

TYPICAL PERFORMANCE

Regulated Output Voltage	15V
Line Regulation (ΔV_{IN} = 3V)	1.5 mV
Load Regulation (ΔI_L = 50 mA)	4.5 mV

FIGURE 2. Basic High Voltage Regulator (V_{OUT} = 7 to 37 Volts)

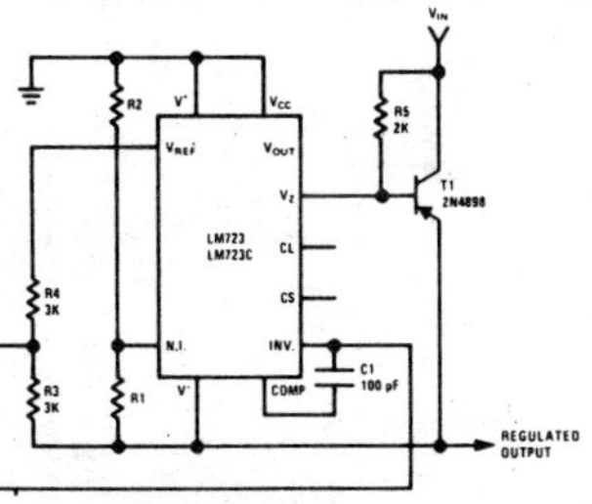

TYPICAL PERFORMANCE

Regulated Output Voltage	-15V
Line Regulation (ΔV_{IN} = 3V)	1 mV
Load Regulation (ΔI_L = 100 mA)	2 mV

FIGURE 3. Negative Voltage Regulator

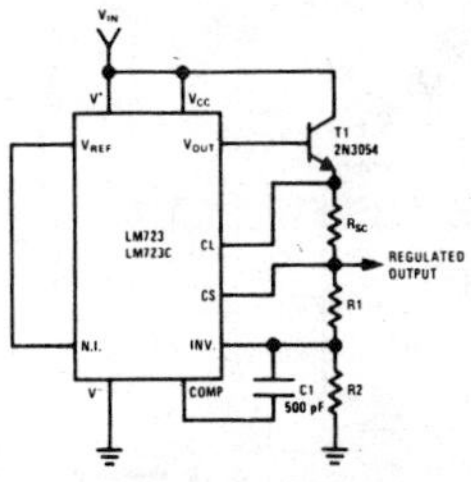

TYPICAL PERFORMANCE

Regulated Output Voltage	+15V
Line Regulation (ΔV_{IN} = 3V)	1.5 mV
Load Regulation (ΔI_L = 1A)	15 mV

FIGURE 4. Positive Voltage Regulator (External NPN Pass Transistor)

Typical Applications (Continued)

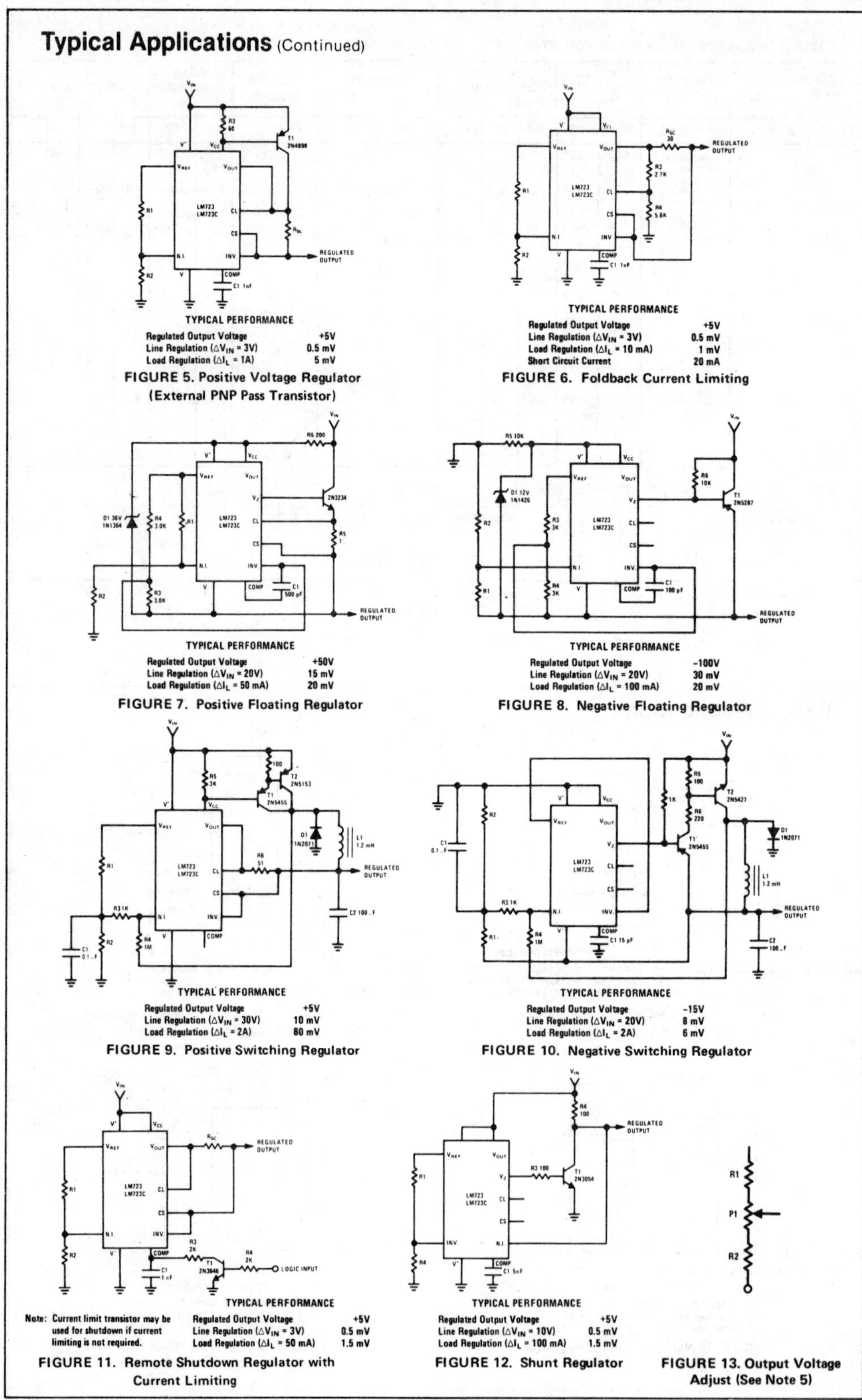

TYPICAL PERFORMANCE

Regulated Output Voltage	+5V
Line Regulation (ΔV_{IN} = 3V)	0.5 mV
Load Regulation (ΔI_L = 1A)	5 mV

FIGURE 5. Positive Voltage Regulator (External PNP Pass Transistor)

TYPICAL PERFORMANCE

Regulated Output Voltage	+5V
Line Regulation (ΔV_{IN} = 3V)	0.5 mV
Load Regulation (ΔI_L = 10 mA)	1 mV
Short Circuit Current	20 mA

FIGURE 6. Foldback Current Limiting

TYPICAL PERFORMANCE

Regulated Output Voltage	+50V
Line Regulation (ΔV_{IN} = 20V)	15 mV
Load Regulation (ΔI_L = 50 mA)	20 mV

FIGURE 7. Positive Floating Regulator

TYPICAL PERFORMANCE

Regulated Output Voltage	−100V
Line Regulation (ΔV_{IN} = 20V)	30 mV
Load Regulation (ΔI_L = 100 mA)	20 mV

FIGURE 8. Negative Floating Regulator

TYPICAL PERFORMANCE

Regulated Output Voltage	+5V
Line Regulation (ΔV_{IN} = 30V)	10 mV
Load Regulation (ΔI_L = 2A)	80 mV

FIGURE 9. Positive Switching Regulator

TYPICAL PERFORMANCE

Regulated Output Voltage	−15V
Line Regulation (ΔV_{IN} = 20V)	8 mV
Load Regulation (ΔI_L = 2A)	6 mV

FIGURE 10. Negative Switching Regulator

Note: Current limit transistor may be used for shutdown if current limiting is not required.

TYPICAL PERFORMANCE

Regulated Output Voltage	+5V
Line Regulation (ΔV_{IN} = 3V)	0.5 mV
Load Regulation (ΔI_L = 50 mA)	1.5 mV

FIGURE 11. Remote Shutdown Regulator with Current Limiting

TYPICAL PERFORMANCE

Regulated Output Voltage	+5V
Line Regulation (ΔV_{IN} = 10V)	0.5 mV
Load Regulation (ΔI_L = 100 mA)	1.5 mV

FIGURE 12. Shunt Regulator

FIGURE 13. Output Voltage Adjust (See Note 5)

LM741/LM741A/LM741C/LM741E Operational Amplifier

General Description

The LM741 series are general purpose operational amplifiers which feature improved performance over industry standards like the LM709. They are direct, plug-in replacements for the 709C, LM201, MC1439 and 748 in most applications.

The amplifiers offer many features which make their application nearly foolproof: overload protection on the input and output, no latch-up when the common mode range is exceeded, as well as freedom from oscillations.

The LM741C/LM741E are identical to the LM741/LM741A except that the LM741C/LM741E have their performance guaranteed over a 0°C to +70°C temperature range, instead of −55°C to +125°C.

Schematic and Connection Diagrams (Top Views)

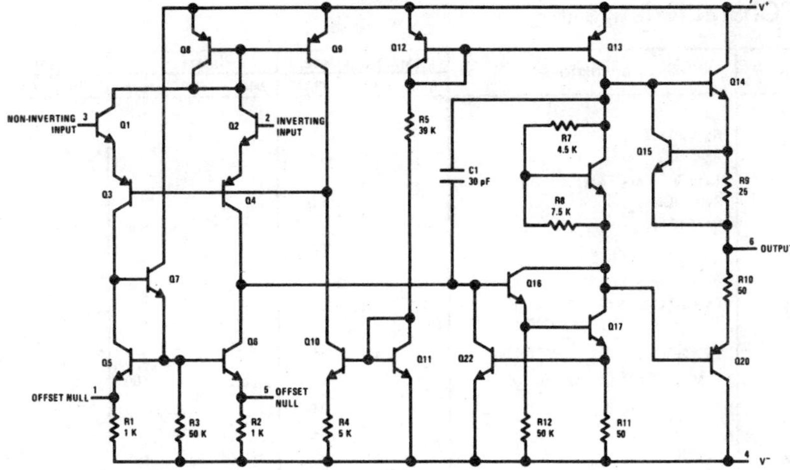

Metal Can Package

NC
OFFSET NULL
V+
INVERTING INPUT
OUTPUT
NON-INVERTING INPUT
OFFSET NULL
V−

Order Number LM741H, LM741AH, LM741CH or LM741EH
See NS Package H08C

Dual-In-Line Package

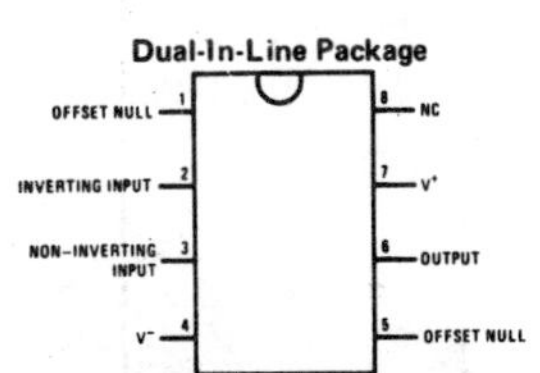

Order Number LM741CN or LM741EN
See NS Package N08B
Order Number LM741CJ
See NS Package J08A

Dual-In-Line Package

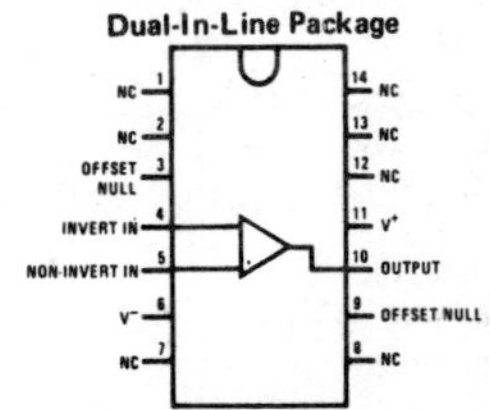

Order Number LM741CN-14
See NS Package N14A
Order Number LM741J-14, LM741AJ-14 or LM741CJ-14
See NS Package J14A

Absolute Maximum Ratings

	LM741A	LM741E	LM741	LM741C
Supply Voltage	±22V	±22V	±22V	±18V
Power Dissipation (Note 1)	500 mW	500 mW	500 mW	500 mW
Differential Input Voltage	±30V	±30V	±30V	±30V
Input Voltage (Note 2)	±15V	±15V	±15V	±15V
Output Short Circuit Duration	Indefinite	Indefinite	Indefinite	Indefinite
Operating Temperature Range	−55°C to +125°C	0°C to +70°C	−55°C to +125°C	0°C to +70°C
Storage Temperature Range	−65°C to +150°C	−65°C to +150°C	−65°C to +150°C	−65°C to +150°C
Lead Temperature (Soldering, 10 seconds)	300°C	300°C	300°C	300°C

Electrical Characteristics (Note 3)

PARAMETER	CONDITIONS	LM741A/LM741E			LM741			LM741C			UNITS
		MIN	TYP	MAX	MIN	TYP	MAX	MIN	TYP	MAX	
Input Offset Voltage	T_A = 25°C										
	$R_S \leq 10\ k\Omega$					1.0	5.0		2.0	6.0	mV
	$R_S \leq 50\Omega$		0.8	3.0							mV
	$T_{AMIN} \leq T_A \leq T_{AMAX}$										
	$R_S \leq 50\Omega$			4.0							mV
	$R_S \leq 10\ k\Omega$						6.0			7.5	mV
Average Input Offset Voltage Drift				15							μV/°C
Input Offset Voltage Adjustment Range	T_A = 25°C, V_S = ±20V	±10				±15			±15		mV
Input Offset Current	T_A = 25°C		3.0	30		20	200		20	200	nA
	$T_{AMIN} \leq T_A \leq T_{AMAX}$			70		85	500			300	nA
Average Input Offset Current Drift				0.5							nA/°C
Input Bias Current	T_A = 25°C		30	80		80	500		80	500	nA
	$T_{AMIN} \leq T_A \leq T_{AMAX}$			0.210			1.5			0.8	μA
Input Resistance	T_A = 25°C, V_S = ±20V	1.0	6.0		0.3	2.0		0.3	2.0		MΩ
	$T_{AMIN} \leq T_A \leq T_{AMAX}$, V_S = ±20V	0.5									MΩ
nput Voltage Range	T_A = 25°C							±12	±13		V
	$T_{AMIN} \leq T_A \leq T_{AMAX}$				±12	±13					V
Large Signal Voltage Gain	T_A = 25°C, $R_L \geq 2\ k\Omega$										
	V_S = ±20V, V_O = ±15V	50									V/mV
	V_S = ±15V, V_O = ±10V				50	200		20	200		V/mV
	$T_{AMIN} \leq T_A \leq T_{AMAX}$, $R_L \geq 2\ k\Omega$,										
	V_S = ±20V, V_O = ±15V	32									V/mV
	V_S = ±15V, V_O = ±10V				25			15			V/mV
	V_S = ±5V, V_O = ±2V	10									V/mV
Output Voltage Swing	V_S = ±20V										
	$R_L \geq 10\ k\Omega$	±16									V
	$R_L \geq 2\ k\Omega$	±15									V
	V_S = ±15V										
	$R_L \geq 10\ k\Omega$				±12	±14		±12	±14		V
	$R_L \geq 2\ k\Omega$				±10	±13		±10	±13		V
Output Short Circuit Current	T_A = 25°C	10	25	35		25			25		mA
	$T_{AMIN} \leq T_A \leq T_{AMAX}$	10		40							mA
Common-Mode Rejection Ratio	$T_{AMIN} \leq T_A \leq T_{AMAX}$										
	$R_S \leq 10\ k\Omega$, V_{CM} = ±12V				70	90		70	90		dB
	$R_S \leq 50\ k\Omega$, V_{CM} = ±12V	80	95								dB

Electrical Characteristics (Continued)

PARAMETER	CONDITIONS	LM741A/LM741E			LM741			LM741C			UNITS
		MIN	TYP	MAX	MIN	TYP	MAX	MIN	TYP	MAX	
Supply Voltage Rejection Ratio	$T_{AMIN} \le T_A \le T_{AMAX}$, $V_S = \pm 20V$ to $V_S = \pm 5V$										
	$R_S \le 50\Omega$	86	96								dB
	$R_S \le 10\ k\Omega$				77	96		77	96		dB
Transient Response	$T_A = 25°C$, Unity Gain										
Rise Time			0.25	0.8		0.3			0.3		μs
Overshoot			6.0	20		5			5		%
Bandwidth (Note 4)	$T_A = 25°C$	0.437	1.5								MHz
Slew Rate	$T_A = 25°C$, Unity Gain	0.3	0.7			0.5			0.5		V/μs
Supply Current	$T_A = 25°C$					1.7	2.8		1.7	2.8	mA
Power Consumption	$T_A = 25°C$										
	$V_S = \pm 20V$		80	150							mW
	$V_S = \pm 15V$					50	85		50	85	mW
LM741A	$V_S = \pm 20V$										
	$T_A = T_{AMIN}$			165							mW
	$T_A = T_{AMAX}$			135							mW
LM741E	$V_S = \pm 20V$			150							mW
	$T_A = T_{AMIN}$			150							mW
	$T_A = T_{AMAX}$			150							mW
LM741	$V_S = \pm 15V$										
	$T_A = T_{AMIN}$					60	100				mW
	$T_A = T_{AMAX}$					45	75				mW

Note 1: The maximum junction temperature of the LM741/LM741A is 150°C, while that of the LM741C/LM741E is 100°C. For operation at elevated temperatures, devices in the TO-5 package must be derated based on a thermal resistance of 150°C/W junction to ambient, or 45°C/W junction to case. The thermal resistance of the dual-in-line package is 100°C/W junction to ambient.

Note 2: For supply voltages less than ±15V, the absolute maximum input voltage is equal to the supply voltage.

Note 3: Unless otherwise specified, these specifications apply for $V_S = \pm 15V$, $-55°C \le T_A \le +125°C$ (LM741/LM741A). For the LM741C/LM741E, these specifications are limited to $0°C \le T_A \le +70°C$.

Note 4: Calculated value from: BW (MHz) = 0.35/Rise Time(μs).

ADC0801, ADC0802, ADC0803, ADC0804, ADC0805 8-Bit μP Compatible A/D Converters

General Description

The ADC0801, ADC0802, ADC0803, ADC0804 and ADC0805 are CMOS 8-bit successive approximation A/D converters which use a differential potentiometric ladder—similar to the 256R products. These converters are designed to allow operation with the NSC800 and INS8080A derivative control bus, and TRI-STATE® output latches directly drive the data bus. These A/Ds appear like memory locations or I/O ports to the microprocessor and no interfacing logic is needed.

A new differential analog voltage input allows increasing the common-mode rejection and offsetting the analog zero input voltage value. In addition, the voltage reference input can be adjusted to allow encoding any smaller analog voltage span to the full 8 bits of resolution.

Features

- Compatible with 8080 μP derivatives—no interfacing logic needed – access time – 135 ns
- Easy interface to all microprocessors, or operates "stand alone"
- Differential analog voltage inputs
- Logic inputs and outputs meet both MOS and T^2L voltage level specifications
- Works with 2.5V (LM336) voltage reference
- On-chip clock generator
- 0V to 5V analog input voltage range with single 5V supply
- No zero adjust required
- 0.3" standard width 20-pin DIP package
- Operates ratiometrically or with 5 V_{DC}, 2.5 V_{DC}, or analog span adjusted voltage reference

Key Specifications

- Resolution: 8 bits
- Total error: ±1/4 LSB, ±1/2 LSB and ±1 LSB
- Conversion time: 100 μs

Typical Applications

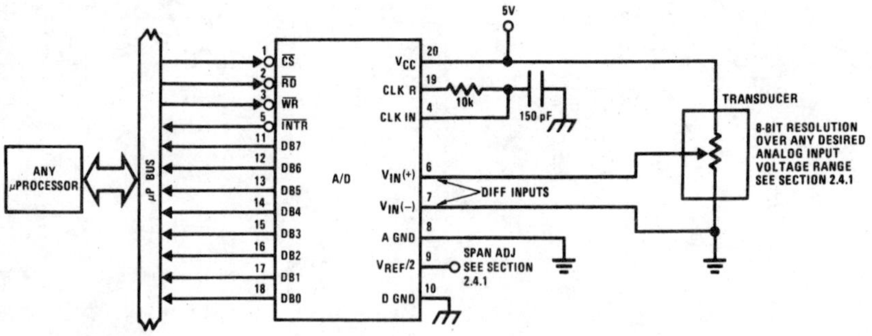

8080 Interface

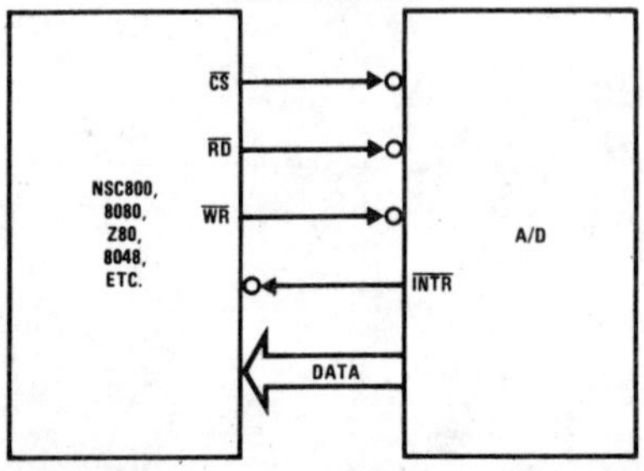

ERROR SPECIFICATION (INCLUDES FULL-SCALE, ZERO ERROR, AND NON-LINEARITY)			
PART NUMBER	FULL-SCALE ADJUSTED	$V_{REF}/2$ = 2.500 V_{DC} (NO ADJUSTMENTS)	$V_{REF}/2$ = NO CONNECTION (NO ADJUSTMENTS)
ADC0801	±1/4 LSB		
ADC0802		±1/2 LSB	
ADC0803	±1/2 LSB		
ADC0804		±1 LSB	
ADC0805			±1 LSB

Absolute Maximum Ratings (Notes 1 and 2)

Supply Voltage (V_{CC}) (Note 3)	6.5V
Voltage	
Logic Control Inputs	−0.3V to +18V
At Other Input and Outputs	−0.3V to (V_{CC} + 0.3V)
Storage Temperature Range	−65°C to +150°C
Package Dissipation at T_A = 25°C	875 mW
Lead Temperature (Soldering, 10 seconds)	300°C

Operating Ratings (Notes 1 and 2)

Temperature Range	$T_{MIN} \leq T_A \leq T_{MAX}$
ADC0801/02LD	$-55°C \leq T_A \leq +125°C$
ADC0801/02/03/04LCD	$-40°C \leq T_A \leq +85°C$
ADC0801/02/03/05LCN	$-40°C \leq T_A \leq +85°C$
ADC0804LCN	$0°C \leq T_A \leq +70°C$
Range of V_{CC}	4.5 V_{DC} to 6.3 V_{DC}

Electrical Characteristics

The following specifications apply for V_{CC} = 5 V_{DC}, $T_{MIN} \leq T_A \leq T_{MAX}$ and f_{CLK} = 640 kHz unless otherwise specified.

PARAMETER	CONDITIONS	MIN	TYP	MAX	UNITS
ADC0801: Total Adjusted Error (Note 8)	With Full-Scale Adj. (See Section 2.5.2)			±1/4	LSB
ADC0802: Total Unadjusted Error (Note 8)	$V_{REF}/2$ = 2.500 V_{DC}			±1/2	LSB
ADC0803: Total Adjusted Error (Note 8)	With Full-Scale Adj. (See Section 2.5.2)			±1/2	LSB
ADC0804: Total Unadjusted Error (Note 8)	$V_{REF}/2$ = 2.500 V_{DC}			±1	LSB
ADC0805: Total Unadjusted Error (Note 8)	$V_{REF}/2$–No Connection			±1	LSB
$V_{REF}/2$ Input Resistance (Pin 9)	ADC0801/02/03/05	2.5	8.0		kΩ
	ADC0804 (Note 9)	1.0	1.3		kΩ
Analog Input Voltage Range	(Note 4) V(+) or V(−)	Gnd−0.05		V_{CC}+0.05	V_{DC}
DC Common-Mode Error	Over Analog Input Voltage Range		±1/16	±1/8	LSB
Power Supply Sensitivity	V_{CC} = 5 V_{DC} ±10% Over Allowed $V_{IN}(+)$ and $V_{IN}(-)$ Voltage Range (Note 4)		±1/16	±1/8	LSB

AC Electrical Characteristics

The following specifications apply for V_{CC} = 5 V_{DC} and T_A = 25°C unless otherwise specified.

PARAMETER		CONDITIONS	MIN	TYP	MAX	UNITS
T_c	Conversion Time	f_{CLK} = 640 kHz (Note 6)	103		114	μs
T_c	Conversion Time	(Note 5, 6)	66		73	1/f_{CLK}
f_{CLK}	Clock Frequency	V_{CC} = 5V, (Note 5)	100	640	1460	kHz
	Clock Duty Cycle	(Note 5)	40		60	%
CR	Conversion Rate In Free-Running Mode	$\overline{INTR}$ tied to $\overline{WR}$ with $\overline{CS}$ = 0 V_{DC}, f_{CLK} = 640 kHz			8770	conv/s
$t_{W(\overline{WR})L}$	Width of $\overline{WR}$ Input (Start Pulse Width)	$\overline{CS}$ = 0 V_{DC} (Note 7)	100			ns
t_{ACC}	Access Time (Delay from Falling Edge of $\overline{RD}$ to Output Data Valid)	C_L = 100 pF		135	200	ns
t_{1H}, t_{0H}	TRI-STATE Control (Delay from Rising Edge of $\overline{RD}$ to Hi-Z State)	C_L = 10 pF, R_L = 10k (See TRI-STATE Test Circuits)		125	200	ns
t_{WI}, t_{RI}	Delay from Falling Edge of $\overline{WR}$ or $\overline{RD}$ to Reset of $\overline{INTR}$			300	450	ns
C_{IN}	Input Capacitance of Logic Control Inputs			5	7.5	pF
C_{OUT}	TRI-STATE Output Capacitance (Data Buffers)			5	7.5	pF

Electrical Characteristics

The following specifications apply for $V_{CC} = 5\ V_{DC}$ and $T_{MIN} \leq T_A \leq T_{MAX}$, unless otherwise specified.

PARAMETER		CONDITIONS	MIN	TYP	MAX	UNITS
CONTROL INPUTS [Note: CLK IN (Pin 4) is the input of a Schmitt trigger circuit and is therefore specified separately]						
V_{IN} (1)	Logical "1" Input Voltage (Except Pin 4 CLK IN)	$V_{CC} = 5.25\ V_{DC}$	2.0		15	V_{DC}
V_{IN} (0)	Logical "0" Input Voltage (Except Pin 4 CLK IN)	$V_{CC} = 4.75\ V_{DC}$			0.8	V_{DC}
I_{IN} (1)	Logical "1" Input Current (All Inputs)	$V_{IN} = 5\ V_{DC}$		0.005	1	μA_{DC}
I_{IN} (0)	Logical "0" Input Current (All Inputs)	$V_{IN} = 0\ V_{DC}$	−1	−0.005		μA_{DC}
CLOCK IN AND CLOCK R						
V_{T+}	CLK IN (Pin 4) Positive Going Threshold Voltage		2.7	3.1	3.5	V_{DC}
V_{T-}	CLK IN (Pin 4) Negative Going Threshold Voltage		1.5	1.8	2.1	V_{DC}
V_H	CLK IN (Pin 4) Hysteresis $(V_{T+}) - (V_{T-})$		0.6	1.3	2.0	V_{DC}
V_{OUT} (0)	Logical "0" CLK R Output Voltage	$I_O = 360\ \mu A$ $V_{CC} = 4.75\ V_{DC}$			0.4	V_{DC}
V_{OUT} (1)	Logical "1" CLK R Output Voltage	$I_O = -360\ \mu A$ $V_{CC} = 4.75\ V_{DC}$	2.4			V_{DC}
DATA OUTPUTS AND $\overline{INTR}$						
V_{OUT}(0)	Logical "0" Output Voltage					
	Data Outputs	$I_{OUT} = 1.6$ mA, $V_{CC} = 4.75\ V_{DC}$			0.4	V_{DC}
	$\overline{INTR}$ Output	$I_{OUT} = 1.0$ mA, $V_{CC} = 4.75\ V_{DC}$			0.4	V_{DC}
V_{OUT} (1)	Logical "1" Output Voltage	$I_O = -360\ \mu A$, $V_{CC} = 4.75\ V_{DC}$	2.4			V_{DC}
V_{OUT} (1)	Logical "1" Output Voltage	$I_O = -10\ \mu A$, $V_{CC} = 4.75\ V_{DC}$	4.5			V_{DC}
I_{OUT}	TRI-STATE Disabled Output Leakage (All Data Buffers)	$V_{OUT} = 0\ V_{DC}$	−3			μA_{DC}
		$V_{OUT} = 5\ V_{DC}$			3	μA_{DC}
I_{SOURCE}		V_{OUT} Short to Gnd, $T_A = 25°C$	4.5	6		mA_{DC}
I_{SINK}		V_{OUT} Short to V_{CC}, $T_A = 25°C$	9.0	16		mA_{DC}
POWER SUPPLY						
I_{CC}	Supply Current (Includes Ladder Current)	$f_{CLK} = 640$ kHz, $V_{REF}/2 = NC$, $T_A = 25°C$ and $\overline{CS}$ = "1"				
		ADC0801/02/03/05		1.1	1.8	mA
		ADC0804 (Note 9)		1.9	2.5	mA

Note 1: Absolute maximum ratings are those values beyond which the life of the device may be impaired.

Note 2: All voltages are measured with respect to Gnd, unless otherwise specified. The separate A Gnd point should always be wired to the D Gnd.

Note 3: A zener diode exists, internally, from V_{CC} to Gnd and has a typical breakdown voltage of 7 V_{DC}.

Note 4: For $V_{IN}(-) \geq V_{IN}(+)$ the digital output code will be 0000 0000. Two on-chip diodes are tied to each analog input (see block diagram) which will forward conduct for analog input voltages one diode drop below ground or one diode drop greater than the V_{CC} supply. Be careful, during testing at low V_{CC} levels (4.5V), as high level analog inputs (5V) can cause this input diode to conduct—especially at elevated temperatures, and cause errors for analog inputs near full-scale. The spec allows 50 mV forward bias of either diode. This means that as long as the analog V_{IN} does not exceed the supply voltage by more than 50 mV, the output code will be correct. To achieve an absolute 0 V_{DC} to 5 V_{DC} input voltage range will therefore require a minimum supply voltage of 4.950 V_{DC} over temperature variations, initial tolerance and loading.

Note 5: Accuracy is guaranteed at f_{CLK} = 640 kHz. At higher clock frequencies accuracy can degrade. For lower clock frequencies, the duty cycle limits can be extended so long as the minimum clock high time interval or minimum clock low time interval is no less than 275 ns.

Note 6: With an asynchronous start pulse, up to 8 clock periods may be required before the internal clock phases are proper to start the conversion process. The start request is internally latched, see *Figure 2* and section 2.0.

Note 7: The $\overline{CS}$ input is assumed to bracket the $\overline{WR}$ strobe input and therefore timing is dependent on the $\overline{WR}$ pulse width. An arbitrarily wide pulse width will hold the converter in a reset mode and the start of conversion is initiated by the low to high transition of the $\overline{WR}$ pulse (see timing diagrams).

Note 8: None of these A/Ds requires a zero adjust (see section 2.5.1). To obtain zero code at other analog input voltages see section 2.5 and *Figure 5.*

Note 9: For ADC0804LCD typical value of $V_{REF}/2$ input resistance is 8 kΩ and of I_{CC} is 1.1 mA.

Typical Performance Characteristics

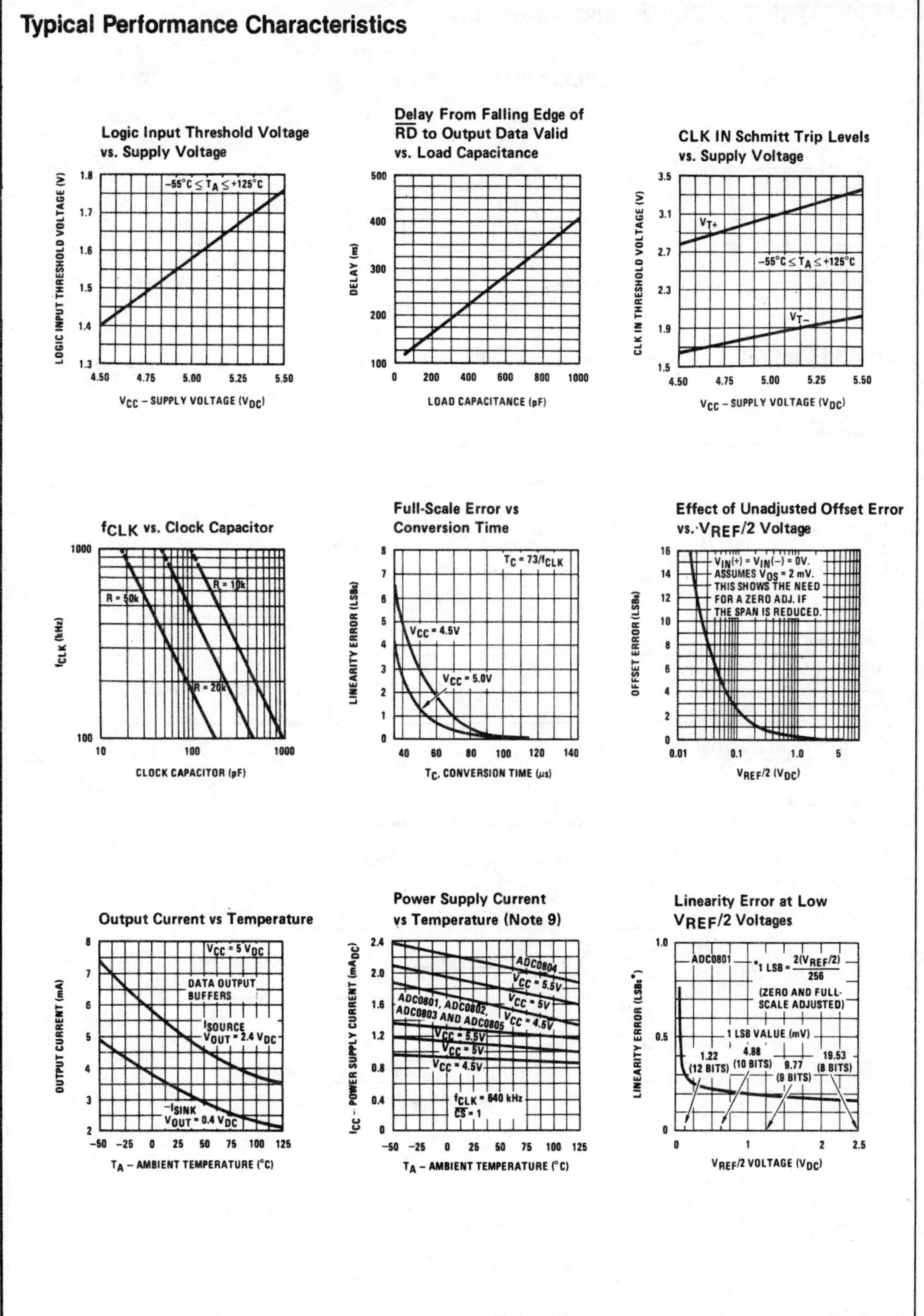

TRI-STATE® Test Circuits and Waveforms

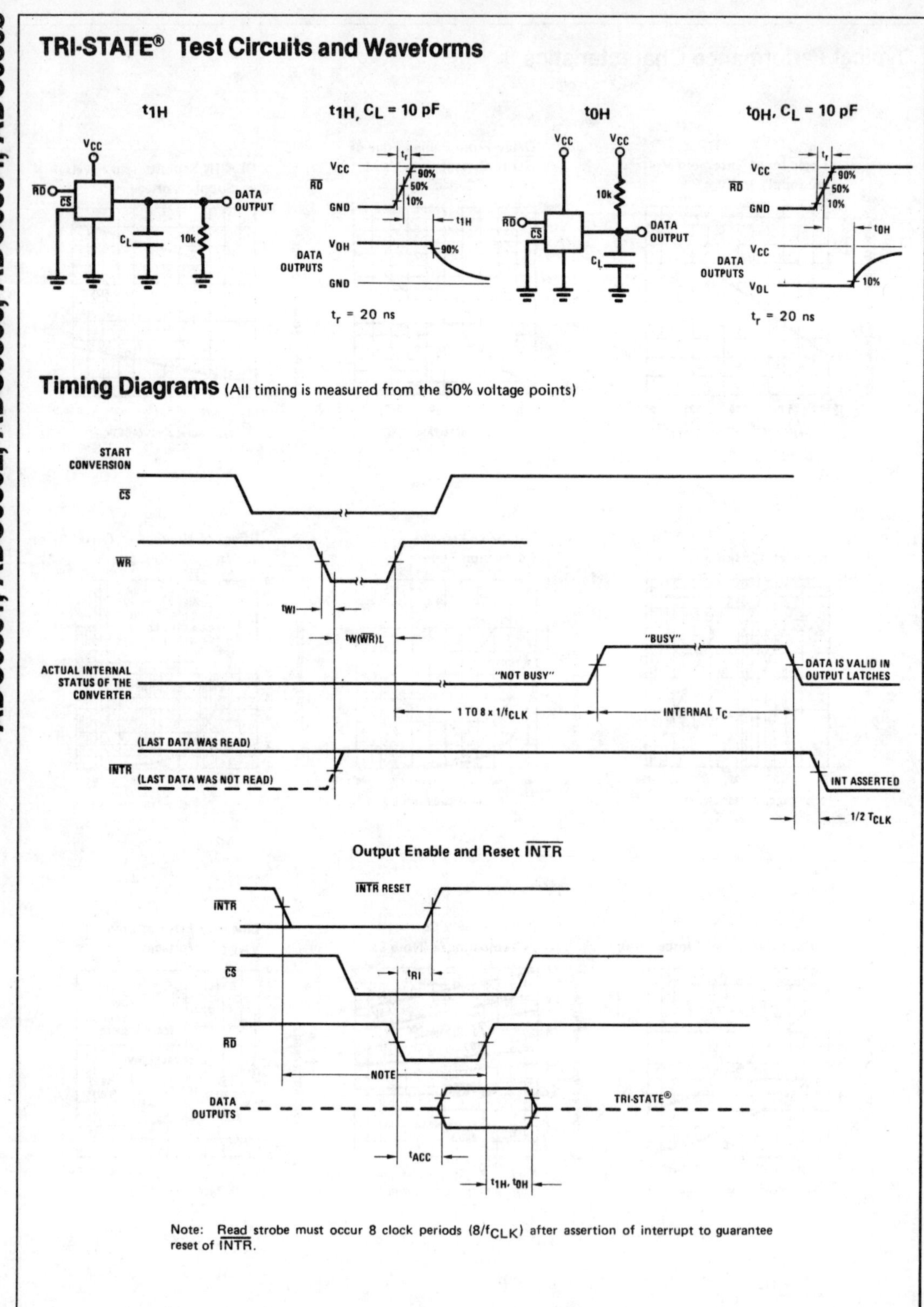

Note: Read strobe must occur 8 clock periods ($8/f_{CLK}$) after assertion of interrupt to guarantee reset of $\overline{INTR}$.

Typical Applications (Continued)

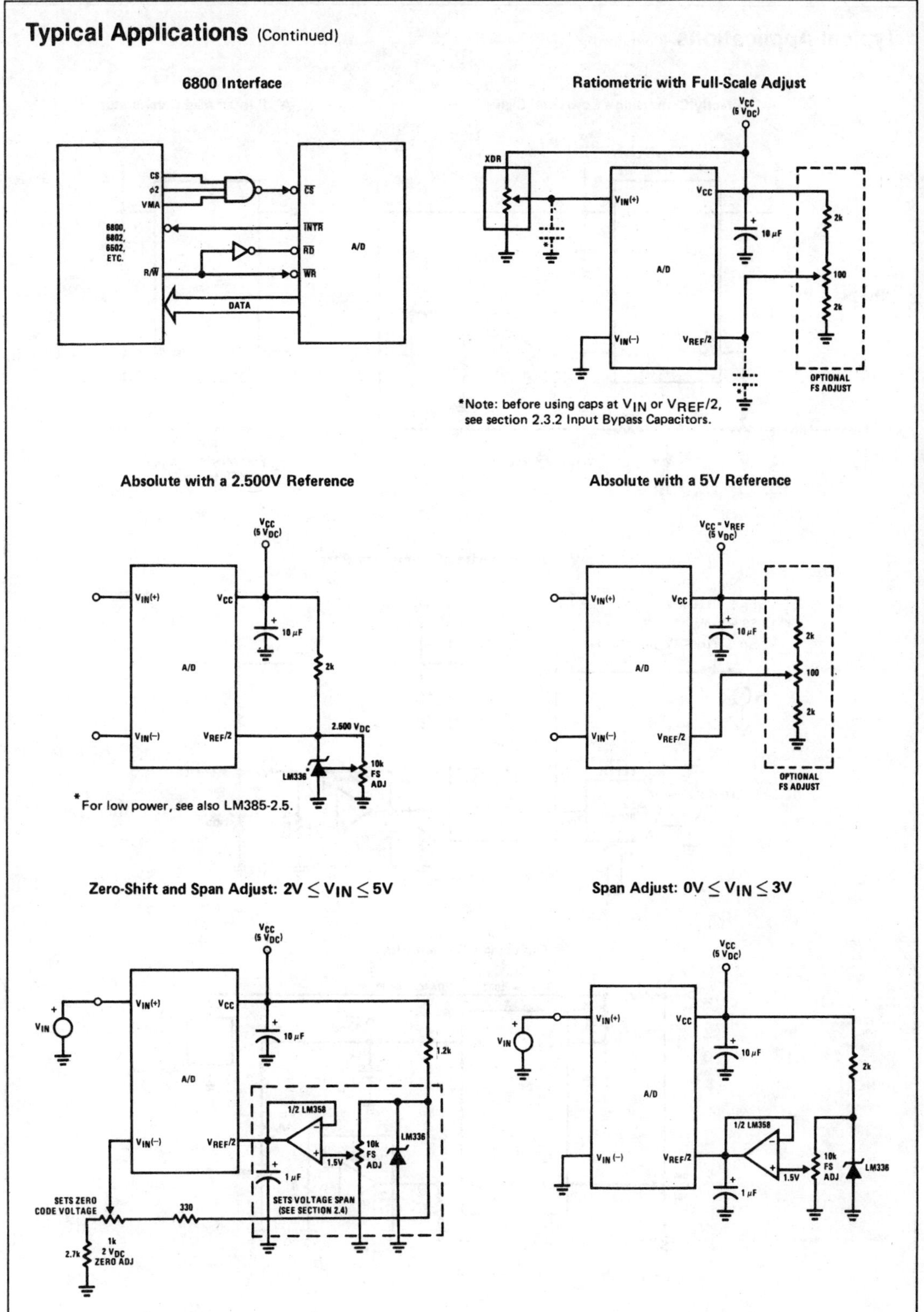

Typical Applications (Continued)

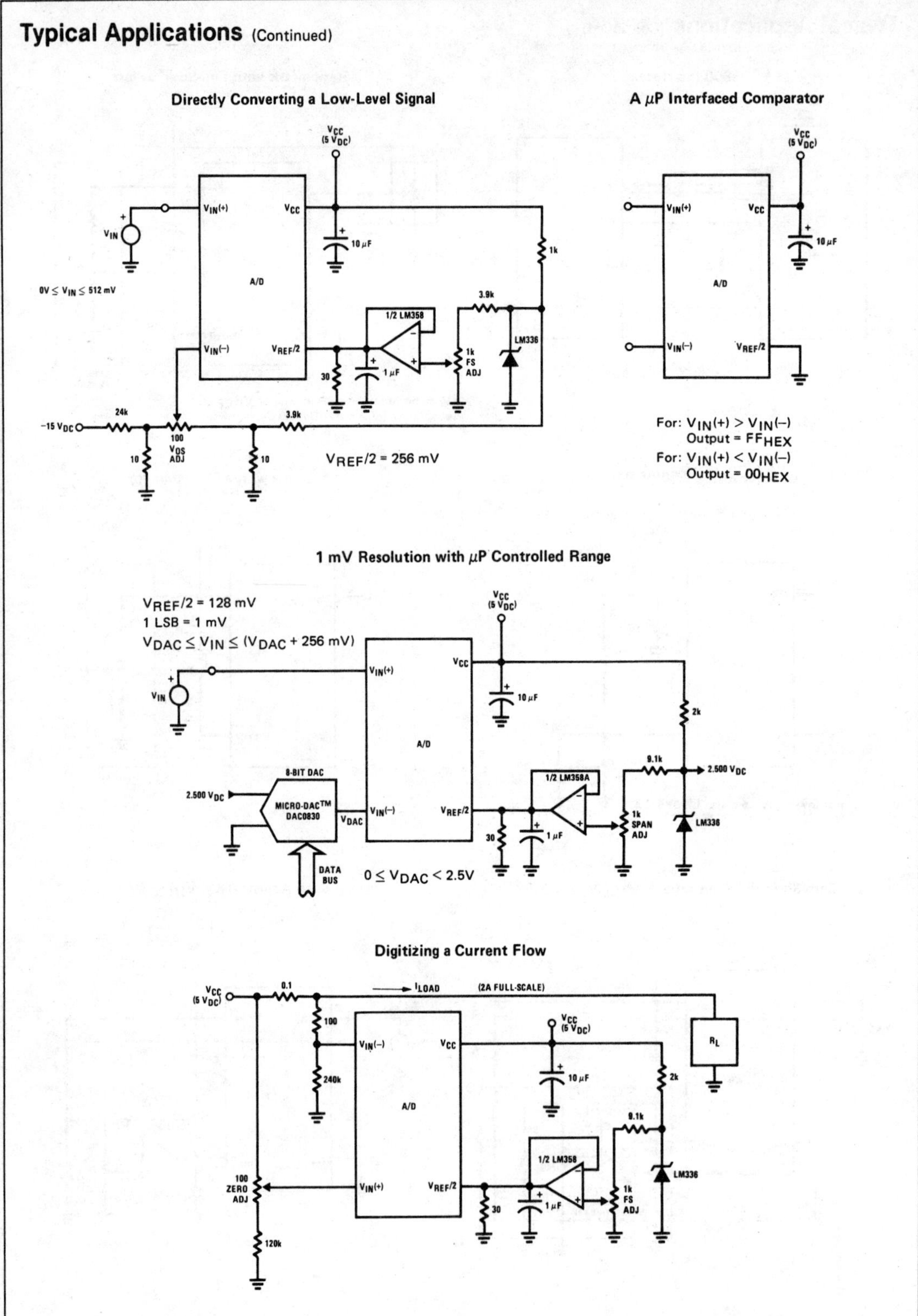

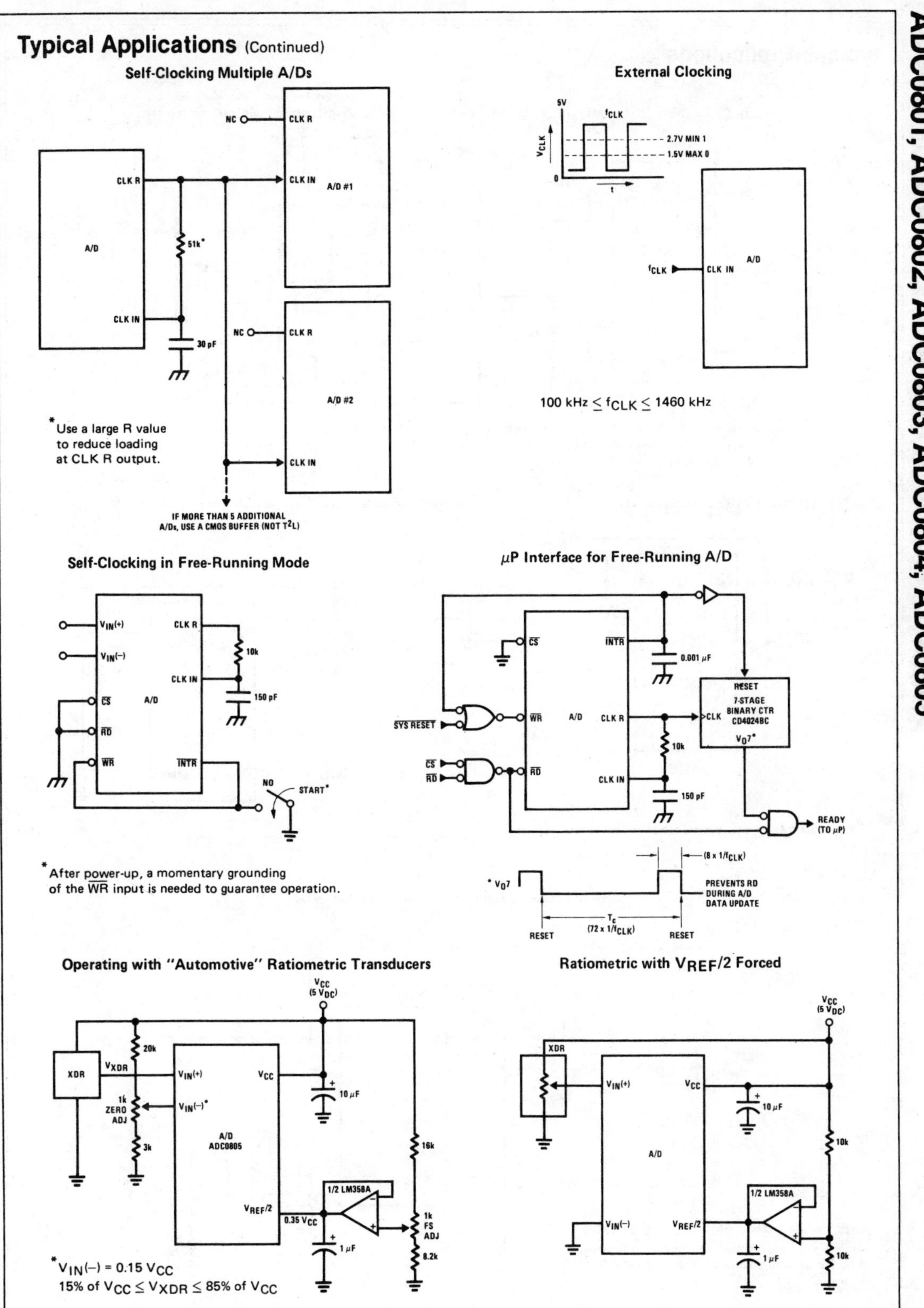
Typical Applications (Continued)
Self-Clocking Multiple A/Ds
NC
CLK R
CLK IN
A/D #1
A/D
51k*
30 pF
A/D #2
*Use a large R value to reduce loading at CLK R output.
IF MORE THAN 5 ADDITIONAL A/Ds, USE A CMOS BUFFER (NOT T²L)
External Clocking
5V
fCLK
VCLK
2.7V MIN 1
1.5V MAX 0
t
100 kHz ≤ fCLK ≤ 1460 kHz
Self-Clocking in Free-Running Mode
VIN(+)
VIN(−)
CS
RD
WR
INTR
10k
150 pF
NO
START*
*After power-up, a momentary grounding of the WR input is needed to guarantee operation.
µP Interface for Free-Running A/D
SYS RESET
0.001 µF
RESET
7-STAGE
BINARY CTR
CD4024BC
VQ7*
READY (TO µP)
(8 x 1/fCLK)
PREVENTS RD DURING A/D DATA UPDATE
Tc (72 x 1/fCLK)
Operating with "Automotive" Ratiometric Transducers
VCC (5 VDC)
XDR
VXDR
20k
1k ZERO ADJ
3k
A/D ADC0805
VREF/2
0.35 VCC
10 µF
1 µF
16k
1/2 LM358A
1k FS ADJ
8.2k
*VIN(−) = 0.15 VCC
15% of VCC ≤ VXDR ≤ 85% of VCC
Ratiometric with VREF/2 Forced

Typical Applications (Continued)

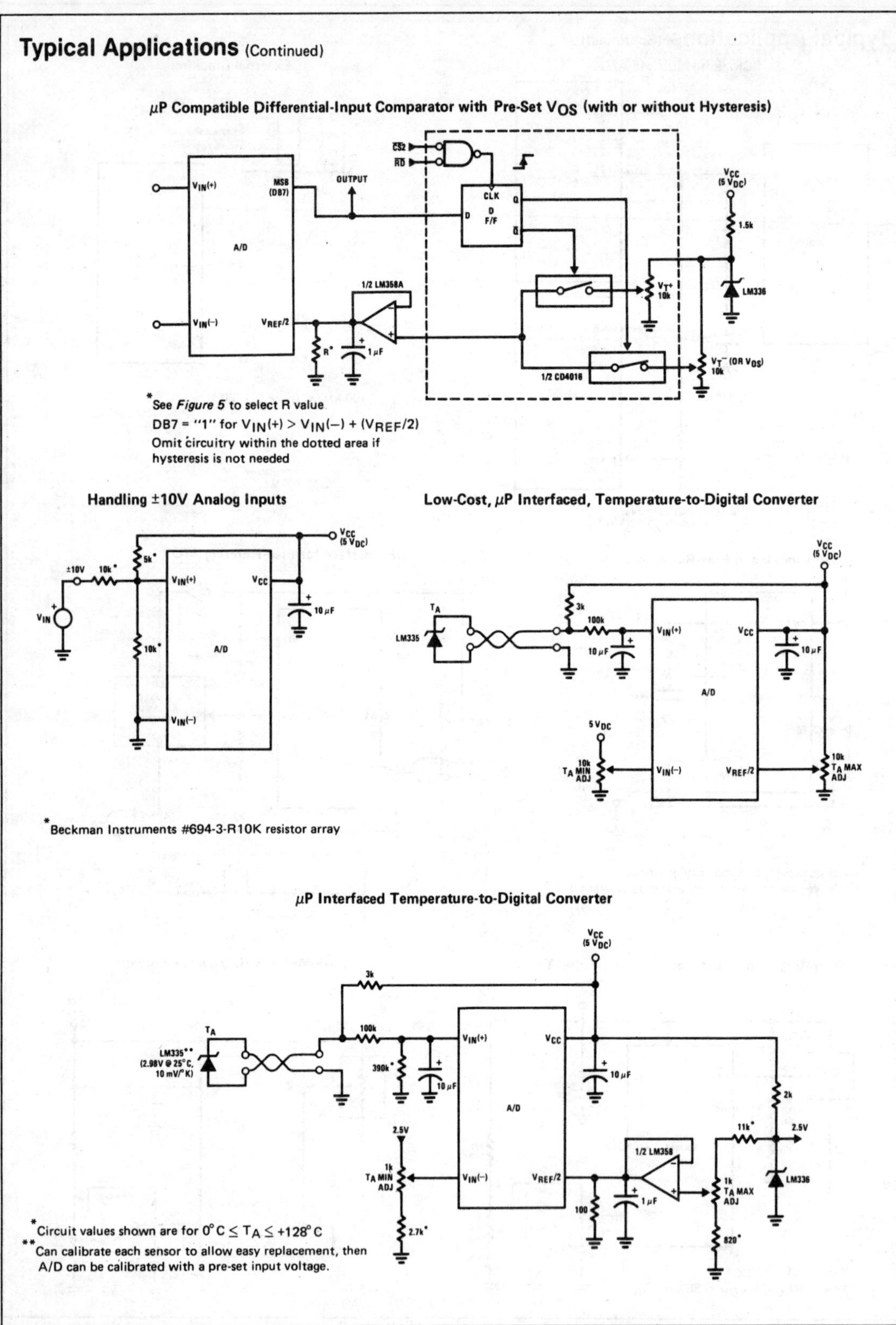

µP Compatible Differential-Input Comparator with Pre-Set V_{OS} (with or without Hysteresis)

*See *Figure 5* to select R value

DB7 = "1" for $V_{IN}(+) > V_{IN}(-) + (V_{REF}/2)$

Omit circuitry within the dotted area if hysteresis is not needed

Handling ±10V Analog Inputs

*Beckman Instruments #694-3-R10K resistor array

Low-Cost, µP Interfaced, Temperature-to-Digital Converter

µP Interfaced Temperature-to-Digital Converter

*Circuit values shown are for $0°C \le T_A \le +128°C$

**Can calibrate each sensor to allow easy replacement, then A/D can be calibrated with a pre-set input voltage.

ADC0801, ADC0802, ADC0803, ADC0804, ADC0805

Typical Applications (Continued)

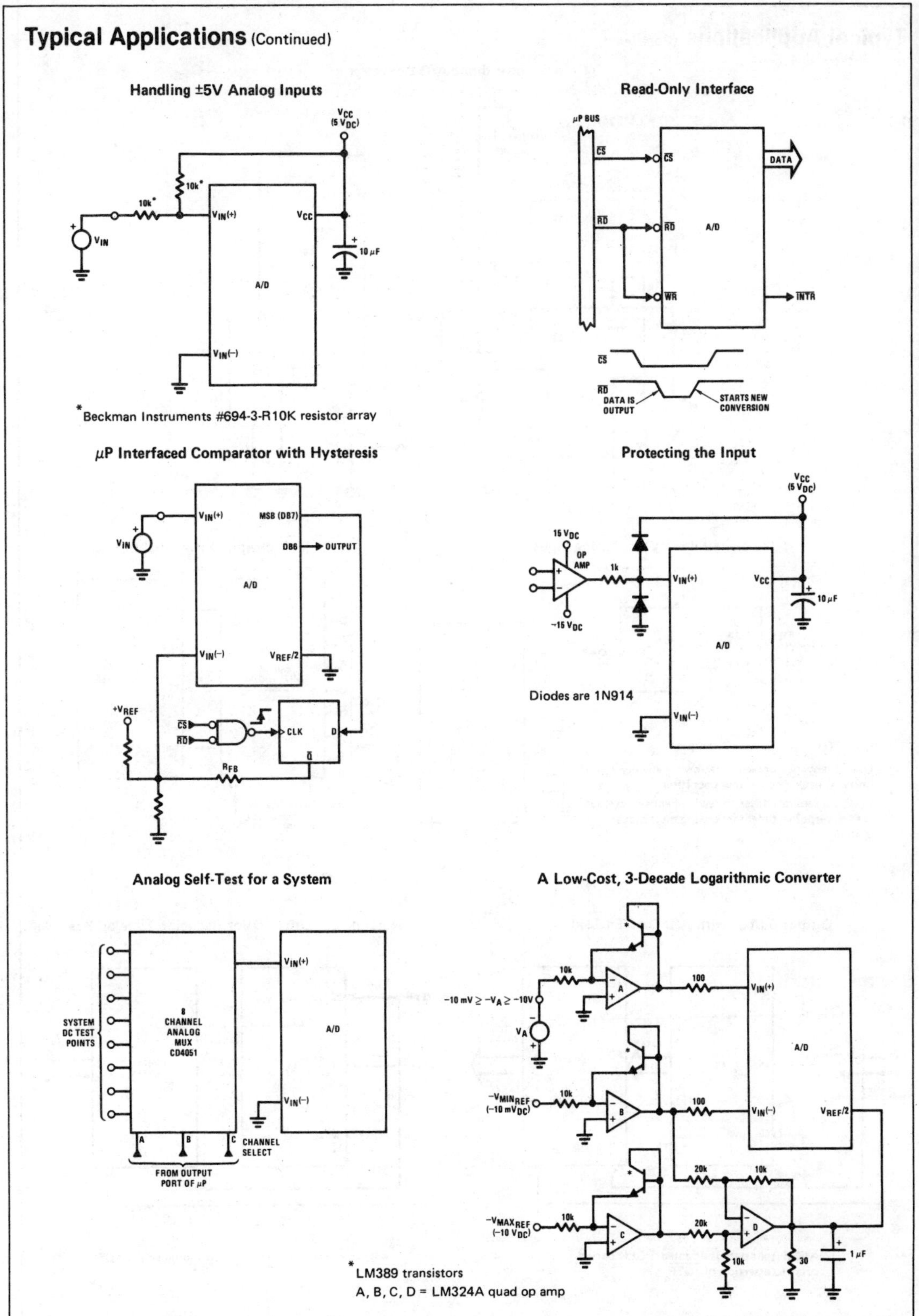

Typical Applications (Continued)

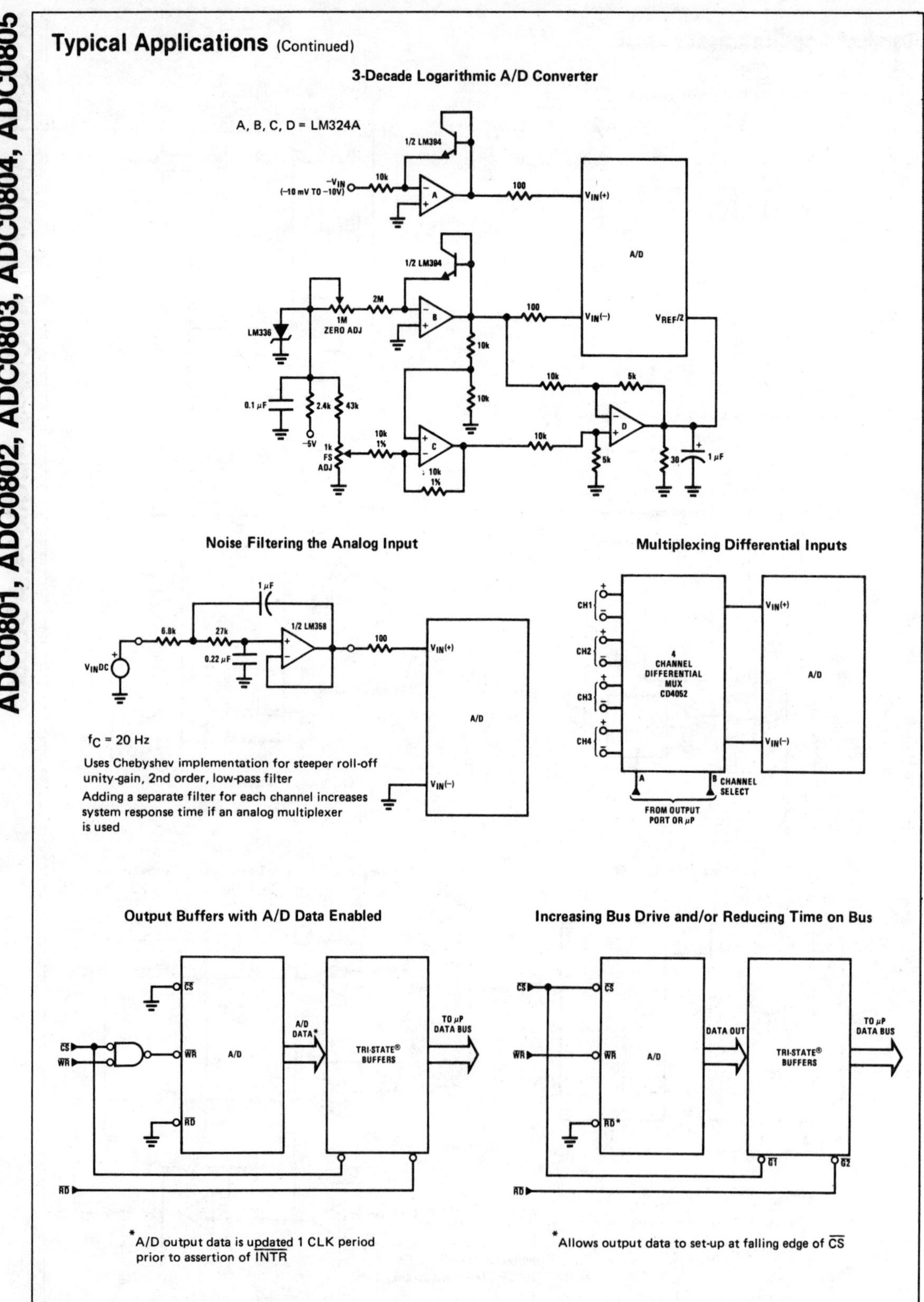

Typical Applications (Continued)

Sampling an AC Input Signal

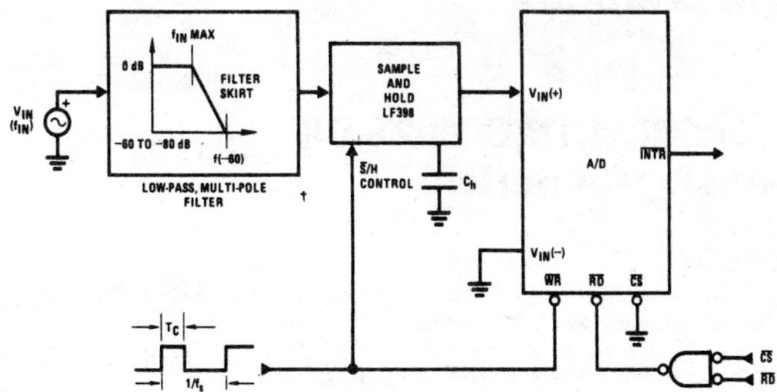

Note 1: Oversample whenever possible [keep fs > 2f(−60)] to eliminate input frequency folding (aliasing) and to allow for the skirt response of the filter.

Note 2: Consider the amplitude errors which are introduced within the passband of the filter.

70% Power Savings by Clock Gating

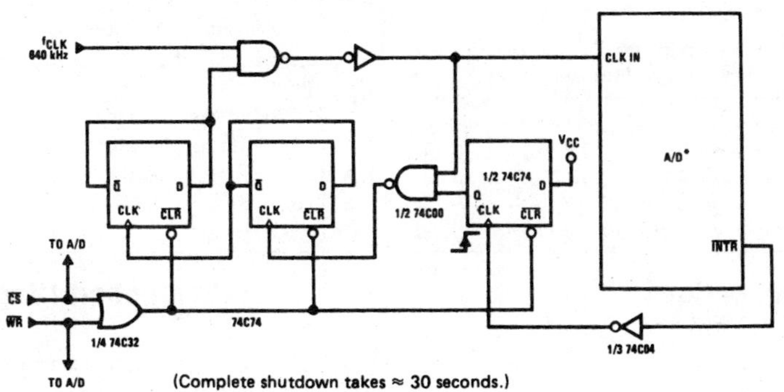

(Complete shutdown takes ≈ 30 seconds.)

Power Savings by A/D and V_{REF} Shutdown

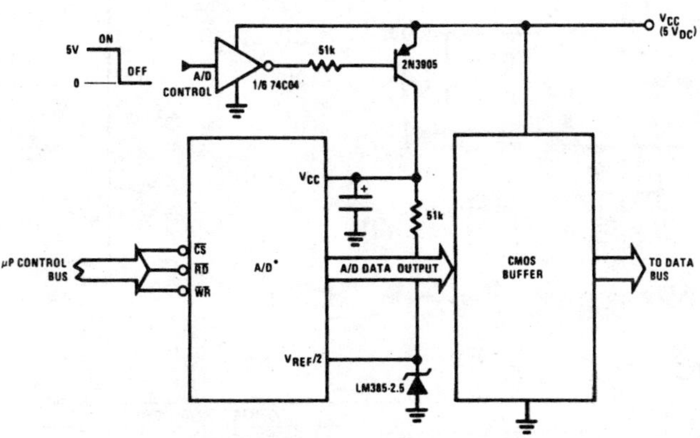

*Use ADC0801, 02, 03 or 05 for lowest power consumption.

Note: Logic inputs can be driven to V_{CC} with A/D supply at zero volts.

Buffer prevents data bus from overdriving outputs of A/D when in shutdown mode.

DAC0800, DAC0801, DAC0802 8-Bit Digital-to-Analog Converters

General Description

The DAC0800 series are monolithic 8-bit high-speed current-output digital-to-analog converters (DAC) featuring typical settling times of 100 ns. When used as a multiplying DAC, monotonic performance over a 40 to 1 reference current range is possible. The DAC0800 series also features high compliance complementary current outputs to allow differential output voltages of 20 Vp-p with simple resistor loads as shown in *Figure 1*. The reference-to-full-scale current matching of better than ±1 LSB eliminates the need for full-scale trims in most applications while the nonlinearities of better than ±0.1% over temperature minimizes system error accumulations.

The noise immune inputs of the DAC0800 series will accept TTL levels with the logic threshold pin, V_{LC}, pin 1 grounded. Simple adjustments of the V_{LC} potential allow direct interface to all logic families. The performance and characteristics of the device are essentially unchanged over the full ± 4.5V to ±18V power supply range; power dissipation is only 33 mW with ± 5V supplies and is independent of the logic input states.

The DAC0800, DAC0802, DAC0800C, DAC0801C and DAC0802C are a direct replacement for the DAC-08, DAC-08A, DAC-08C, DAC-08E and DAC-08H, respectively.

Features

- Fast settling output current 100 ns
- Full scale error ±1 LSB
- Nonlinearity over temperature ±0.1%
- Full scale current drift ±10 ppm/°C
- High output compliance −10V to +18V
- Complementary current outputs
- Interface directly with TTL, CMOS, PMOS and others
- 2 quadrant wide range multiplying capability
- Wide power supply range ±4.5V to ±18V
- Low power consumption 33 mW at ±5V
- Low cost

Typical Applications

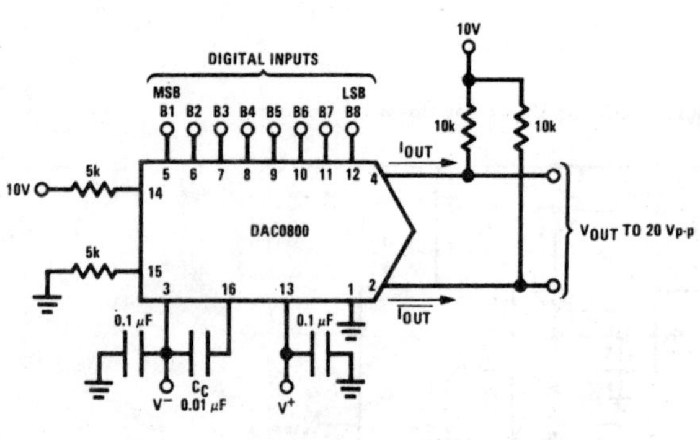

FIGURE 1. ±20 Vp-p Output Digital-to-Analog Converter

Connection Diagram

Dual-In-Line Package

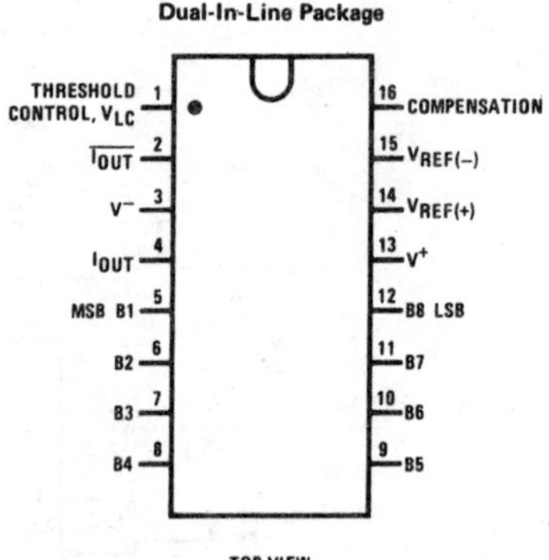

TOP VIEW

Ordering Information

NON LINEARITY	TEMPERATURE RANGE	ORDER NUMBERS*					
		D PACKAGE (D16C)		J PACKAGE (J16A)		N PACKAGE (N16A)	
±0.1% FS	$-55°C \le T_A \le +125°C$	DAC0802LD	DAC-08AQ				
±0.1% FS	$0°C \le T_A \le +70°C$			DAC0802LCJ	DAC-08HQ	DAC0802LCN	DAC-08HP
±0.19% FS	$-55°C \le T_A \le +125°C$	DAC0800LD	DAC-08Q				
±0.19% FS	$0°C \le T_A \le +70°C$			DAC0800LCJ	DAC-08EQ	DAC0800LCN	DAC-08EP
±0.39% FS	$0°C \le T_A \le +70°C$			DAC0801LCJ	DAC-08CQ	DAC0801LCN	DAC-08CP

*Note. Devices may be ordered by using either order number.

Absolute Maximum Ratings

Supply Voltage	±18V or 36V
Power Dissipation (Note 1)	500 mW
Reference Input Differential Voltage (V14 to V15)	V^- to V^+
Reference Input Common-Mode Range (V14, V15)	V^- to V^+
Reference Input Current	5 mA
Logic Inputs	V^- to V^- plus 36V
Analog Current Outputs	*Figure 24*
Storage Temperature	−65°C to +150°C
Lead Temperature (Soldering, 10 seconds)	300°C

Operating Conditions

	MIN	MAX	UNITS
Temperature (T_A)			
DAC0802L	−55	+125	°C
DAC0800L	−55	+125	°C
DAC0800LC	0	+70	°C
DAC0801LC	0	+70	°C
DAC0802LC	0	+70	°C

Electrical Characteristics

(V_S = ±15V, I_{REF} = 2 mA, $T_{MIN} \le T_A \le T_{MAX}$ unless otherwise specified. Output characteristics refer to both I_{OUT} and $\overline{I_{OUT}}$.)

	PARAMETER	CONDITIONS	DAC0802L/ DAC0802LC			DAC0800L/ DAC0800LC			DAC0801LC			UNITS
			MIN	TYP	MAX	MIN	TYP	MAX	MIN	TYP	MAX	
	Resolution		8	8	8	8	8	8	8	8	8	Bits
	Monotonicity		8	8	8	8	8	8	8	6	8	Bits
	Nonlinearity				±0.1			±0.19			±0.39	%FS
t_s	Settling Time	To ±1/2 LSB, All Bits Switched "ON" or "OFF", T_A = 25°C		100	135					100	150	ns
		DAC0800L					100	135				ns
		DAC0800LC					100	150				ns
t_{PLH}, t_{PHL}	Propagation Delay	T_A = 25°C										
	Each Bit			35	60		35	60		35	60	ns
	All Bits Switched			35	60		35	60		35	60	ns
TCI_{FS}	Full Scale Tempco			±10	±50		±10	±50		±10	±80	ppm/°C
V_{OC}	Output Voltage Compliance	Full Scale Current Change < 1/2 LSB, R_{OUT} > 20 MΩ Typ	−10		18	−10		18	−10		18	V
I_{FS4}	Full Scale Current	V_{REF} = 10.000V, R14 = 5.000 kΩ R15 = 5.000 kΩ, T_A = 25°C	1.984	1.992	2.000	1.94	1,99	2.04	1.94	1.99	2.04	mA
I_{FSS}	Full Scale Symmetry	$I_{FS4} - I_{FS2}$		±0.5	±4.0		±1	±8.0		±2	±16	μA
I_{ZS}	Zero Scale Current			0.1	1.0		0.2	2.0		0.2	4.0	μA
I_{FSR}	Output Current Range	V^- = −5V	0	2.0	2.1	0	2.0	2.1	0	2.0	2.1	mA
		V^- = −8V to −18V	0	2.0	4.2	0	2.0	4.2	0	2.0	4.2	mA
	Logic Input Levels											
V_{IL}	Logic "0"	V_{LC} = 0V			0.8			0.8			0.8	V
V_{IH}	Logic "1"		2.0			2.0			2.0			V
	Logic Input Current	V_{LC} = 0V										
I_{IL}	Logic "0"	$-10V \le V_{IN} \le +0.8V$		−2.0	−10		−2.0	−10		−2.0	−10	μA
I_{IH}	Logic "1"	$2V \le V_{IN} \le +18V$		0.002	10		0.002	10		0.002	10	μA
V_{IS}	Logic Input Swing	V^- = −15V	−10		18	−10		18	−10		18	V
V_{THR}	Logic Threshold Range	V_S = ±15V	−10		13.5	−10		13.5	−10		13.5	V
I_{15}	Reference Bias Current			−1.0	−3.0		−1.0	−3.0		−1.0	−3.0	μA
dI/dt	Reference Input Slew Rate	*(Figure 24)*	4.0	8.0		4.0	8.0		4.0	8.0		mA/μs
$PSSI_{FS+}$	Power Supply Sensitivity	$4.5V \le V^+ \le 18V$		0.0001	0.01		0.0001	0.01		0.0001	0.01	%/%
$PSSI_{FS-}$		$-4.5V \le V^- \le 18V$		0.0001	0.01		0.0001	0.01		0.0001	0.01	%/%
		I_{REF} = 1 mA										
	Power Supply Current	V_S = ±5V, I_{REF} = 1 mA										
I+				2.3	3.8		2.3	3.8		2.3	3.8	mA
I−				−4.3	−5.8		−4.3	−5.8		−4.3	−5.8	mA
		V_S = 5V, −15V, I_{REF} = 2 mA										
I+				2.4	3.8		2.4	3.8		2.4	3.8	mA
I−				−6.4	−7.8		−6.4	−7.8		−6.4	−7.8	mA
		V_S = ±15V, I_{REF} = 2 mA										
I+				2.5	3.8		2.5	3.8		2.5	3.8	mA
I−				−6.5	−7.8		−6.5	−7.8		−6.5	−7.8	mA
P_D	Power Dissipation	±5V, I_{REF} = 1 mA		33	48		33	48		33	48	mW
		5V, −15V, I_{REF} = 2 mA		108	136		108	136		108	136	mW
		±15V, I_{REF} = 2 mA		135	174		135	174		135	174	mW

Note 1: The maximum junction temperature of the DAC0800, DAC0801 and DAC0802 is 125°C. For operating at elevated temperatures, devices in the dual-in-line J or D package must be derated based on a thermal resistance of 100°C/W, junction to ambient, 175°C/W for the molded dual-in-line N package.

Block Diagram

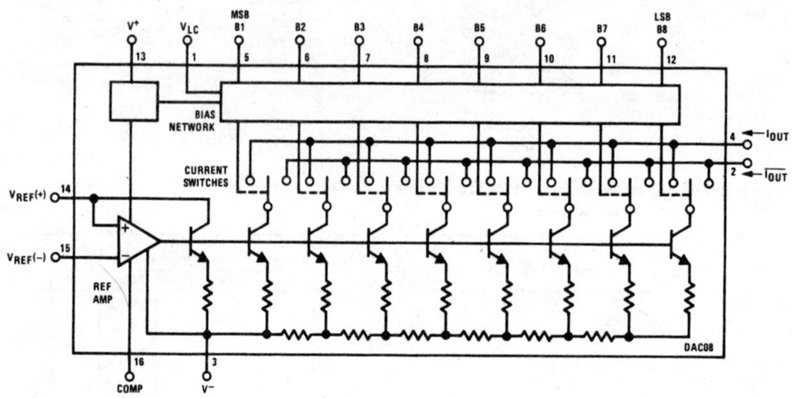

Equivalent Circuit

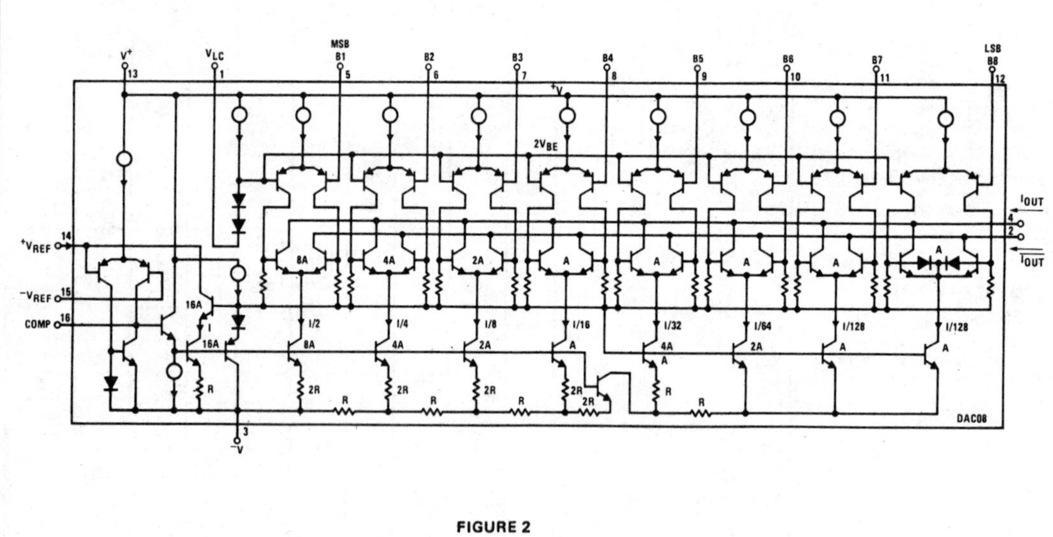

FIGURE 2

Typical Performance Characteristics

Full Scale Current vs Reference Current

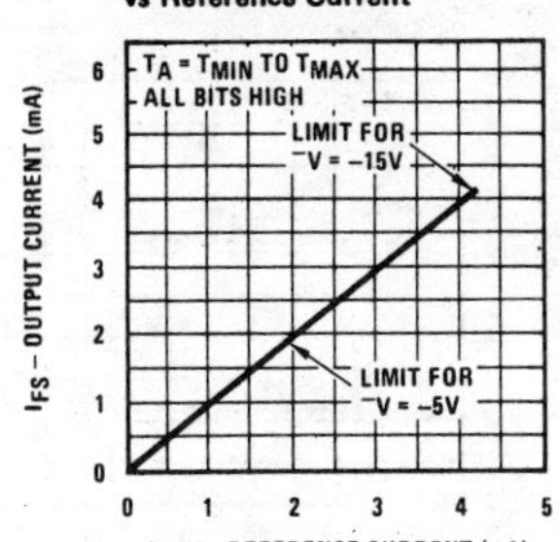

FIGURE 3

LSB Propagation Delay vs I_{FS}

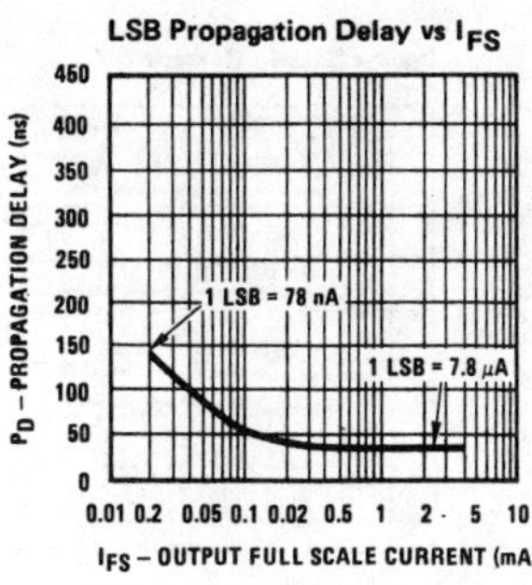

FIGURE 4

Reference Input Frequency Response

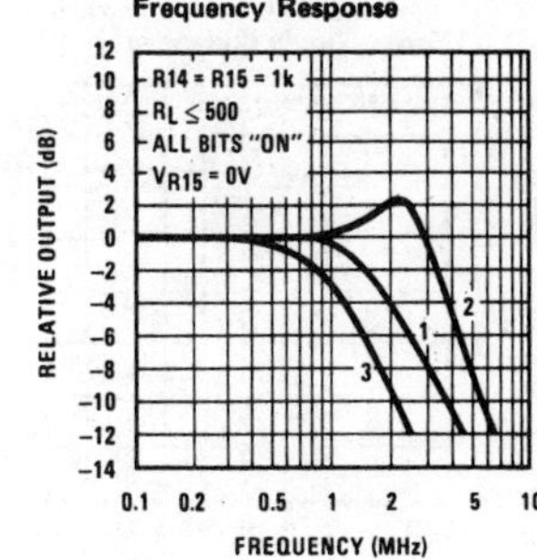

Curve 1: C_C = 15 pF, V_{IN} = 2 Vp-p centered at 1V.

Curve 2: C_C = 15 pF, V_{IN} = 50 mVp-p centered at 200 mV.

Curve 3: C_C = 0 pF, V_{IN} = 100 mVp-p at 0V and applied through 50 Ω connected to pin 14. 2V applied to R14.

FIGURE 5

Reference Amp Common-Mode Range

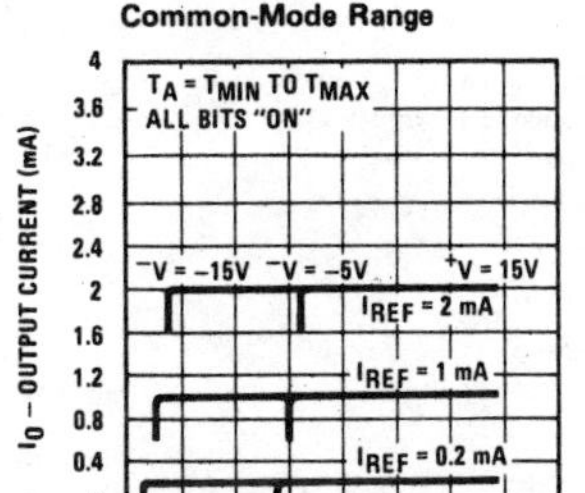

Note. Positive common-mode range is always (V+) − 1.5V.

FIGURE 6

Logic Input Current vs Input Voltage

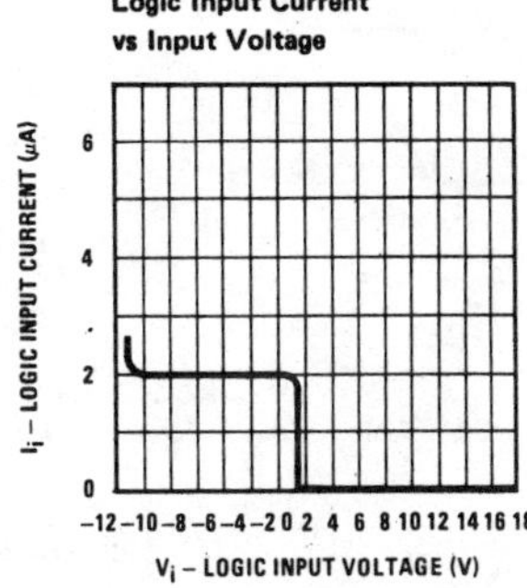

FIGURE 7

$V_{TH} - V_{LC}$ vs Temperature

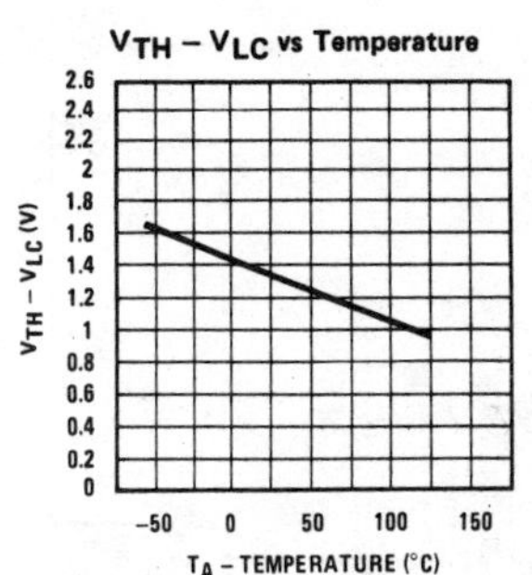

FIGURE 8

Output Current vs Output Voltage (Output Voltage Compliance)

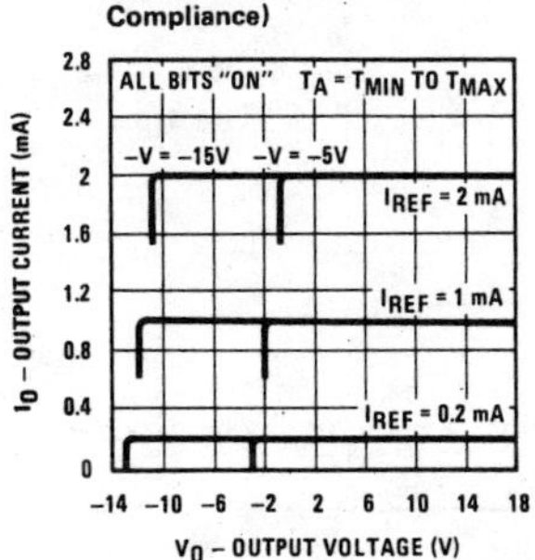

FIGURE 9

Output Voltage Compliance vs Temperature

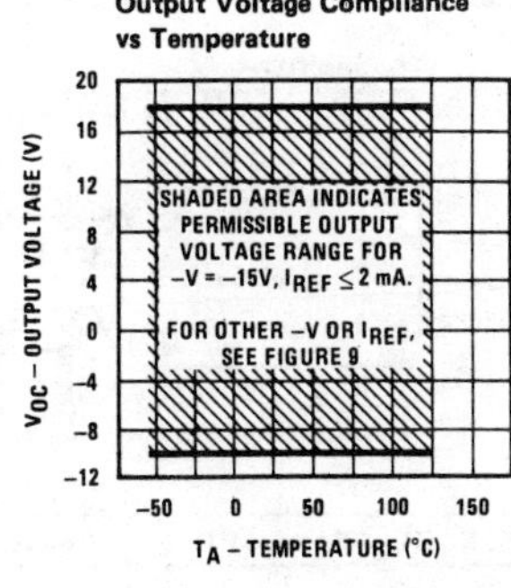

FIGURE 10

Bit Transfer Characteristics

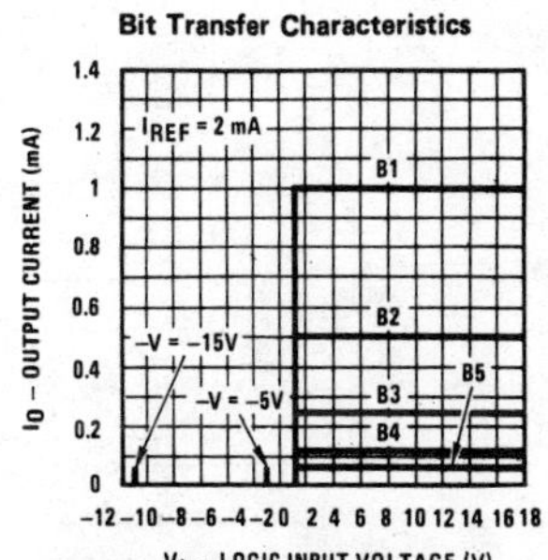

Note. B1–B8 have identical transfer characteristics. Bits are fully switched with less than 1/2 LSB error, at less than ±100 mV from actual threshold. These switching points are guaranteed to lie between 0.8 and 2V over the operating temperature range (V_{LC} = 0V).

FIGURE 11

Typical Performance Characteristics (Continued)

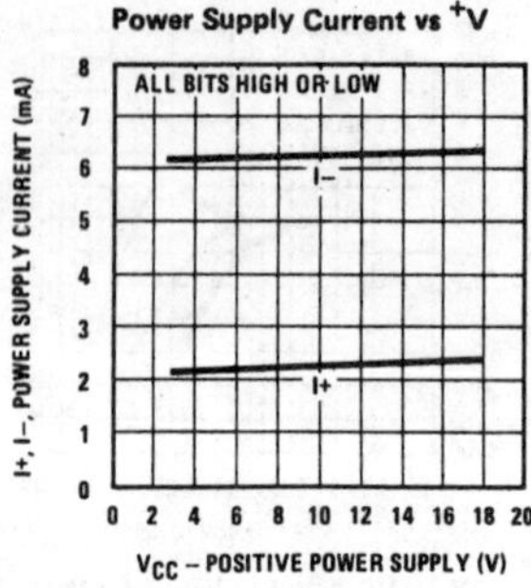

FIGURE 12

Power Supply Current vs −V

ALL BITS MAY BE HIGH OR LOW

I− WITH I_{REF} = 2 mA

I− WITH I_{REF} = 1 mA

I− WITH I_{REF} = 0.2 mA

I+

POWER SUPPLY CURRENT (mA)

V – NEGATIVE POWER SUPPLY (V)

FIGURE 13

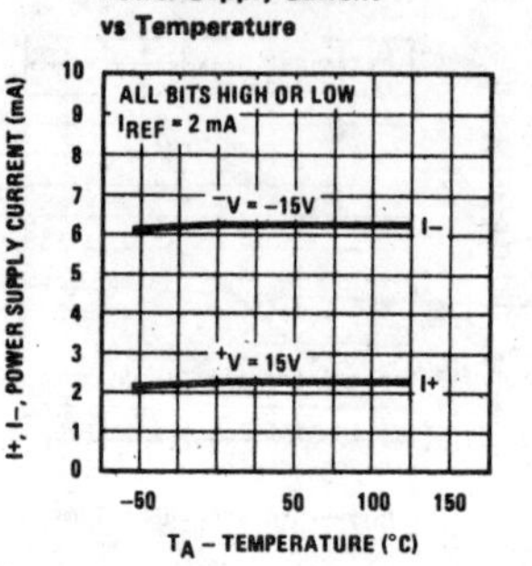

FIGURE 14

Typical Applications (Continued)

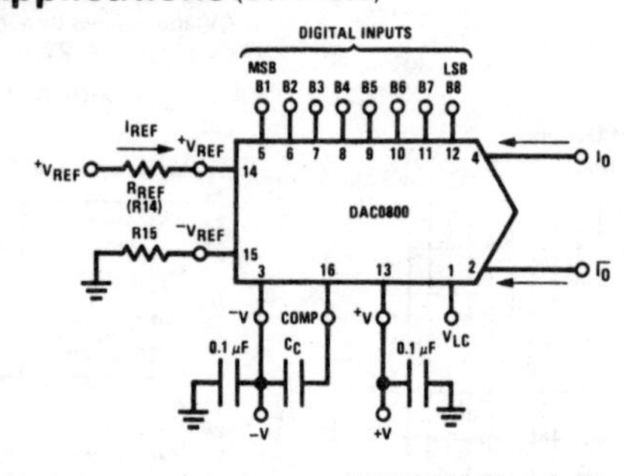

$$I_{FS} \approx \frac{+V_{REF}}{R_{REF}} \times \frac{255}{256}$$

$I_O + \overline{I_O} = I_{FS}$ for all logic states

For fixed reference, TTL operation, typical values are:

V_{REF} = 10.000V
R_{REF} = 5.000k
R15 ≈ R_{REF}
C_C = 0.01 µF
V_{LC} = 0V (Ground)

FIGURE 15. Basic Positive Reference Operation

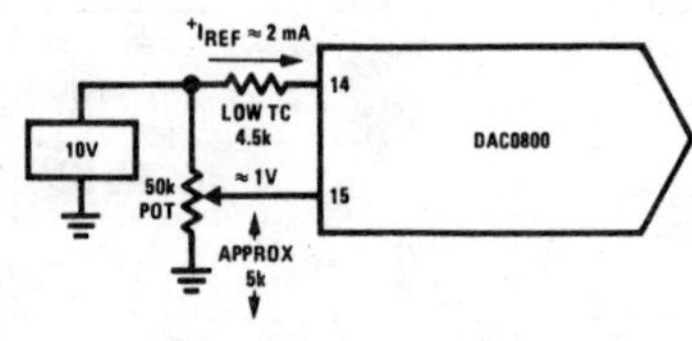

FIGURE 16. Recommended Full Scale Adjustment Circuit

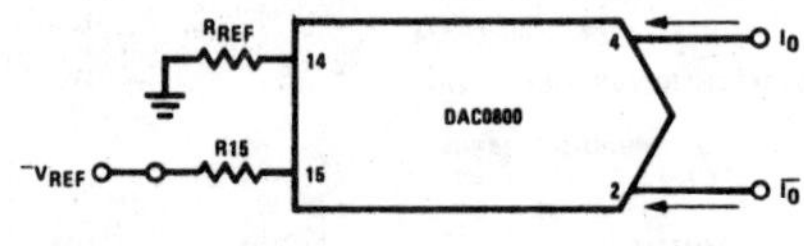

$$I_{FS} \approx \frac{-V_{REF}}{R_{REF}} \times \frac{255}{256}$$

Note. R_{REF} sets I_{FS}; R15 is for bias current cancellation

FIGURE 17. Basic Negative Reference Operation

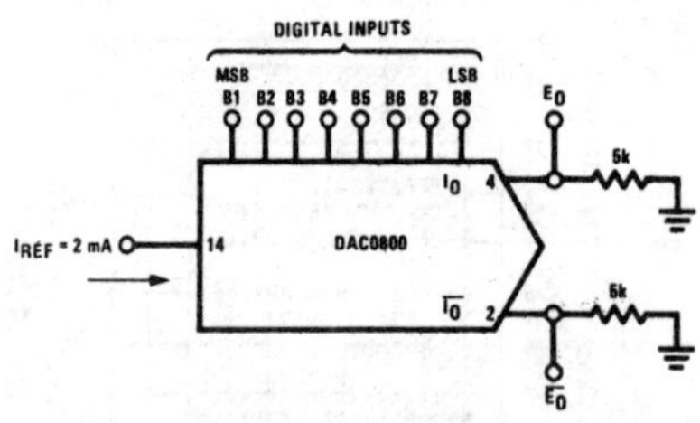

	B1	B2	B3	B4	B5	B6	B7	B8	I_O mA	$\overline{I_O}$ mA	E_O	$\overline{E_O}$
Full Scale	1	1	1	1	1	1	1	1	1.992	0.000	−9.960	0.000
Full Scale−LSB	1	1	1	1	1	1	1	0	1.984	0.008	−9.920	−0.040
Half Scale+LSB	1	0	0	0	0	0	0	1	1.008	0.984	−5.040	−4.920
Half Scale	1	0	0	0	0	0	0	0	1.000	0.992	−5.000	−4.960
Half Scale−LSB	0	1	1	1	1	1	1	1	0.992	1.000	−4.960	−5.000
Zero Scale+LSB	0	0	0	0	0	0	0	1	0.008	1.984	−0.040	−9.920
Zero Scale	0	0	0	0	0	0	0	0	0.000	1.992	0.000	−9.960

FIGURE 18. Basic Unipolar Negative Operation

Typical Applications (Continued)

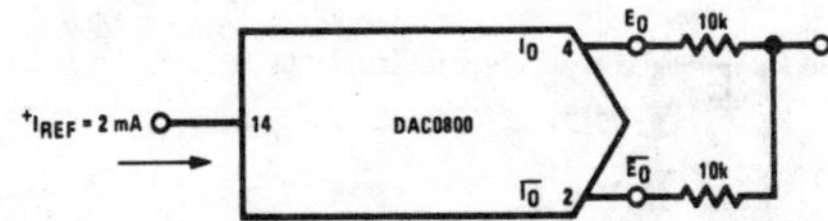

	B1	B2	B3	B4	B5	B6	B7	B8	E_O	$\overline{E_O}$
Pos. Full Scale	1	1	1	1	1	1	1	1	−9.920	+10.000
Pos. Full Scale–LSB	1	1	1	1	1	1	1	0	−9.840	+9.920
Zero Scale+LSB	1	0	0	0	0	0	0	1	−0.080	+0.160
Zero Scale	1	0	0	0	0	0	0	0	0.000	+0.080
Zero Scale–LSB	0	1	1	1	1	1	1	1	+0.080	0.000
Neg. Full Scale+LSB	0	0	0	0	0	0	0	1	+9.920	−9.840
Neg. Full Scale	0	0	0	0	0	0	0	0	+10.000	−9.920

FIGURE 19. Basic Bipolar Output Operation

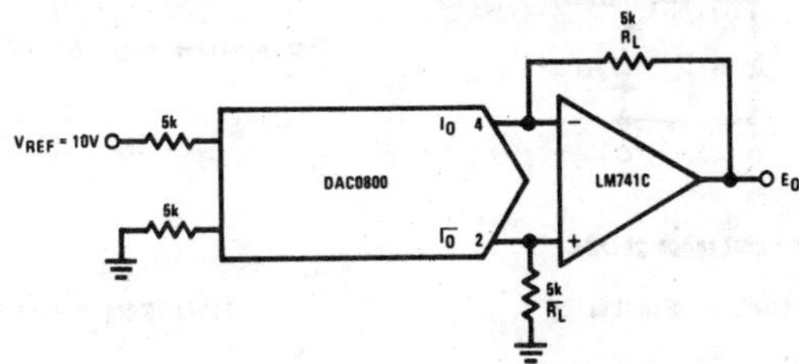

If $R_L = \overline{R_L}$ within ±0.05%, output is symmetrical about ground

	B1	B2	B3	B4	B5	B6	B7	B8	E_O
Pos. Full Scale	1	1	1	1	1	1	1	1	+9.920
Pos. Full Scale–LSB	1	1	1	1	1	1	1	0	+9.840
(+) Zero Scale	1	0	0	0	0	0	0	0	+0.040
(−) Zero Scale	0	1	1	1	1	1	1	1	−0.040
Neg. Full Scale+LSB	0	0	0	0	0	0	0	1	−9.840
Neg. Full Scale	0	0	0	0	0	0	0	0	−9.920

FIGURE 20. Symmetrical Offset Binary Operation

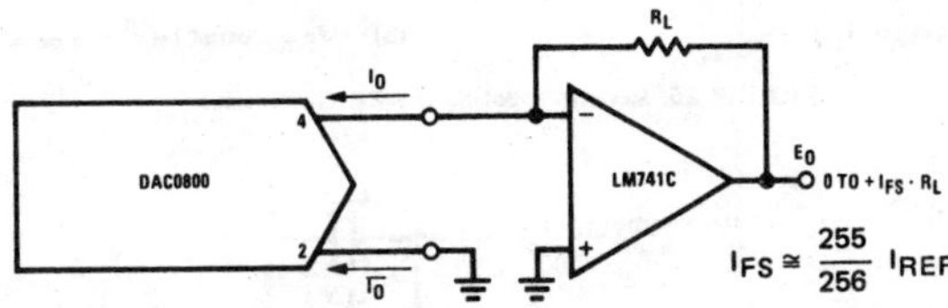

For complementary output (operation as negative logic DAC), connect inverting input of op amp to $\overline{I_O}$ (pin 2), connect I_O (pin 4) to ground.

FIGURE 21. Positive Low Impedance Output Operation

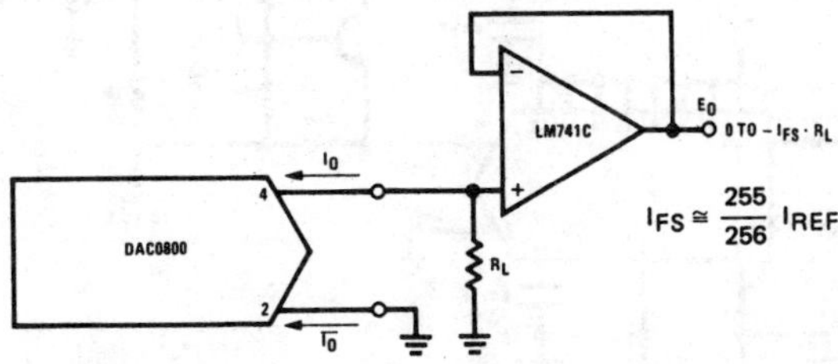

For complementary output (operation as a negative logic DAC) connect non-inverting input of op amp to $\overline{I_O}$ (pin 2); connect I_O (pin 4) to ground.

FIGURE 22. Negative Low Impedance Output Operation

Typical Applications (Continued)

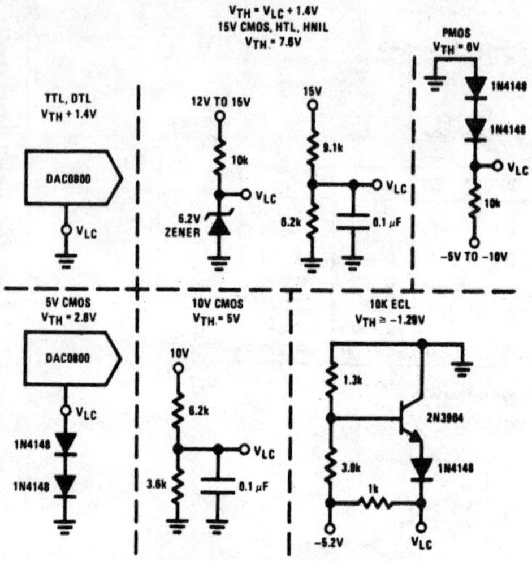

Note. Do not exceed negative logic input range of DAC.

FIGURE 23. Interfacing with Various Logic Families

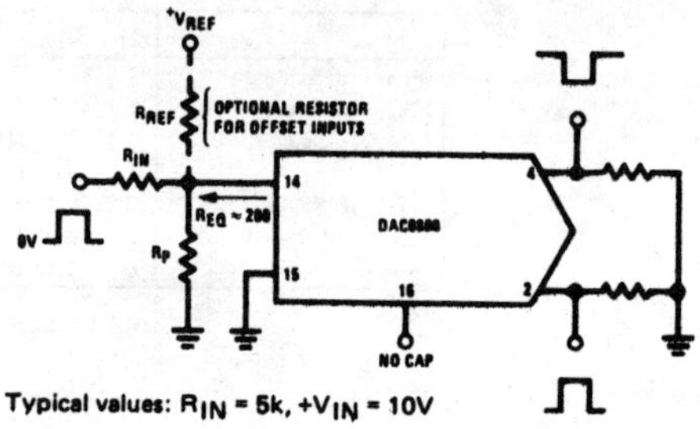

Typical values: R_{IN} = 5k, $+V_{IN}$ = 10V

FIGURE 24. Pulsed Reference Operation

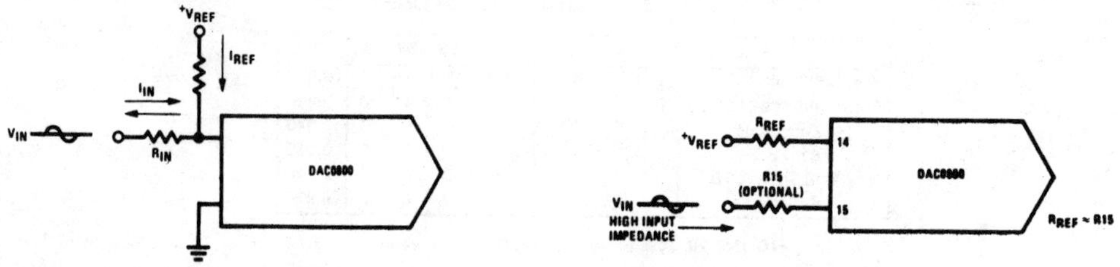

(a) $I_{REF} \geq$ peak negative swing of I_{IN}

(b) $+V_{REF}$ must be above peak positive swing of V_{IN}

FIGURE 25. Accommodating Bipolar References

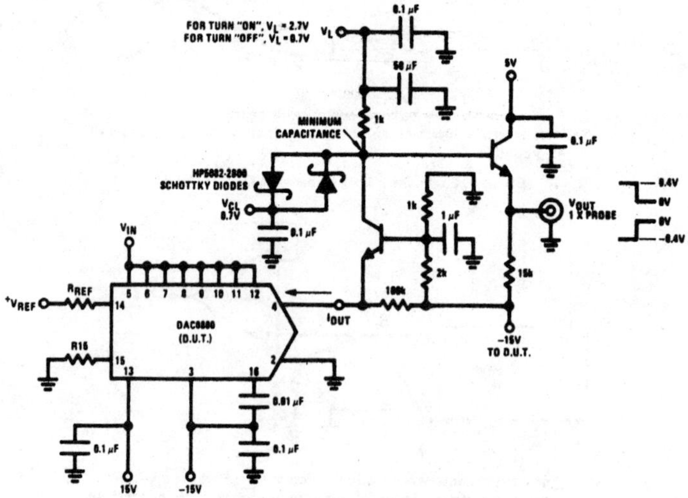

FIGURE 26. Settling Time Measurement

Typical Applications (Continued)

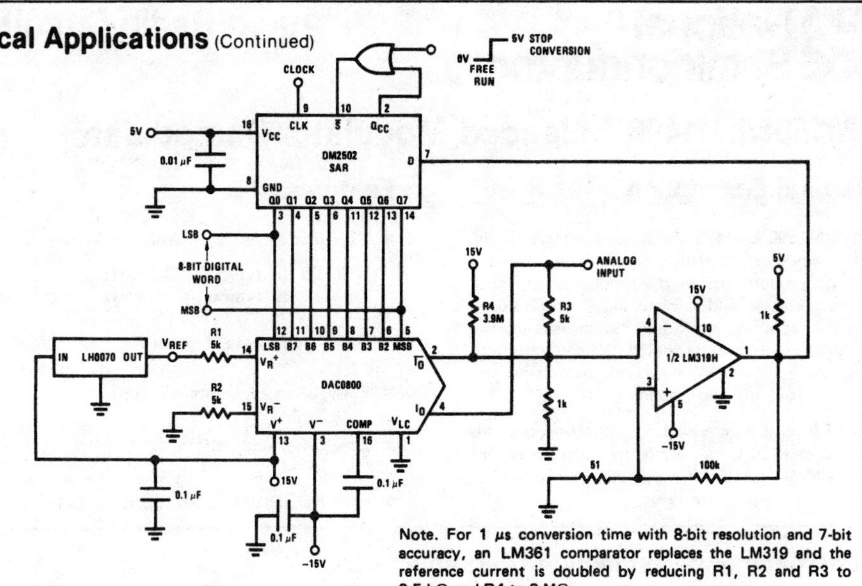

Note. For 1 μs conversion time with 8-bit resolution and 7-bit accuracy, an LM361 comparator replaces the LM319 and the reference current is doubled by reducing R1, R2 and R3 to 2.5 kΩ and R4 to 2 MΩ.

FIGURE 27. A Complete 2 μs Conversion Time, 8-Bit A/D Converter

LM1596/LM1496 Balanced Modulator-Demodulator

General Description

The LM1596/LM1496 are double balanced modulator-demodulators which produce an output voltage proportional to the product of an input (signal) voltage and a switching (carrier) signal. Typical applications include suppressed carrier modulation, amplitude modulation, synchronous detection, FM or PM detection, broadband frequency doubling and chopping.

The LM1596 is specified for operation over the −55°C to +125°C military temperature range. The LM1496 is specified for operation over the 0°C to +70°C temperature range.

Features

- Excellent carrier suppression

  65 dB typical at 0.5 MHz
  50 dB typical at 10 MHz
- Adjustable gain and signal handling
- Fully balanced inputs and outputs
- Low offset and drift
- Wide frequency response up to 100 MHz

Schematic and Connection Diagrams

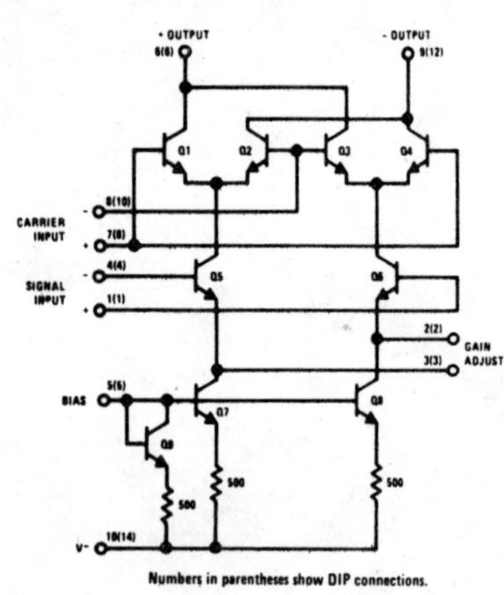

Numbers in parentheses show DIP connections.

Metal Can Package

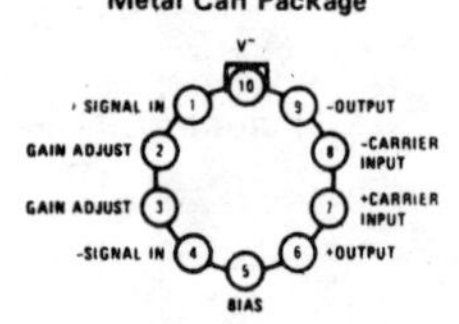

TOP VIEW

Note: Pin 10 is connected electrically to the case through the device substrate.

Order Number LM1496H or LM1596H
See NS Package H08C

Dual-In-Line Package

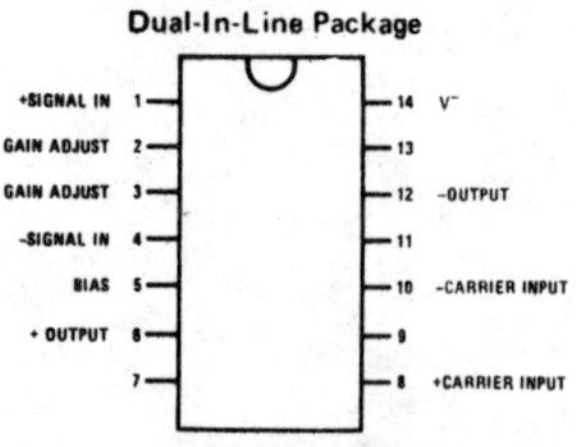

TOP VIEW

Order Number LM1496N
See NS Package N14A

Typical Application and Test Circuit

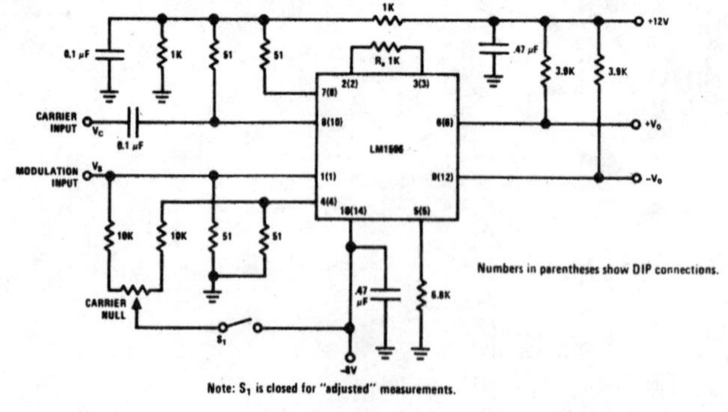

Numbers in parentheses show DIP connections.

Note: S_1 is closed for "adjusted" measurements.

Suppressed Carrier Modulator

Absolute Maximum Ratings

Internal Power Dissipation (Note 1)	500 mW
Applied Voltage (Note 2)	30V
Differential Input Signal ($V_7 - V_8$)	±5.0V
Differential Input Signal ($V_4 - V_1$)	$\pm(5+I_5R_e)$ V
Input Signal ($V_2 - V_1$, $V_3 - V_4$)	5.0V
Bias Current (I_5)	12 mA
Operating Temperature Range LM1596	−55°C to +125°C
LM1496	0°C to +70°C
Storage Temperature Range	−65°C to +150°C
Lead Temperature (Soldering, 10 sec)	300°C

Electrical Characteristics (T_A = 25°C, unless otherwise specified, see test circuit)

PARAMETER	CONDITIONS	LM1596			LM1496			UNITS
		MIN	TYP	MAX	MIN	TYP	MAX	
Carrier Feedthrough	V_C = 60 mVrms sine wave f_C = 1.0 kHz, offset adjusted		40			40		μVrms
	V_C = 60 mVrms sine wave f_C = 10 MHz, offset adjusted		140			140		μVrms
	V_C = 300 mV_{pp} square wave f_C = 1.0 kHz, offset adjusted		0.04	0.2		0.04	0.2	mVrms
	V_C = 300 mV_{pp} square wave f_C = 1.0 kHz, offset not adjusted		20	100		20	150	mVrms
Carrier Suppression	f_S = 10 kHz, 300 mVrms f_C = 500 kHz, 60 mVrms sine wave offset adjusted	50	65		50	65		dB
	f_S = 10 kHz, 300 mVrms f_C = 10 MHz, 60 mVrms sine wave offset adjusted		50			50		dB
Transadmittance Bandwidth	R_L = 50Ω Carrier Input Port, V_C = 60 mVrms sine wave f_S = 1.0 kHz, 300 mVrms sine wave		300			300		MHz
	Signal Input Port, V_S = 300 mVrms sine wave $V_7 - V_8$ = 0.5Vdc		80			80		MHz
Voltage Gain, Signal Channel	V_S = 100 mVrms, f = 1.0 kHz $V_7 - V_8$ = 0.5Vdc	2.5	3.5		2.5	3.5		V/V
Input Resistance, Signal Port	f = 5.0 MHz $V_7 - V_8$ = 0.5 Vdc		200			200		kΩ
Input Capacitance, Signal Port	f = 5.0 MHz $V_7 - V_8$ = 0.5 Vdc		2.0			2.0		pF
Single Ended Output Resistance	f = 10 MHz		40			40		kΩ
Single Ended Output Capacitance	f = 10 MHz		5.0			5.0		pF
Input Bias Current	$(I_1 + I_4)/2$		12	25		12	30	μA
Input Bias Current	$(I_7 + I_8)/2$		12	25		12	30	μA
Input Offset Current	$(I_1 - I_4)$		0.7	5.0		0.7	5.0	μA
Input Offset Current	$(I_7 - I_8)$		0.7	5.0		5.0	5.0	μA
Average Temperature Coefficient of Input Offset Current	$(-55°C < T_A < +125°C)$		2.0					nA/°C
	$(0°C < T_A < +70°C$					2.0		nA/°C
Output Offset Current	$(I_6 - I_9)$		14	50		14	60	μA
Average Temperature Coefficient of Output Offset Current	$(-55°C < T_A < +125°C)$		90					nA/°C
	$(0°C < T_A < +70°C)$					90		nA/°C
Signal Port Common Mode Input Voltage Range	f_S = 1.0 kHz		5.0			5.0		V_{p-p}
Signal Port Common Mode Rejection Ratio	$V_7 - V_8$ = 0.5 Vdc		−85			−85		dB
Common Mode Quiescent Output Voltage			8.0			8.0		Vdc
Differential Output Swing Capability			8.0			8.0		V_{p-p}
Positive Supply Current	$(I_6 + I_9)$		2.0	3.0		2.0	3.0	mA
Negative Supply Current	(I_{10})		3.0	4.0		3.0	4.0	mA
Power Dissipation			33			33		mW

Note 1: LM1596 rating applies to case temperatures to +125°C; derate linearly at 6.5 mW/°C for ambient temperature above 75°C. LM1496 rating applies to case temperatures to +70°C.

Note 2: Voltage applied between pins 6-7, 8-1, 9-7, 9-8, 7-4, 7-1, 8-4, 6-8, 2-5, 3-5.

Typical Performance Characteristics

Carrier Suppression vs Carrier Input Level

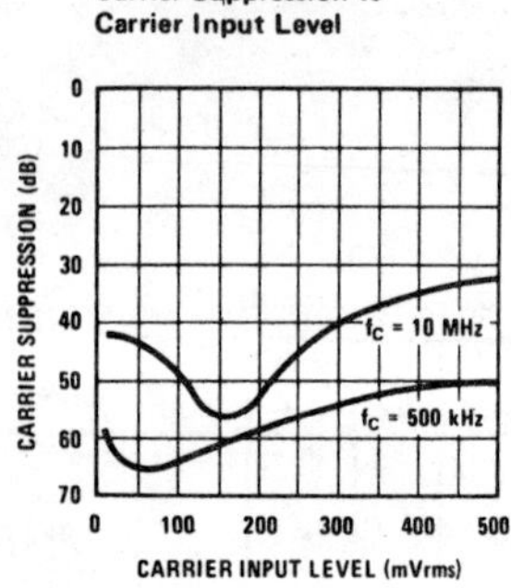

Carrier Suppression vs Frequency

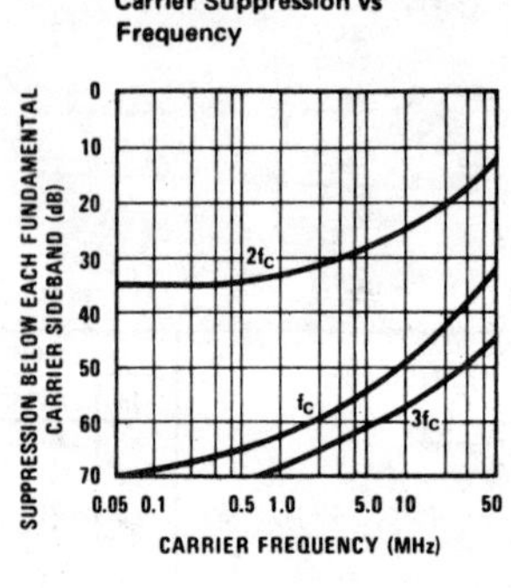

Carrier Feedthrough vs Frequency

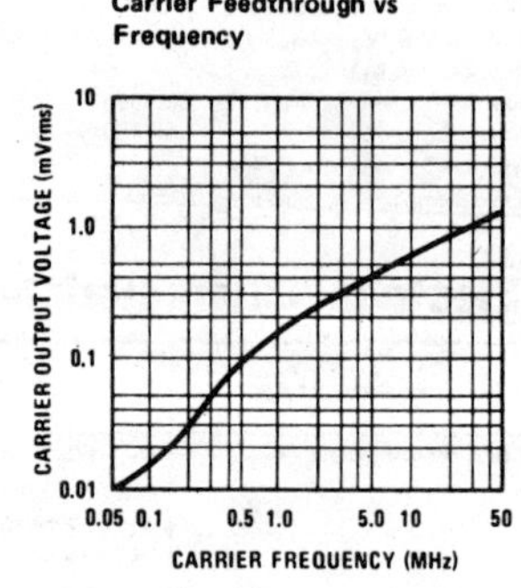

Sideband Output vs Carrier Levels

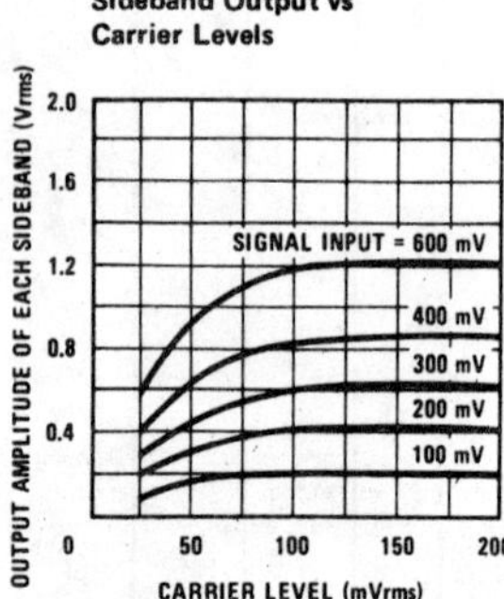

Sideband and Signal Port Transadmittances vs Frequency

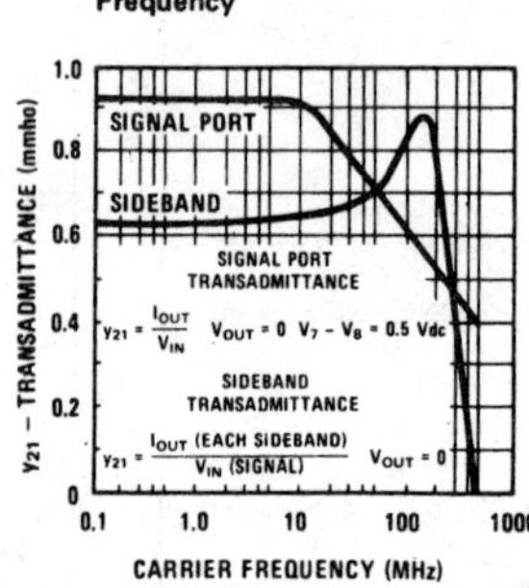

Signal-Port Frequency Response

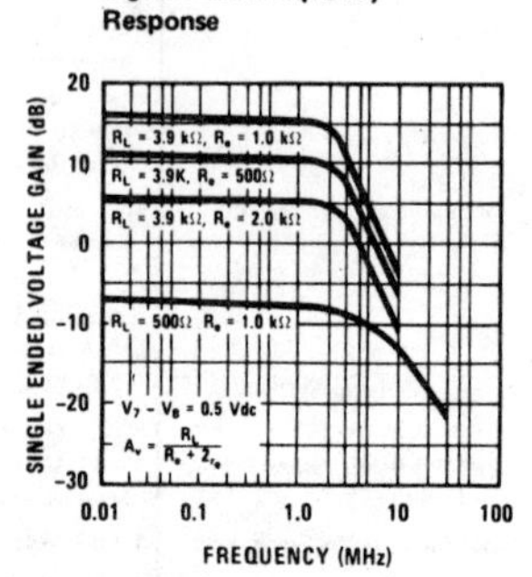

Typical Applications (Continued)

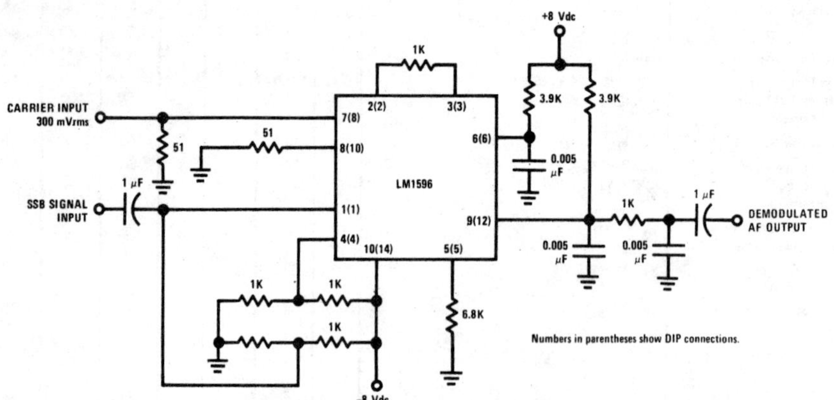

SSB Product Detector

This figure shows the LM1596 used as a single sideband (SSB) suppressed carrier demodulator (product detector). The carrier signal is applied to the carrier input port with sufficient amplitude for switching operation. A carrier input level of 300 mVrms is optimum. The composite SSB signal is applied to the signal input port with an amplitude of 5.0 to 500 mVrms. All output signal components except the desired demodulated audio are filtered out, so that an offset adjustment is not required. This circuit may also be used as an AM detector by applying composite and carrier signals in the same manner as described for product detector operation.

Typical Applications (Continued)

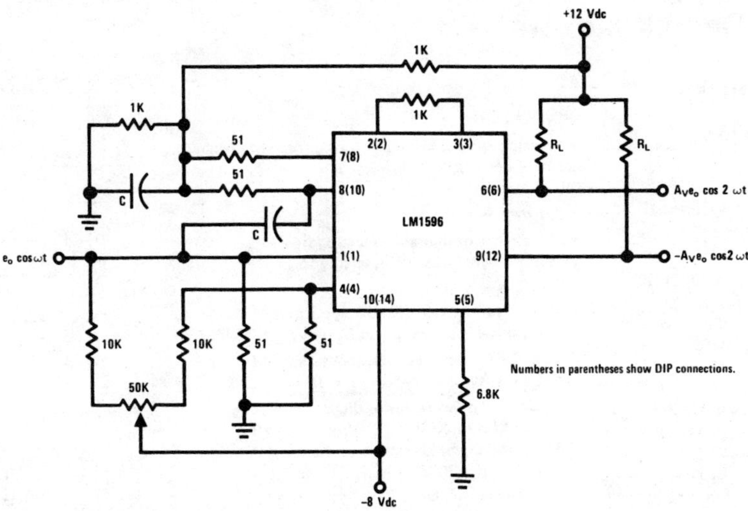

Broadband Frequency Doubler

The frequency doubler circuit shown will double low-level signals with low distortion. The value of C should be chosen for low reactance at the operating frequency.

Signal level at the carrier input must be less than 25 mV peak to maintain operation in the linear region of the switching differential amplifier. Levels to 50 mV peak may be used with some distortion of the output waveform. If a larger input signal is available a resistive divider may be used at the carrier input, with full signal applied to the signal input.

CD4051B, CD4052B, CD4053B Types

COS/MOS Analog Multiplexers/Demultiplexers*

With Logic-Level Conversion

High-Voltages Types (20-Volt Rating)

CD4051B – Single 8-Channel
CD4052B – Differential 4-Channel
CD4053B – Triple 2-Channel

RCA-CD4051B, CD4052B, and CD4053B analog multiplexers/demultiplexers are digitally controlled analog switches having low ON impedance and very low OFF leakage current. Control of analog signals up to 20 V peak-to-peak can be achieved by digital signal amplitudes of 4.5 to 20 V (if V_{DD}-V_{SS} = 3 V, a V_{DD}-V_{EE} of up to 13 V can be controlled; for V_{DD}-V_{EE} level differences above 13 V, a V_{DD}-V_{SS} of at least 4.5 V is required). For example, if V_{DD} = +4.5V, V_{SS} = 0, and V_{EE} = −13.5 V, analog signals from −13.5 V to +4.5 V can be controlled by digital inputs of 0 to 5 V. These multiplexer circuits dissipate extremely low quiescent power over the full V_{DD}-V_{SS} and V_{DD}-V_{EE} supply-voltage ranges, independent of the logic state of the control signals. When a logic "1" is present at the inhibit input terminal all channels are off.

The CD4051B is a single 8-channel multiplexer having three binary control inputs, A, B, and C, and an inhibit input. The three binary signals select 1 of 8 channels to be turned on, and connect one of the 8 inputs to the output.

The CD4052B is a differential 4-channel multiplexer having two binary control inputs, A and B, and an inhibit input. The two binary input signals select 1 of 4 pairs of channels to be turned on and connect the analog inputs to the outputs.

The CD4053B is a triple 2-channel multiplexer having three separate digital control inputs, A, B, and C, and an inhibit input. Each control input selects one of a pair of channels which are connected in a single-pole double-throw configuration.

The CD4051B, CD4052B, and CD4053B are supplied in 16-lead ceramic dual-in-line packages (D and F suffixes), 16-lead plastic dual-in-line packages (E suffix), and in chip form (H suffix).

* When these devices are used as demultiplexers, the "CHANNEL IN/OUT" terminals are the outputs and the "COMMON OUT/IN" terminals are the inputs.

Applications:

- **Analog and digital multiplexing and demultiplexing**
- **A/D and D/A conversion**
- **Signal gating**

Features:

- **Wide range of digital and analog signal levels: digital 3 to 20 V, analog to 20 V_{p-p}**
- **Low ON resistance: 125 Ω (typ.) over 15 V_{p-p} signal-input range for V_{DD}-V_{EE} = 15 V**
- **High OFF resistance: channel leakage of ±100 pA (typ.) @ V_{DD}-V_{EE} = 18 V**
- **Logic-level conversion for digital addressing signals of 3 to 20 V (V_{DD}-V_{SS} = 3 to 20 V) to switch analog signals to 20 V p-p (V_{DD}-V_{EE} = 20 V); see introductory text**
- **Matched switch characteristics: R_{ON} = 5 Ω (typ.) for V_{DD}-V_{EE} = 15 V**
- **Very low quiescent power dissipation under under all digital-control input and supply conditions: 0.2 μW (typ.) @ V_{DD}-V_{SS} = V_{DD}-V_{EE} = 10 V**
- **Binary address decoding on chip**
- **5-, 10-, and 15-V parametric ratings**
- **100% tested for quiescent current at 20 V**
- **Maximum input current of 1 μA at 18 V over full package temperature range; 100 nA at 18 V and 25°C**

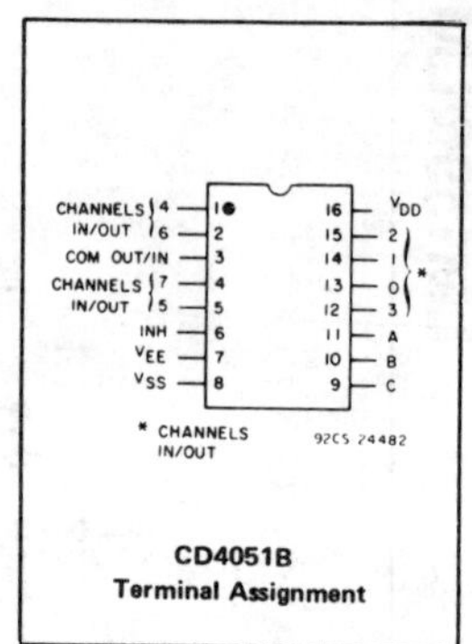

CD4051B
Terminal Assignment

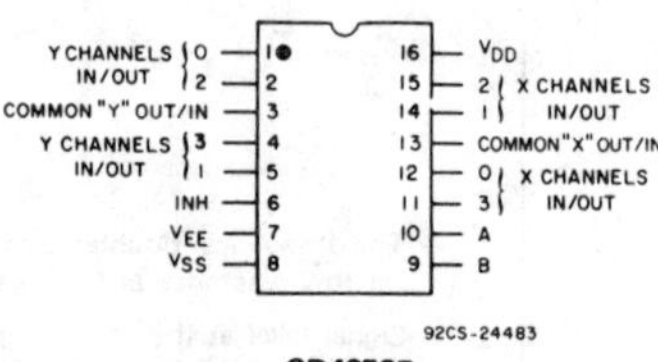

CD4052B
Terminal Assignment

RECOMMENDED OPERATING CONDITIONS AT T_A =25°C (Unless Otherwise Specified)

For maximum reliability, nominal operating conditions should be selected so that operation is always within the following ranges. Values shown apply to all types except as noted.

CHARACTERISTIC	V_{DD}	Min.	Max.	Units
Supply-Voltage Range (T_A = Full Package-Temp. Range)	–	3	18	V
Multiplexer Switch Input Current Capability*	–	–	25	mA
Output Load Resistance	–	100	–	Ω

* In certain applications, the external load-resistor current may include both V_{DD} and signal-line components. To avoid drawing V_{DD} current when switch current flows into the transmission gate inputs, the voltage drop across the bidirectional switch must not exceed 0.8 volt (calculated from R_{ON} values shown in ELECTRICAL CHARACTERISTICS CHART). No V_{DD} current will flow through R_L if the switch current flows into terminal 3 on the CD4051; terminals 3 and 13 on the CD4052; terminals 4,14, and 15 on the CD4053.

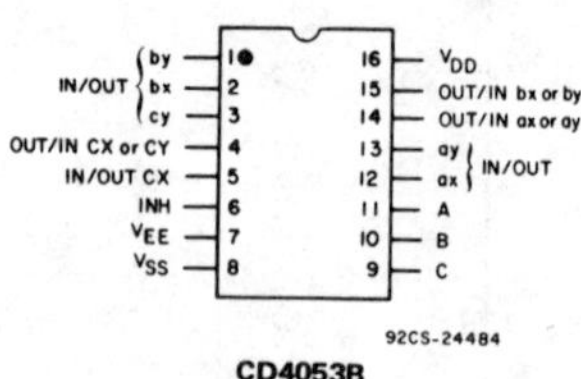

CD4053B
Terminal Assignment

MAXIMUM RATINGS, *Absolute-Maximum Values:*

DC SUPPLY-VOLTAGE RANGE, (V_{DD})
(Voltages referenced to V_{SS} or V_{EE}, whichever is more negative) . . . . −0.5 to +20 V
INPUT VOLTAGE RANGE, ALL INPUTS . . . . −0.5 to V_{DD} +0.5 V
DC INPUT CURRENT, ANY ONE INPUT . . . . ±10 mA
POWER DISSIPATION PER PACKAGE (P_D):
For T_A = −40 to +60°C (PACKAGE TYPE E) . . . . 500 mW
For T_A = +60 to +85°C (PACKAGE TYPE E) . . . . . Derate Linearly at 12 mW/°C to 200 mW
For T_A = −55 to +100°C (PACKAGE TYPES D,F) . . . . 500 mW
For T_A = +100 to +125°C (PACKAGE TYPES D, F) . . . Derate Linearly at 12 mW/°C to 200 mW
DEVICE DISSIPATION PER OUTPUT TRANSISTOR
FOR T_A = FULL PACKAGE-TEMPERATURE RANGE (All Package Types) . . . . . 100 mW
OPERATING-TEMPERATURE RANGE (T_A):
PACKAGE TYPES D, F, H . . . . −55 to +125°C
PACKAGE TYPE E . . . . −40 to +85°C
STORAGE TEMPERATURE RANGE (T_{stg}) . . . . −65 to +150°C
LEAD TEMPERATURE (DURING SOLDERING):
At distance 1/16 ± 1/32 inch (1.59 ± 0.79 mm) from case for 10 s max. . . . . +265°C

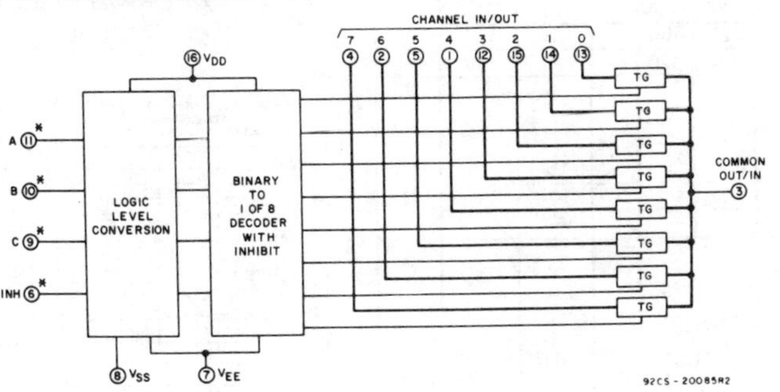

Fig. 1 – Functional diagram of CD4051B.

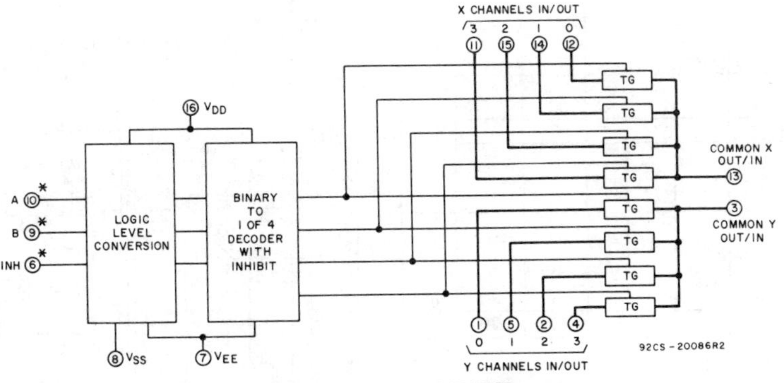

Fig. 2 – Functional diagram of CD4052B.

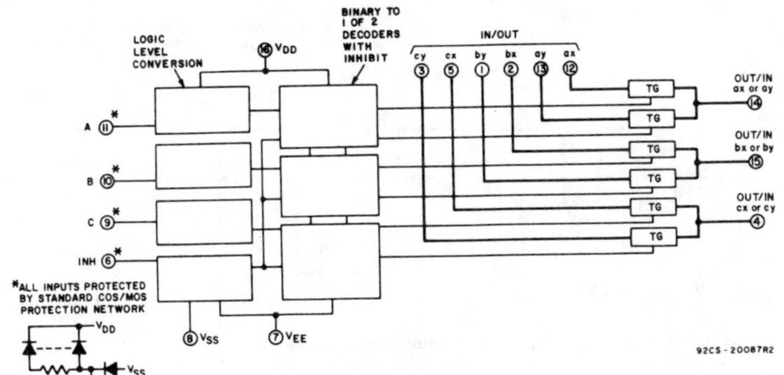

Fig. 3 – Functional diagram of CD4053B.

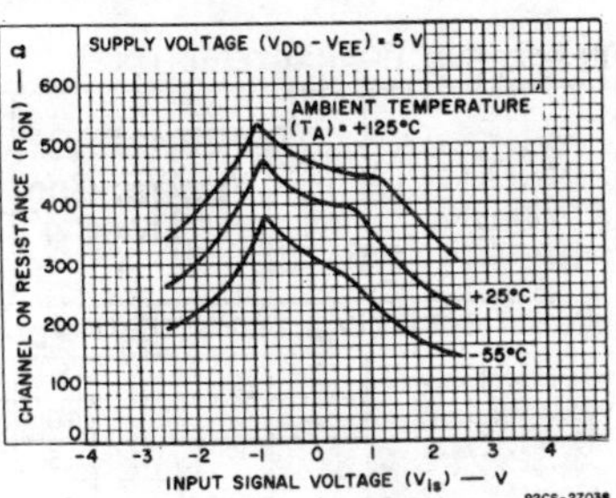

Fig.4 – Typical channel ON resistance vs input signal voltage (all types).

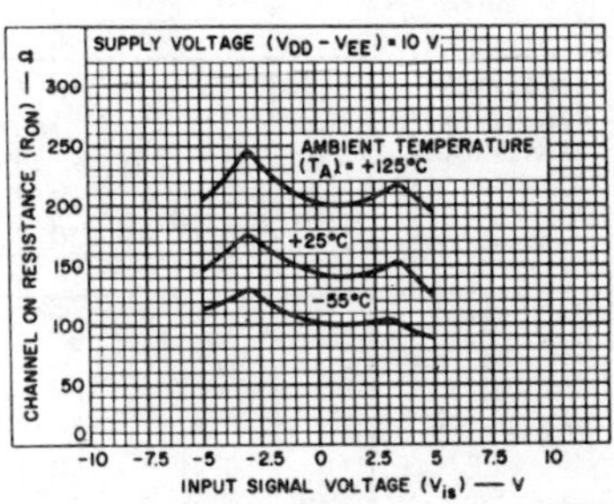

Fig.5 – Typical channel ON resistance vs. input signal voltage (all types).

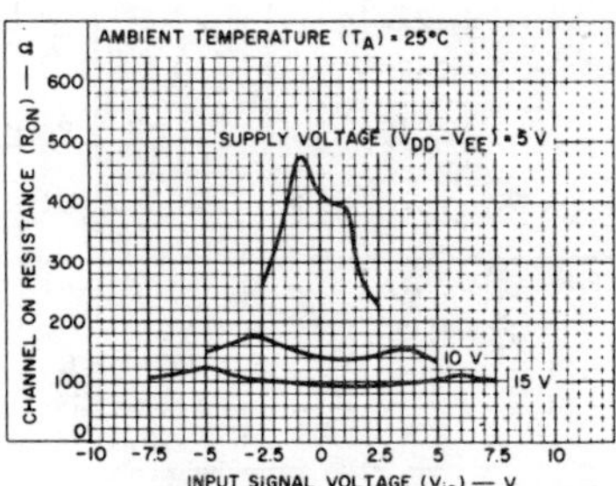

Fig.6 – Typical channel ON resistance vs. input signal voltage (all types).

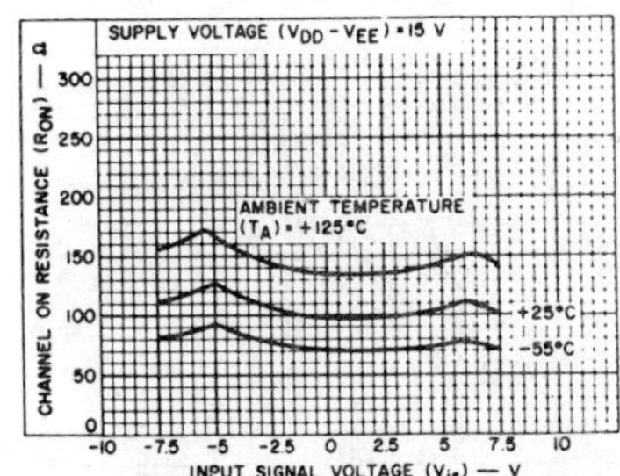

Fig.7 – Typical channel ON resistance vs. input signal voltage (all types).

CD4051B, CD4052B, CD4053B Types

ELECTRICAL CHARACTERISTICS

CHARACTERISTIC	CONDITIONS V_{is} (V)	V_{EE} (V)	V_{SS} (V)	V_{DD} (V)	LIMITS at Indicated Temperature (°C) Values at −55,+25,+125, apply to D,F,H pkg Values at −40,+25,+85, apply to E pkgs −55	−40	+85	+125	+25 Min.	Typ.	Max.	Units
SIGNAL INPUTS (V_{is}) AND OUTPUTS (V_{OS})												
Quiescent Device Current, I_{DD} Max.				5	5	5	150	150	–	0.04	5	μA
				10	10	10	300	300	–	0.04	10	
				15	20	20	600	600	–	0.04	20	
				20	100	100	3000	3000	–	0.08	100	
On-State Resistance $0 \leq V_{is} \leq V_{DD}$ r_{on} Max.		0	0	5	800	850	1200	1300	–	470	1050	Ω
		0	0	10	310	330	520	550	–	180	400	
		0	0	15	200	210	300	320	–	125	240	
Change in On-State Resistance (Between Any Two Channels) Δr_{on}		0	0	5	–	–	–	–	–	15	–	Ω
		0	0	10	–	–	–	–	–	10	–	
		0	0	15	–	–	–	–	–	5	–	
OFF Channel Leakage Current: Any Channel OFF Max. or All Channels OFF (Common OUT/IN) Max.		0	0	18	±100*		±1000*		–	±0.01	±100*	nA
Capacitance: Input, C_{is}		−5	−5	5	–	–	–	–	–	5	–	pF
Output, C_{os} CD4051					–	–	–	–	–	30	–	
CD4052					–	–	–	–	–	18	–	
CD4053					–	–	–	–	–	9	–	
Feedthrough, C_{ios}					–	–	–	–	–	0.2	–	
Propagation Delay Time (Signal Input to Output)	V_{DD} ⎍	R_L = 200 kΩ		5	–	–	–	–	–	30	60	ns
		C_L = 50 pF		10	–	–	–	–	–	15	30	
		t_r, t_f = 20 ns		15	–	–	–	–	–	10	20	

* Determined by minimum feasible leakage measurement for automatic testing.

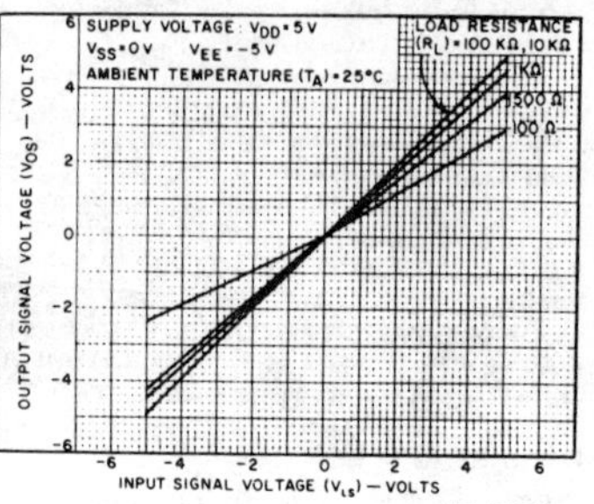

Fig.8 – Typical ON characteristics for 1 of 8 channels (CD4051B).

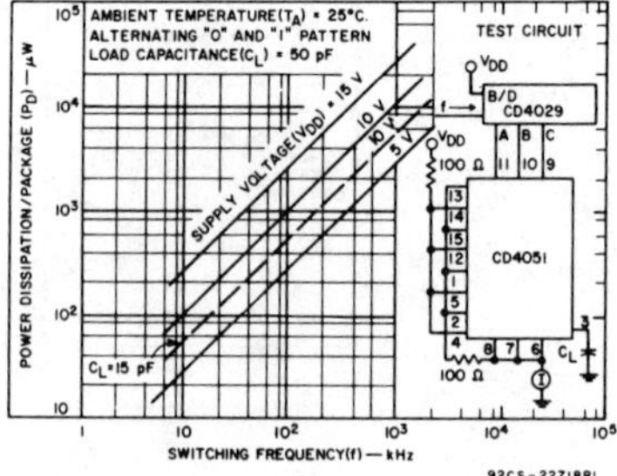

Fig.9 – Typical dynamic power dissipation vs. switching frequency (CD4051B).

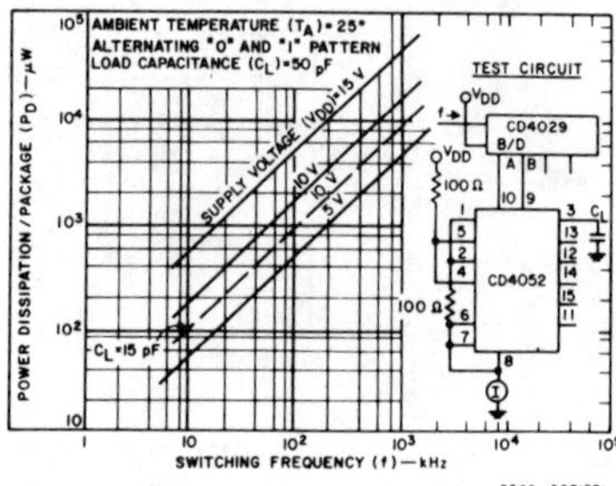

Fig.10 – Typical dynamic power dissipation vs. switching frequency (CD4052B).

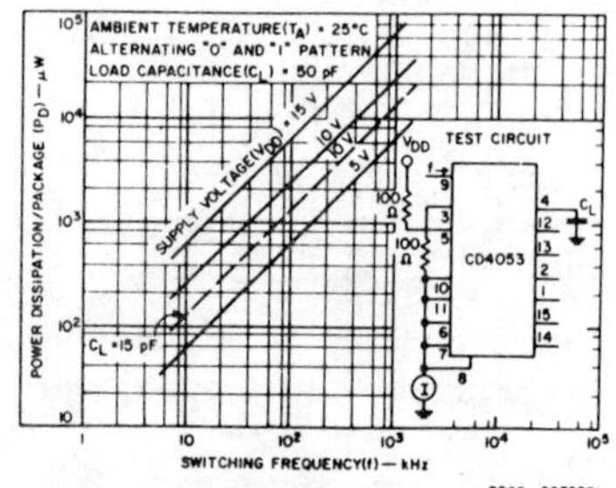

Fig.11 – Typical dynamic power dissipation vs. switching frequency (CD4053B).

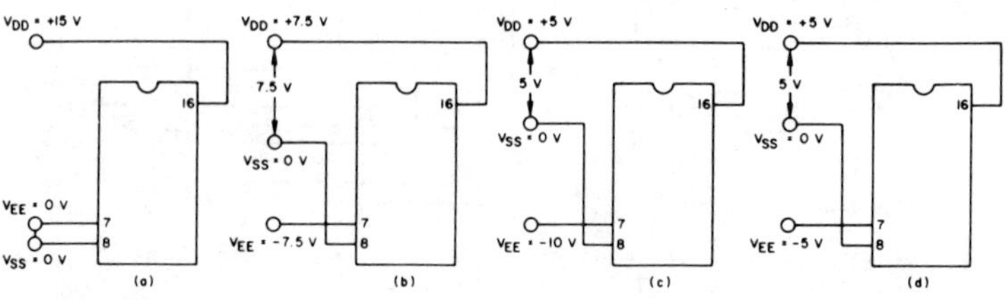

The ADDRESS (digital-control inputs) and INHIBIT logic levels are: "0" = V_{SS} and "1" = V_{DD}. The analog signal (through the TG) may swing from V_{EE} to V_{DD}.

Fig.12 – Typical bias voltages.

ELECTRICAL CHARACTERISTICS (Cont'd)

CHARACTERISTIC	CONDITIONS V_{is} (V)	V_{EE} (V)	V_{SS} (V)	V_{DD} (V)	LIMITS at Indicated Temperature (°C) Values at −55,+25,+125, apply to D,F,H pkg; Values at −40,+25,+85, apply to E pkgs: −55	−40	+85	+125	+25 Min.	+25 Typ.	+25 Max.	Units
CONTROL (ADDRESS or INHIBIT) V_C												
Input Low Voltage, V_{IL} Max.	=V_{DD} thru 1 kΩ	V_{EE}=V_{SS}, R_L=1 kΩ to V_{SS}, I_{IS}< 2 µA on all OFF Channels		5	1.5				–	–	1.5	V
				10	3				–	–	3	
				15	4				–	–	4	
Input High Voltage, V_{IH} Min.				5	3.5				3.5	–	–	
				10	7				7	–	–	
				15	11				11	–	–	
Input Current, I_{IN} Max.	V_{IN} = 0,18			18	±0.1	±0.1	±1	±1	–	±10^{-5}	±0.1	µA
Propagation Delay Time: Address-to-Signal OUT (Channels ON or OFF) See Figs.14,15,18	t_r,t_f = 20 ns, C_L = 50 pF	0	0	5	–	–	–	–	–	360	720	ns
		0	0	10	–	–	–	–	–	160	320	
		0	0	15	–	–	–	–	–	120	240	
		−5	0	5	–	–	–	–	–	225	450	
Inhibit-to-Signal OUT (Channel turning ON)	R_L=10 kΩ, C_L=50 pF, t_r, t_f = 20 ns	0	0	5	–	–	–	–	–	360	720	ns
		0	0	10	–	–	–	–	–	160	320	
		0	0	15	–	–	–	–	–	120	240	
		−10	0	5	–	–	–	–	–	200	400	
Inhibit-to-Signal OUT (Channel turning OFF)	R_L=300Ω, C_L=50 pF, t_r,t_f = 20 ns	0	0	5	–	–	–	–	–	200	450	ns
		0	0	10	–	–	–	–	–	90	210	
		0	0	15	–	–	–	–	–	70	160	
		−10	0	5	–	–	–	–	–	130	300	
Input Capacitance, C_{IN} (Any Address or Inhibit Input)					–	–	–	–	–	5	7.5	pF

CD4051B

INHIBIT	C	B	A	"ON" CHANNEL(S)
0	0	0	0	0
0	0	0	1	1
0	0	1	0	2
0	0	1	1	3
0	1	0	0	4
0	1	0	1	5
0	1	1	0	6
0	1	1	1	7
1	X	X	X	NONE

CD4052B

INHIBIT	B	A	"ON" CHANNEL(S)
0	0	0	0x, 0y
0	0	1	1x, 1y
0	1	0	2x, 2y
0	1	1	3x, 3y
1	X	X	NONE

CD4053B

INHIBIT	A or B or C	"ON" CHANNEL(S)
0	0	ax or bx or cx
0	1	ay or by or cy
1	X	NONE

X = Don't care

Fig. 13 – Truth tables.

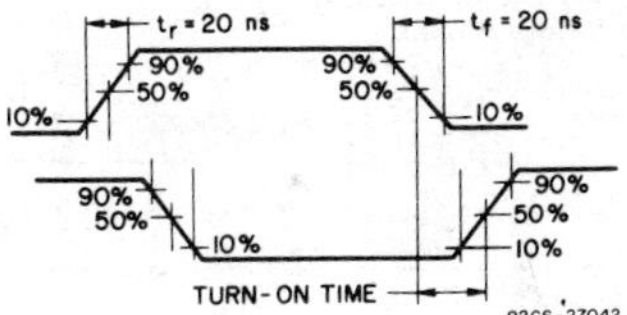

Fig. 14 – Waveforms, channel being turned ON (R_L = 10 kΩ).

Fig. 15 – Waveforms, channel being turned OFF (R_L = 300 Ω).

TEST CIRCUITS

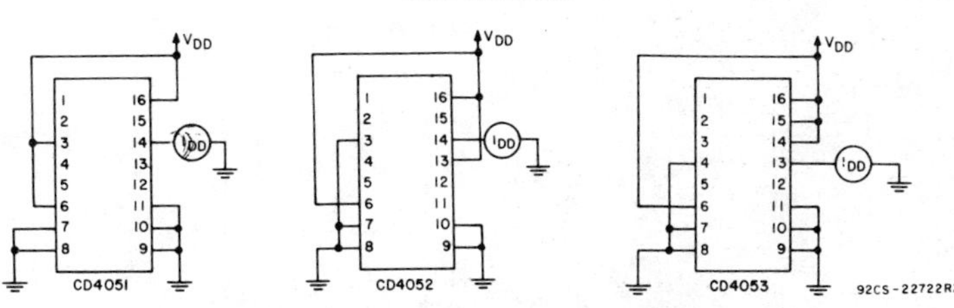

Fig. 16 – OFF channel leakage current – any channel OFF.

ELECTRICAL CHARACTERISTICS (Cont'd)

CHARACTERISTIC	TEST CONDITIONS V_{is} (V)	V_{DD} (V)	R_L (kΩ)			LIMITS TYPICAL VALUE	UNITS
Cutoff (−3-dB) Frequency Channel ON (Sine Wave Input)	5•	10	1	V_{OS} at Common OUT/IN	CD4053	30	MHz
	$V_{EE} = V_{SS}$, $20 \log \frac{V_{OS}}{V_{is}} = -3$ dB				CD4052	25	
					CD4051	20	
				V_{OS} at Any Channel		60	
Total Harmonic Distortion, THD	2•	5	10			0.3	%
	3•	10				0.2	
	5•	15				0.12	
	$V_{EE} = V_{SS}$, f_{is} = 1 kHz sine wave						
−40-dB Feedthrough Frequency (All Channels OFF)	5•	10	1	V_{OS} at Common OUT/IN	CD4053	8	MHz
	$V_{EE} = V_{SS}$, $20 \log \frac{V_{OS}}{V_{is}} = -40$dB				CD4052	10	
					CD4051	12	
				V_{OS} at Any Channel		8	
−40-dB Signal Crosstalk Frequency	5•	10	1	Between Any 2 Channels		3	MHz
	$V_{EE} = V_{SS}$, $20 \log \frac{V_{OS}}{V_{is}} = -40$dB			Between Sections CD4052 Only	Measured on Common	6	
					Measured on Any Channel	10	
				Between Any 2 Sections CD4053 Only	In Pin 2, Out Pin 14	2.5	
					In Pin 15, Out Pin 14	6	
Address-or-Inhibit-to Signal Crosstalk	–	10	10#			65	mV (Peak)
	$V_{EE}=0, V_{SS}=0$, t_r, t_f = 20 ns, $V_C = V_{DD} - V_{SS}$ (Square Wave)						

• Peak-to-peak voltage symmetrical about $\frac{V_{DD} - V_{EE}}{2}$

Both ends of channel

TEST CIRCUITS (Cont'd)

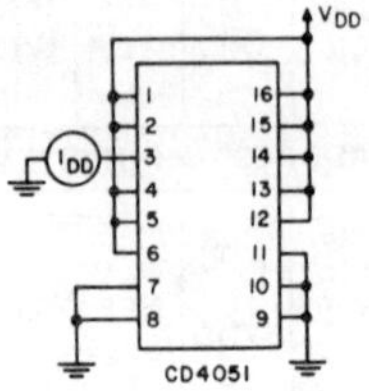

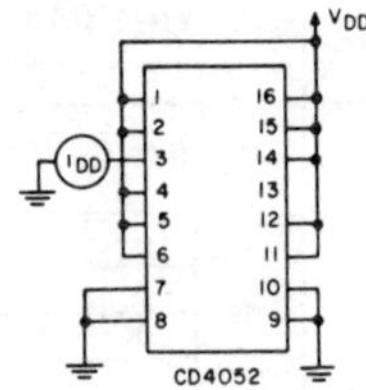

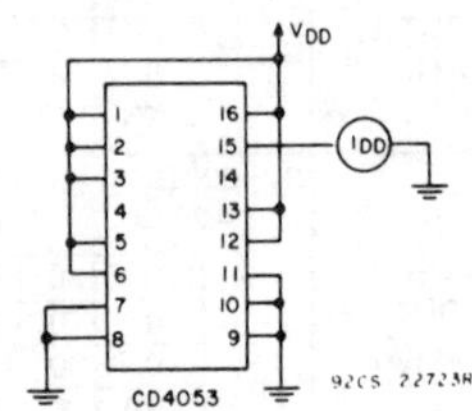

Fig.17 – OFF channel leakage current – all channels OFF.

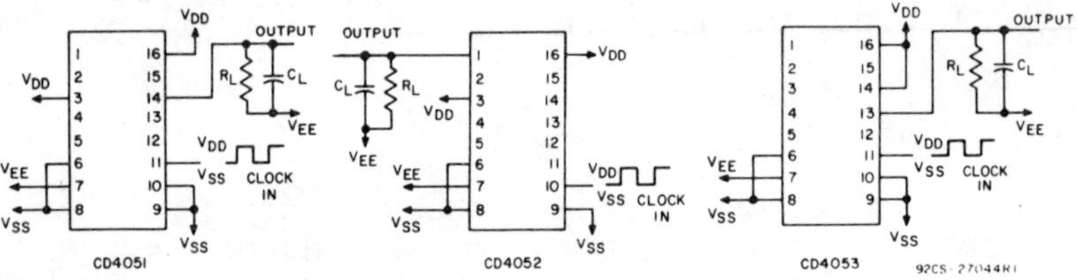

Fig.18 – Propagation delay – address input to signal output.

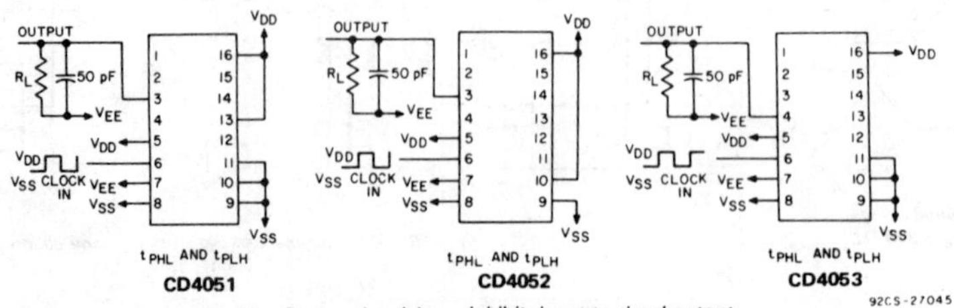

Fig.19 – Propagation delay – inhibit input to signal output.

COUNTERS

54/7493, LS93

4-Bit Binary Ripple Counter

DESCRIPTION

The '93 is a 4-bit, ripple-type Binary Counter. The device consists of four master-slave flip-flops internally connected to provide a divide-by-two section and a divide-by-eight section. Each section has a separate Clock input to initiate state changes of the counter on the HIGH-to-LOW clock transition. State changes of the Q outputs do not occur simultaneously because of internal ripple delays. Therefore, decoded output signals are subject to decoding spikes and should not be used for clocks or strobes.

A gated AND asynchronous Master Reset ($MR_1 \cdot MR_2$) is provided which overrides both clocks and resets (clears) all the flip-flops.

Since the output from the divide-by-two section is not internally connected to the succeeding stages, the device may be operated in various counting modes. In a 4-bit ripple counter the output Q_0 must be connected externally to input $\overline{CP}_1$. The input count pulses are applied to input $\overline{CP}_0$. Simultaneous divisions of 2, 4, 8 and 16 are performed at the Q_0, Q_1, Q_2 and Q_3 outputs as shown in the Function Table. As a 3-bit ripple counter the input count pulses are applied to input $\overline{CP}_1$. Simultaneous frequency divisions of 2, 4 and 8 are available at the Q_1, Q_2 and Q_3 outputs. Independent use of the first flip-flop is available if the reset function coincides with reset of the 3-bit ripple-through counter.

TYPE	TYPICAL f_{MAX}	TYPICAL SUPPLY CURRENT (Total)
7493	40MHz	28mA
74LS93	42MHz	9mA

ORDERING CODE

PACKAGES	COMMERCIAL RANGES $V_{CC}=5V \pm 5\%$; $T_A = 0°C$ to $+70°C$	MILITARY RANGES $V_{CC}=5V \pm 10\%$; $T_A = -55°C$ to $+125°C$
Plastic DIP	N7493N • N74LS93N	
Ceramic DIP	N7493F • N74LS93F	S5493F • S54LS93F
Flatpack		S5493W • S54LS93W

INPUT AND OUTPUT LOADING AND FAN-OUT TABLE

PINS	DESCRIPTION	54/74	54/74LS
MR	Master Reset Inputs	1ul	1LSul
$\overline{CP}_0$	Input	2ul	6LSul
$\overline{CP}_1$	Input	2ul	4LSul
Q_0-Q_3	Outputs	10ul	10LSul

NOTE
Where a 54/74 unit load (ul) is understood to be 40μA I_{IH} and −1.6mA I_{IL}, and a 54/74LS unit load (LSul) is 20μA I_{IH} and −0.4mA I_{IL}.

PIN CONFIGURATION

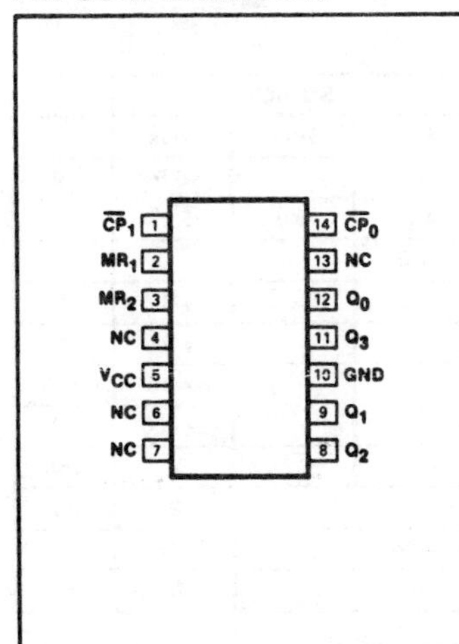

LOGIC SYMBOL

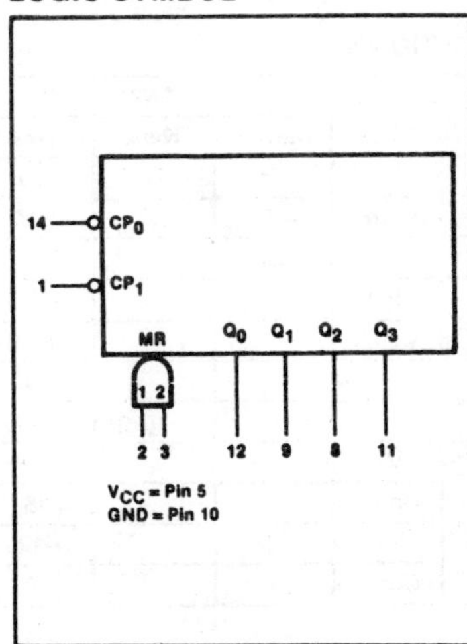

LOGIC SYMBOL (IEEE/IEC)

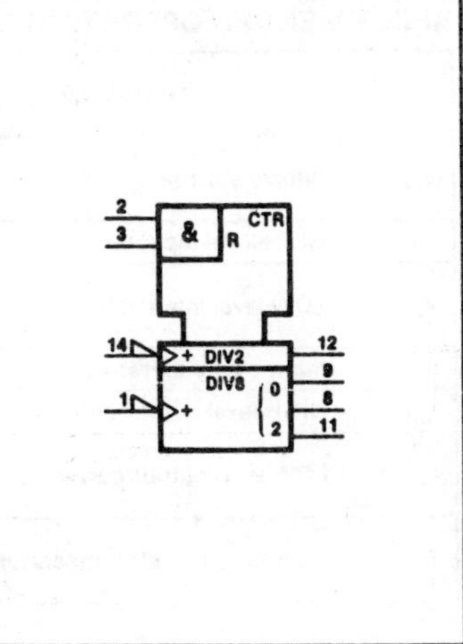

LOGIC DIAGRAM

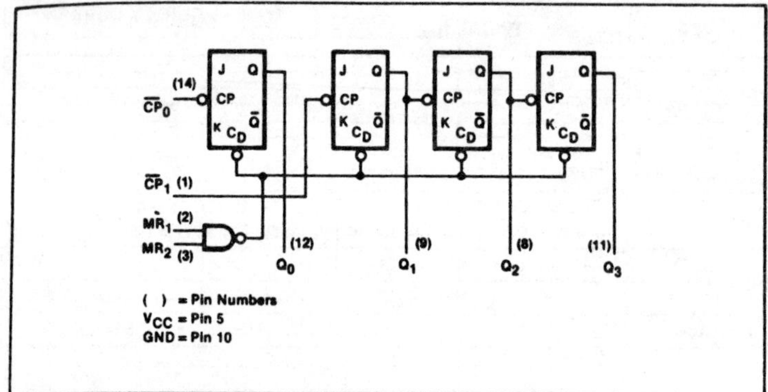

FUNCTION TABLE

COUNT	OUTPUTS			
	Q_0	Q_1	Q_2	Q_3
0	L	L	L	L
1	H	L	L	L
2	L	H	L	L
3	H	H	L	L
4	L	L	H	L
5	H	L	H	L
6	L	H	H	L
7	H	H	H	L
8	L	L	L	H
9	H	L	L	H
10	L	H	L	H
11	H	H	L	H
12	L	L	H	H
13	H	L	H	H
14	L	H	H	H
15	H	H	H	H

NOTE
Output Q_0 connected to input $\overline{CP}_1$.

3

MODE SELECTION

RESET INPUTS		OUTPUTS			
MR_1	MR_2	Q_0	Q_1	Q_2	Q_3
H	H	L	L	L	L
L	H	Count			
H	L	Count			
L	L	Count			

H = HIGH voltage level
L = LOW voltage level
X = Don't care

ABSOLUTE MAXIMUM RATINGS (Over operating free-air temperature range unless otherwise noted.)

PARAMETER		54	54LS	74	74LS	UNIT
V_{CC}	Supply voltage	7.0	7.0	7.0	7.0	V
V_{IN}	Input voltage	−0.5 to +5.5	−0.5 to +7.0	−0.5 to +5.5	−0.5 to +7.0	V
I_{IN}	Input current	−30 to +5	−30 to +1	−30 to +5	−30 to +1	mA
V_{OUT}	Voltage applied to output in HIGH output state	−0.5 to $+V_{CC}$	−0.5 to $+V_{CC}$	−0.5 to $+V_{CC}$	−0.5 to $+V_{CC}$	V
T_A	Operating free-air temperature range	−55 to +125		0 to 70		°C

NOTE
V_{IN} is limited to 5.5V on $\overline{CP}_0$ and $\overline{CP}_1$ inputs only on the 54/74LS93.

RECOMMENDED OPERATING CONDITIONS

PARAMETER			54/74			54/74LS			UNIT
			Min	Nom	Max	Min	Nom	Max	
V_{CC}	Supply voltage	Mil	4.5	5.0	5.5	4.5	5.0	5.5	V
		Com'l	4.75	5.0	5.25	4.75	5.0	5.25	V
V_{IH}	HIGH-level input voltage		2.0			2.0			V
V_{IL}	LOW-level input voltage	Mil			+0.8			+0.7	V
		Com'l			+0.8			+0.8	V
I_{IK}	Input clamp current				−12			−18	mA
I_{OH}	HIGH-level output current				−800			−400	μA
I_{OL}	LOW-level output current	Mil			16			4	mA
		Com'l			16			8	mA
T_A	Operating free-air temperature	Mil	−55		+125	−55		+125	°C
		Com'l	0		70	0		70	°C

DC ELECTRICAL CHARACTERISTICS (Over recommended operating free-air temperature range unless otherwise noted.)

	PARAMETER	TEST CONDITIONS[1]			54/7493 Min	54/7493 Typ[2]	54/7493 Max	54/74LS93 Min	54/74LS93 Typ[2]	54/74LS93 Max	UNIT
V_{OH}	HIGH-level output voltage	V_{CC} = MIN, V_{IH} = MIN, V_{IL} = MAX, I_{OH} = MAX		Mil	2.4	3.4		2.5	3.4		V
				Com'l	2.4	3.4		2.7	3.4		V
V_{OL}	LOW-level output voltage	V_{CC} = MIN, V_{IH} = MIN, V_{IL} = MAX	I_{OL} = MAX	Mil		0.2	0.4		0.25	0.4	V
				Com'l		0.2	0.4		0.35	0.5	V
			I_{OL} = 4mA	74LS					0.25	0.4	V
V_{IK}	Input clamp voltage	V_{CC} = MIN, $I_I = I_{IK}$					−1.5			−1.5	V
I_I	Input current at maximum input voltage	V_{CC} = MAX	V_I = 5.5V	All inputs '93			1.0				mA
			V_I = 7.0V	MR inputs						0.1	mA
			V_I = 5.5V	$\overline{CP}_0$, $\overline{CP}_1$ inputs						0.2	mA
I_{IH}	HIGH-level input current	V_{CC} = MAX	V_I = 2.4V	MR inputs			40				μA
				$\overline{CP}_0$, $\overline{CP}_1$ inputs			80				μA
			V_I = 2.7V	MR inputs						20	μA
				$\overline{CP}_0$, $\overline{CP}_1$ inputs[5]						40	μA
I_{IL}	LOW-level input current	V_{CC} = MAX	V_I = 0.4V	MR inputs			−1.6			−0.4	mA
				$\overline{CP}_0$ input			−3.2			−2.4	mA
				$\overline{CP}_1$ input			−3.2			−1.6	mA
I_{OS}	Short-circuit output current[3]	V_{CC} = MAX		Mil	−20		−55	−20		−100	mA
				Com'l	−18		−55	−20		−100	mA
I_{CC}	Supply current[4] (total)	V_{CC} = MAX		Mil		28	46		9	15	mA
				Com'l		28	53		9	15	mA

1. For conditions shown as MIN or MAX, use the appropriate value specified under recommended operating conditions for the applicable type.
2. All typical values are at V_{CC} = 5V, T_A = 25°C.
3. I_{OS} is tested with V_{OUT} = +0.5V and V_{CC} = MAX + 0.5V. Not more than one output should be shorted at a time and duration of the short circuit should not exceed one second.
4. I_{CC} is measured with all outputs open, both MR inputs grounded following momentary connection to 4.5V, and all other inputs grounded.
5. The maximum limit for the 54LS93 only is 80μA for $\overline{CP}_0$ and $\overline{CP}_1$ inputs.

AC CHARACTERISTICS T_A = 25°C, V_{CC} = 5.0V

	PARAMETER	TEST CONDITIONS	54/74 C_L = 15pF, R_L = 400Ω Min	54/74 C_L = 15pF, R_L = 400Ω Max	54/74LS C_L = 15pF, R_L = 2kΩ Min	54/74LS C_L = 15pF, R_L = 2kΩ Max	UNIT
f_{MAX} f_{MAX}	$\overline{CP}_0$ input count frequency $\overline{CP}_1$ input count frequency	Waveform 1	10 10		32 16		MHz
t_{PLH} t_{PHL}	Propagation delay $\overline{CP}_0$ input to Q_0 output	Waveform 1				16 18	ns
t_{PLH} t_{PHL}	Propagation delay $\overline{CP}_1$ input to Q_1 output	Waveform 1				16 21	ns
t_{PLH} t_{PHL}	Propagation delay $\overline{CP}_1$ input to Q_2 output	Waveform 1				32 35	ns
t_{PLH} t_{PHL}	Propagation delay $\overline{CP}_1$ input to Q_3 output	Waveform 1				51 51	ns
t_{PLH} t_{PHL}	Propagation delay $\overline{CP}_0$ input to Q_3 output	Waveform 1		135 135		70 70	ns
t_{PHL}	MR input to any output	Waveform 2				40	ns

NOTE
Per industry convention, f_{MAX} is the worst case value of the maximum device operating frequency with no constraints on t_r, t_f, pulse width or duty cycle.

AC SETUP REQUIREMENTS $T_A = 25°C$, $V_{CC} = 5.0V$

PARAMETER		TEST CONDITIONS	54/74		54/74LS		UNIT
			Min	Max	Min	Max	
t_W	$\overline{CP}_0$ pulse width	Waveform 1	50		15		ns
t_W	$\overline{CP}_1$ pulse width	Waveform 1	50		30		ns
t_W	MR pulse width	Waveform 2	50		15		ns
t_{rec}	Recovery time, MR to $\overline{CP}$	Waveform 2			25		ns

3

AC WAVEFORMS

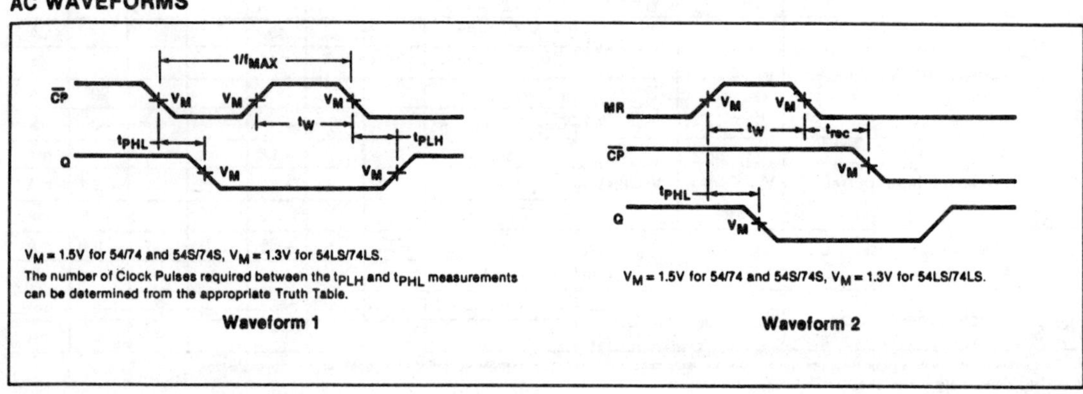

$V_M = 1.5V$ for 54/74 and 54S/74S, $V_M = 1.3V$ for 54LS/74LS.
The number of Clock Pulses required between the t_{PLH} and t_{PHL} measurements can be determined from the appropriate Truth Table.

Waveform 1

$V_M = 1.5V$ for 54/74 and 54S/74S, $V_M = 1.3V$ for 54LS/74LS.

Waveform 2

TEST CIRCUITS AND WAVEFORMS

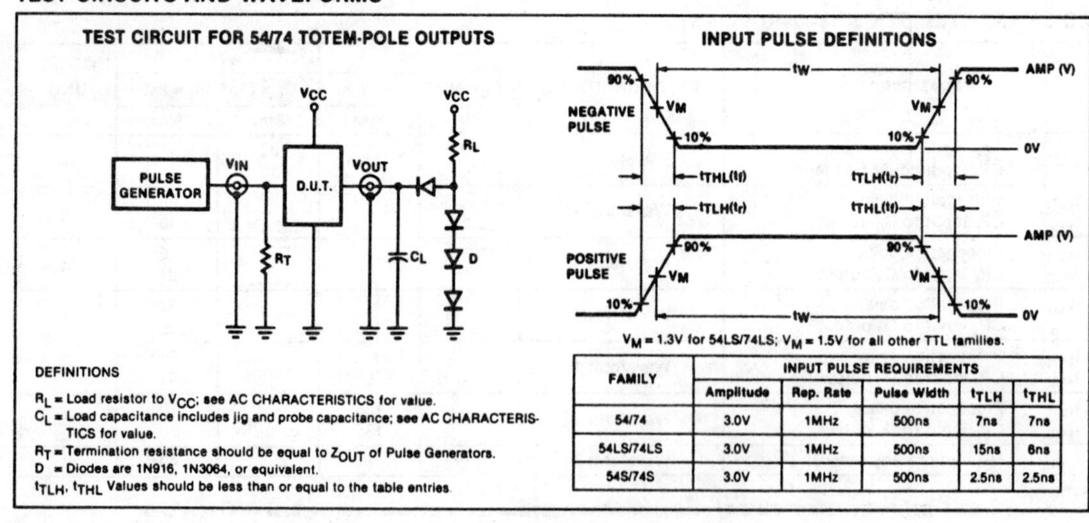

$V_M = 1.3V$ for 54LS/74LS; $V_M = 1.5V$ for all other TTL families.

DEFINITIONS

R_L = Load resistor to V_{CC}; see AC CHARACTERISTICS for value.
C_L = Load capacitance includes jig and probe capacitance; see AC CHARACTERISTICS for value.
R_T = Termination resistance should be equal to Z_{OUT} of Pulse Generators.
D = Diodes are 1N916, 1N3064, or equivalent.
t_{TLH}, t_{THL} Values should be less than or equal to the table entries.

FAMILY	INPUT PULSE REQUIREMENTS				
	Amplitude	Rep. Rate	Pulse Width	t_{TLH}	t_{THL}
54/74	3.0V	1MHz	500ns	7ns	7ns
54LS/74LS	3.0V	1MHz	500ns	15ns	6ns
54S/74S	3.0V	1MHz	500ns	2.5ns	2.5ns

Voltage Regulators

LM78XX Series Voltage Regulators

General Description

The LM78XX series of three terminal regulators is available with several fixed output voltages making them useful in a wide range of applications. One of these is local on card regulation, eliminating the distribution problems associated with single point regulation. The voltages available allow these regulators to be used in logic systems, instrumentation, HiFi, and other solid state electronic equipment. Although designed primarily as fixed voltage regulators these devices can be used with external components to obtain adjustable voltages and currents.

The LM78XX series is available in an aluminum TO-3 package which will allow over 1.0A load current if adequate heat sinking is provided. Current limiting is included to limit the peak output current to a safe value. Safe area protection for the output transistor is provided to limit internal power dissipation. If internal power dissipation becomes too high for the heat sinking provided, the thermal shutdown circuit takes over preventing the IC from overheating.

Considerable effort was expended to make the LM78XX series of regulators easy to use and minimize the number of external components. It is not necessary to bypass the output, although this does improve transient response. Input bypassing is needed only if the regulator is located far from the filter capacitor of the power supply.

For output voltage other than 5V, 12V and 15V the LM117 series provides an output voltage range from 1.2V to 57V.

Features

- Output current in excess of 1A
- Internal thermal overload protection
- No external components required
- Output transistor safe area protection
- Internal short circuit current limit
- Available in the aluminum TO-3 package

Voltage Range

LM7805C	5V
LM7812C	12V
LM7815C	15V

Schematic and Connection Diagrams

Metal Can Package
TO-3 (K)
Aluminum

BOTTOM VIEW

Order Numbers
LM7805CK
LM7812CK
LM7815CK
See Package KC02A

Plastic Package
TO-220 (T)

TOP VIEW

Order Numbers:
LM7805CT
LM7812CT
LM7815CT
See Package T03B

Absolute Maximum Ratings

Input Voltage (V_O = 5V, 12V and 15V)	35V
Internal Power Dissipation (Note 1)	Internally Limited
Operating Temperature Range (T_A)	0°C to +70°C
Maximum Junction Temperature	
(K Package)	150°C
(T Package)	125°C
Storage Temperature Range	−65°C to +150°C
Lead Temperature (Soldering, 10 seconds)	
TO-3 Package K	300°C
TO-220 Package T	230°C

Electrical Characteristics LM78XXC (Note 2) 0°C ≤ Tj ≤ 125°C unless otherwise noted.

OUTPUT VOLTAGE				5V			12V			15V			UNITS
INPUT VOLTAGE (unless otherwise noted)				10V			19V			23V			
	PARAMETER	CONDITIONS		MIN	TYP	MAX	MIN	TYP	MAX	MIN	TYP	MAX	
V_O	Output Voltage	T_j = 25°C, 5 mA ≤ I_O ≤ 1A		4.8	5	5.2	11.5	12	12.5	14.4	15	15.6	V
		P_D ≤ 15W, 5 mA ≤ I_O ≤ 1A		4.75		5.25	11.4		12.6	14.25		15.75	V
		V_{MIN} ≤ V_{IN} ≤ V_{MAX}		(7 ≤ V_{IN} ≤ 20)			(14.5 ≤ V_{IN} ≤ 27)			(17.5 ≤ V_{IN} ≤ 30)			V
ΔV_O	Line Regulation	I_O = 500 mA	T_j = 25°C		3	50		4	120		4	150	mV
			ΔV_{IN}	(7 ≤ V_{IN} ≤ 25)			(14.5 ≤ V_{IN} ≤ 30)			(17.5 ≤ V_{IN} ≤ 30)			V
			0°C ≤ T_j ≤ +125°C			50			120			150	mV
			ΔV_{IN}	(8 ≤ V_{IN} ≤ 20)			(15 ≤ V_{IN} ≤ 27)			(18.5 ≤ V_{IN} ≤ 30)			V
		I_O ≤ 1A	T_j = 25°C			50			120			150	mV
			ΔV_{IN}	(7.3 ≤ V_{IN} ≤ 20)			(14.6 ≤ V_{IN} ≤ 27)			(17.7 ≤ V_{IN} ≤ 30)			V
			0° ≤ T_j ≤ +125°C			25			60			75	mV
			ΔV_{IN}	(8 ≤ IN ≤ 12)			(16 ≤ V_{IN} ≤ 22)			(20 ≤ V_{IN} ≤ 26)			V
ΔV_O	Load Regulation	T_j = 25°C	5 mA ≤ I_O ≤ 1.5A		10	50		12	120		12	150	mV
			250 mA ≤ I_O ≤ 750 mA			25			60			75	mV
		5 mA ≤ I_O ≤ 1A, 0°C ≤ T_j ≤ +125°C				50			120			150	mV
I_Q	Quiescent Current	I_O ≤ 1A	T_j = 25°C			8			8			8	mA
			0°C ≤ T_j ≤ +125°C			8.5			8.5			8.5	mA
ΔI_Q	Quiescent Current Change	5 mA ≤ I_O ≤ 1A				0.5			0.5			0.5	mA
		T_j = 25°C, I_O ≤ 1A				1.0			1.0			1.0	mA
		V_{MIN} ≤ V_{IN} ≤ V_{MAX}		(7.5 ≤ V_{IN} ≤ 20)			(14.8 ≤ V_{IN} ≤ 27)			(17.9 ≤ V_{IN} ≤ 30)			V
		I_O ≤ 500 mA, 0°C ≤ T_j ≤ +125°C				1.0			1.0			1.0	mA
		V_{MIN} ≤ V_{IN} ≤ V_{MAX}		(7 ≤ V_{IN} ≤ 25)			(14.5 ≤ V_{IN} ≤ 30)			(17.5 ≤ V_{IN} ≤ 30)			V
V_N	Output Noise Voltage	T_A = 25°C, 10 Hz ≤ f ≤ 100 kHz			40			75			90		μV
$\frac{\Delta V_{IN}}{\Delta V_{OUT}}$	Ripple Rejection	f = 120 Hz	I_O ≤ 1A, T_j = 25°C or	62	80		55	72		54	70		dB
			I_O ≤ 500 mA	62			55			54			dB
			0°C ≤ T_j ≤ +125°C										
		V_{MIN} ≤ V_{IN} ≤ V_{MAX}		(8 ≤ V_{IN} ≤ 18)			(15 ≤ V_{IN} ≤ 25)			(18.5 ≤ V_{IN} ≤ 28.5)			V
R_O	Dropout Voltage	T_j = 25°C, I_{OUT} = 1A			2.0			2.0			2.0		V
	Output Resistance	f = 1 kHz			8			18			19		mΩ
	Short-Circuit Current	T_j = 25°C			2.1			1.5			1.2		A
	Peak Output Current	T_j = 25°C			2.4			2.4			2.4		A
	Average TC of V_{OUT}	0°C ≤ T_j ≤ +125°C, I_O = 5 mA			0.6			1.5			1.8		mV/°C
V_{IN}	Input Voltage Required to Maintain Line Regulation	T_j = 25°C, I_O ≤ 1A		7.3			14.6			17.7			V

NOTE 1: Thermal resistance of the TO-3 package (K, KC) is typically 4°C/W junction to case and 35°C/W case to ambient. Thermal resistance of the TO-220 package (T) is typically 4°C/W junction to case and 50°C/W case to ambient.

NOTE 2: All characteristics are measured with capacitor across the inut of 0.22 μF, and a capacitor across the output of 0.1 μF. All characteristics except noise voltage and ripple rejection ratio are measured using pulse techniques (t_W ≤ 10 ms, duty cycle ≤ 5%). Output voltage changes due to changes in internal temperature must be taken into account separately.